Springer Series on Atomic, Optical, and Plasma Physics

Volume 116

The Springer Series on Atomic, Optical, and Plasma Physics covers in a comprehensive manner theory and experiment in the entire field of atoms and molecules and their interaction with electromagnetic radiation. Books in the series provide a rich source of new ideas and techniques with wide applications in fields such as chemistry, materials science, astrophysics, surface science, plasma technology, advanced optics, aeronomy, and engineering. Laser physics is a particular connecting theme that has provided much of the continuing impetus for new developments in the field, such as quantum computation and Bose-Einstein condensation. The purpose of the series is to cover the gap between standard undergraduate textbooks and the research literature with emphasis on the fundamental ideas, methods, techniques, and results in the field.

More information about this series at http://www.springer.com/series/411

Yuri A. Rezunkov

High Power Laser Propulsion

Springer

Yuri A. Rezunkov
Research Institute for Optoelectronic Instrument
Engineering (NII OEP JSC)
Sosnovy Bor, Russia

ISSN 1615-5653 ISSN 2197-6791 (electronic)
Springer Series on Atomic, Optical, and Plasma Physics
ISBN 978-3-030-79692-1 ISBN 978-3-030-79693-8 (eBook)
https://doi.org/10.1007/978-3-030-79693-8

This Springer imprint is published by the registered company Springer Nature Switzerland AG
The registered company address is: Gewerbestrasse 11, 6330 Cham, Switzerland

Preface

> At last, there is a third, the most enticing method of obtaining speed. This is energy transfer to a missile from without, from the earth. The missile itself could not be supplied with material (i.e., weight, in the form of explosives or fuel) energy. It would be transmitted from the planet in the form of a parallel bundle of electromagnetic beams, with small wavelength.
>
> K. E. Tsiolkovsky,
> *Spacecraft*, 1924

More than 60 years ago, on October 4, 1957, the first artificial object—Sputnik—was launched into a low Earth orbit by a rocket booster. Since then, huge space-vehicle launch sites have been created in many countries, including Russia, the USA, France, China, and Japan; space-vehicle missions to planets in the solar system have been performed; and astronauts have visited the Moon. Many prospective projects in the fields of manned and unmanned missions are assumed to be implemented in the near future, including vehicle missions to nearby stars.

Approximately the same numbers of years have passed since the first laser, a highly directional source of "a parallel bundle of electromagnetic beams," was first demonstrated in 1960. Currently, high-power lasers of various types (including gases, liquid, solid states, and free electrons) operate in a wide range of radiation wavelengths, from X rays up to the middle-infrared spectral range. Lasers achieved widespread use both for civil and naval applications. Global energetic systems based on power delivery via high-power laser radiation have been developed. Some of these projects involve space-transportation systems that produce thrust with high-power lasers.

In the 1970s and 1980s, several projects on space missions involving laser orbital transfer vehicles (LOTVs) propelled by ground-based or space-borne high-power lasers were developed. A LOTV of 30 tons in a mass mission from LEOs to geostationary orbit were assumed to be achieved by using a 500-MW laser. A high-power laser propulsion (HPLP) system had to include the following equipment:

(a) laser equipment consisting of 6–10 large lasers
(b) laser power transmitter that has an adaptive telescope to control the laser beam direction

(c) orbital re-transmitter to redirect the laser beam to the orbiting LOTV
(d) LOTV with a receiver telescope system to collect the laser power and redirect the power into a laser-propulsion engine
(e) land-based launching space center to control the space missions

Moreover, in the 1980s, the Apollo Lightcraft project was developed under the guidance of NASA as a new space-transportation system. The project included all design components to develop the spacecraft, from the general concept of the vehicle device to the electronic systems that guide the launching and landing of the craft. But one of the problems that had to be solved was mounting laser-propulsion engine on the craft board. The Lightcraft was assumed to be propelled by high-power space-borne lasers with gigawatt-power emission. The laser-propulsion engine had to operate in two regimes of thrust production: until an altitude of 70 km, it operates via air-breathing propulsion, and then it operates in a rocket mode with onboard propellant.

Modern lasers can generate power of only a few MW. This limitation is caused by the technical and technological problems of developing a laser as a whole engineering device.

Recently, the idea of interstellar missions featuring small vehicles that use high-power laser radiation met with "sudden" approval from Yuri Milner, a Russian businessman, and Stephen Hawking, an English theoretical physicist. In 2016, they supported the Breakthrough Starshot project with the goal of launching a nano-vehicle with a 1-g mass to the Alpha Centauri star system by using sails irradiated by lasers. It is assumed that the vehicle would reach Alpha Centauri in 20 years to take photos of planets in this star system and then send this information to the Earth observers.

The following principal technologies defining the mission success will likely be developed in the near future:

- progress in the region of micro-electronic concerning all micro-units of the space vehicle, namely, the photo camera, power sources, navigation electronics, and so on
- creation of solar sails with nanotechnology using materials with a thickness of a few hundreds of atoms
- development of a multi-component laser system with a total power of 50–70 GW with adaptive phasing of a number of lasers with less power

These projects seem fantastic right now, but they can be considered as scientific forecasting of the future of space-transportation system development, which is based on HPLP.

The theoretical basis of laser propulsion was developed by Russian Academician A. M. Prokhorov and American Professor Anthony Pirri in the early 1970s. Many experiments on laser propulsion were performed by American and Russian scientists during the 1970s and 1980s. At this time, Russian Prof. Yu. P. Raizer developed the basic theory of plasma ignition via high-power laser radiation. Laser plasma and chemical reactions running through the plasma under the laser radiation play an important role in producing thrust.

At the beginning of the twenty-first century, a second wave of interest in laser propulsion was observed, during which a few International Symposium on Beamed Energy Propulsion (ISBEP) meetings took place in different countries every 2 years, where researchers addressed the practical concerns of laser propulsion, starting from the achievements made in the USA, Germany, and Russia up to the development of new laser-propulsion applications.

Simultaneously, these meetings and discussions displayed both the advantages of applying lasers to produce thrust as compared with rocket technology and the principal problems of such practical adaptations. The laser propulsion could be used to develop space-transportation systems devoted to the creation of space-communications systems, internet, and other energetic and communicatory infrastructures based on small space-vehicle systems.

Unfortunately, all of these concepts have not been realized until now! And the direction of the research into laser-propulsion transportation systems remains undefined. To that end, this book considers the principal achievements in the development of HPLP as well as the theoretical and technological problems that have appeared.

In particular, the theory of laser radiation interaction with plasma ignited under the laser radiation, developed by Raizer, is considered as applied to the efficiency of the laser power input into the plasma. Basic physical phenomena of thrust production by laser power are analyzed to determine the optimal type of laser propulsion. The principles of propulsion engine designs and assembly of the vehicles with laser propulsion are considered, including the following:

- designs of laser-propulsion engines, operating as both air-breathing and ramjet engines to launch vehicles into LEOs
- laser-optics assemblies of vehicles with laser propulsion, whereby power is collected from a remote laser and directed into a propulsion engine
- laser adaptive systems that control the laser beam to propel vehicles into Earth orbits by delivering the laser power through the Earth's atmosphere

The development of HPLP systems is not progressing because of problems with efficient thrust production in subsonic and supersonic modes of engine operation and because of the lack of an optimal design for the laser-propulsion engine. One of the complex problems is the lack of consequent experimental equipment that could permit real testing of the laser-propulsion engines operating with high-power laser radiation.

This book is one of the first real attempts to generalize the accumulated experience of investigations into laser propulsion, considering laser-propulsion phenomena, laser-propulsion techniques, vehicle designs with laser propulsion, and high-power laser systems that provide movement for the vehicles.

Sosnovy Bor, Russia Yuri Rezunkov
January 2021

Acknowledgments

Publishing this book wouldn't have been possible without the assistance and support of many people.

First of all, I would like to thank my loving wife, Galina Razina, for taking over all daily chores, thereby allowing me more time to work on the manuscript. Her patience and support have been truly endless.

I benefited greatly from the work of my colleagues from the Scientific Research Institute for Optoelectronic Instrument Engineering (NII OEP) and professional acquaintances over the years. They are too many to mention by name in this short paragraph, but I included the titles of their publications in the References where it was relevant.

I'm grateful to SPIE Press science editors for their detailed review of the first draft of the book. Their comments were invaluable. Whatever errors remain are my responsibility.

Also I would like to thank Svetlana Naidenkova, the Chief Librarian at the NII OEP's Scientific Library, who greatly assisted me in searching databases for the required scientific papers and books, Irina Arkanova—for producing a fair amount of the figures presented here, and my niece, Nataliia Pogudina—for prior designing the book layout.

And last but not least, I'm particularly grateful to Dr. A.A. Schmidt for his friendship and constant support and attention to this work.

Contents

1 A Brief History of Laser Propulsion 1
1.1 Introduction 2
1.2 Main Stages of Laser-Propulsion Developments 3
1.3 Physical Processes Underlying Laser Propulsion 15
1.3.1 General Classification of the Laser-Propulsion Phenomena 15
1.3.2 Basic Thrust Characteristics of Laser-Propulsion Engines 18
1.4 General Concepts of Laser Propulsion 23
1.4.1 Launching Space Vehicles into Low Earth Orbits with Laser Propulsion 24
1.4.2 Laser Propulsion for the Correction of LEO Satellites 25
1.4.3 Interorbital Missions of Space Vehicles with the Laser Propulsion 30
1.5 Original Concepts of High-Power Laser Propulsion 34
1.5.1 The "4P" Vehicles 34
1.5.2 Lightcraft Technology Demonstrator (LTD) 35
1.5.3 Laser Impulse Space Propulsion—LISP 36
1.5.4 Principal Concept Design of the High-power Laser-Propulsion Systems 38
References 40

2 Basic Gas-Dynamic Theories of the Laser Air-Breathing and Rocket Propulsion 43
2.1 Introduction 43
2.2 Gas-dynamic Theory of Laser Propulsion 46
2.2.1 Specific Properties of Pulsejet Laser Propulsion 46
2.2.2 Rocket Laser Propulsion at Space Conditions 54
2.3 Physics of Laser Plasma Ignited in Gases as Applied to Laser Propulsion 62

2.3.1 Model of Multi-Ionized Plasma Ignited by Laser Pulses in Gases 63
2.3.2 Conversion Efficiency of Laser Power into Plasma Temperature 72
2.4 Numerical Calculations of Non-stationary and Non-isentropic Gas Flows as Applied to Laser Propulsion 78
2.4.1 Perfect Gas Flow Models and Numerical Algorithms to Calculate Gas Flow of Pulsejet Laser Propulsion 79
2.4.2 Model of Equilibrium (Thermal) Plasma 82
2.4.3 Model of Non-equilibrium Plasma as Applied to Pulsejet Laser Propulsion 85
2.4.4 Discussion on the Applicability of Various Models of Plasma Ignited 87
References 89

3 Laser Ablation of Solid Materials, Laser Ablation Propulsion 93
3.1 Introduction 93
3.2 Physical Phenomena Underlying of Laser Ablation Propulsion 95
3.2.1 Basic Concept of Developed Evaporation of High-Melting and Low-Melting Materials 95
3.2.2 Simplified Gas-Dynamics Model of Laser ablation Propulsion 97
3.2.3 "Absorption Explosion" Model of Plasma Ignition at Laser Ablation of Solid Targets 99
3.2.4 Gas-Dynamic Models of the Laser Radiation Interaction with Ionized Gas (Gaseous Plasma) 102
3.3 Effects of Solid Target Structure on Laser Ablation Propulsion 105
3.3.1 Direct Laser Ablation Propulsion 105
3.3.2 Combined Laser Ablation Propulsion 107
3.3.3 Confined Laser Ablation of Multilayer Structured Targets 108
3.4 Laser Ablation Propulsion Based on Ablation of High-Energy Polymers 111
3.4.1 Basic Plasma-chemical Reactions Proceeding in the CHO-Polymer Vapor Under Laser Radiation 112
3.4.2 Similarity Laws of Laser Ablation Propulsion Based on Polymer Propellants 119
3.5 Semi-empirical Models of Laser ablation Propulsion Based on CHO-Polymers 121
3.5.1 Gas-Dynamics of the Laser Ablation Propulsion 121
3.5.2 Vapor and Plasma Models of the Laser ablation Propulsion Using Critical Laser Power Flux 125

3.6 Efficiency of the Laser Ablation Propulsion Based on CHO-Polymers 127
References 130

4 Aerospace Laser-Propulsion Engine 133
4.1 Introduction 133
4.2 The Aerospace Laser-Propulsion Engine Conception 135
4.2.1 Designing of Two-Mirror Beam Concentrator 136
4.2.2 Optical Model of the Two-Mirror Beam Concentrator 140
4.2.3 Numerical Techniques to Develop the Two-Mirror Beam Concentrator 143
4.3 ASLPE Thrust Characteristics in a Pulsed Mode of Operation 149
4.4 Adaptation of ASLPE for Continuous Wave (CW) Laser Propulsion 156
4.4.1 Principles of CW Laser Propulsion 156
4.4.2 CW ASLPE Thrust Characteristics 158
4.5 Analysis of Available Technologies as Applied to ASLPE Development and its Engineering Constraints 162
4.5.1 Effects of Slit on Thrust Production 162
4.5.2 Thermo-physical Model of the ASLPE Device 164
4.6 Preliminary Conclusion 169
References 169

5 Supersonic Laser Propulsion 171
5.1 Introduction 171
5.2 Lightcraft Engineering Version Adapted to the Pulsejet Supersonic Laser Propulsion 173
5.2.1 Perspective Designs of the Lightcraft 181
5.2.2 Intermediate Conclusion 183
5.3 Physical Phenomena Going with Ramjet Supersonic Laser Propulsion 183
5.3.1 Gas-Dynamics Effects Induced by Lasers in a Supersonic Gas Flow 183
5.4 Merging of Individual Shock Waves into a Quasi-Stationary Integrated Shock Wave 188
5.5 Supersonic Laser Ablation Propulsion 190
5.5.1 The Effects of Gas Jet Injection into Supersonic Gas Flows 190
5.5.2 Theoretical Model of Supersonic Laser Ablation Propulsion 192
5.5.3 Thrust Characteristics of Supersonic Laser Ablation Propulsion 194

5.5.4 Peculiar Properties of Thrust Production at the Supersonic Laser Ablation Propulsion 199
5.6 Conclusion 204
References 204

6 Space Mini-vehicles with Laser Propulsion 207
6.1 Introduction to the Problem 207
6.2 Scenario of the SMV Orbital Maneuvers 211
6.3 Space Debris Removal Out of Geosynchronous Earth Orbit (GEO) by Using Laser-Propelled Space Mini-vehicles 222
6.4 Onboard Laser-Propulsion System as Applied to SMV 229
6.4.1 Receiver Telescope 230
6.4.2 Optical Turret 231
6.4.3 Optical Switch 233
6.4.4 The Unit of Laser-Propulsion Engines 234
6.4.5 Requirements to Optical Elements of the Onboard Laser-Propulsion System 235
6.5 Brief Outcome 237
References 238

7 Laser Power Transfer to Space Vehicles with Laser Propulsion 241
7.1 Introduction into the Problem 242
7.2 Models of the Aerosols and Gases Attenuation, Absorption, and Scattering of Laser Radiation in the Upper Atmosphere 243
7.2.1 Models of the Atmospheric Aerosols and Gases 243
7.2.2 Nonlinear Effects Developed During Propagation of High-Power Laser Radiation in the Upper Atmosphere 247
7.3 Self-Empirical Models of the Upper Atmosphere Turbulence 251
7.4 Phase and Intensity Profiles of the Laser Beam That Passed Through a Turbulent Atmosphere 258
7.4.1 Tentative Conclusion 260
7.5 Basic Atmospheric Effects Limiting Delivery of the Airborne Laser Power to Space Vehicle 260
7.5.1 Scenario of Laser Power Delivery to a Space Vehicle 260
7.5.2 Turbulence Effects on a Laser Beam as Applied to High-Power Laser Propulsion 264
7.6 Adaptive Laser Systems for the High-Power Laser Propulsion 268
7.6.1 Statement of the Problem 268
7.6.2 Adaptive Optical Laser Circuits and Special Equipment 270
7.6.3 Laser Adaptive Optical Systems as Applied to Beaming a Remote Target 273
7.6.4 Principal Outcomes 280
References 280

Conclusion 283

Index 287

About the Author

Yu. A. Rezunkov (born in 1950) is a leading scientist of the Scientific Research Institute for Optoelectronic Instrument Engineering (NII OEP JSC) in Sosnovy Bor (Russia). His research interests lie in the fields of laser physics, laser optics, and interaction of laser radiation with matter.

He graduated from Leningrad Polytechnic Institute (now Peter the Great St. Petersburg Polytechnic University) with diploma in aerodynamics and thermodynamics in 1974. In 1977, he earned a PhD degree and joined Sosnovy Bor Branch of Vavilov State Optical Institute (now NII OEP JSC).

Yu. A. Rezunkov has published a number of papers on various scientific subjects, including propagation of high-power laser radiation in the Earth's atmosphere, development of laser lightning protection systems, phase conjugation techniques to compensate for the laser beam wave front distortions caused by an atmospheric turbulence as well as development of the high-peak power picosecond CO_2 lasers. Special attention in his papers has been paid to the development of the laser-propulsion theory and applied techniques. The laser ablation propulsion based on the CHO-polymers was the subject of his Doctor of Science thesis, successfully defended in 2006.

At present, Dr. Yu. A. Rezunkov works on a study of laser effects on optic-electronic equipment.

Abbreviations

ABL	Air-borne laser
ABLP	Air-breathing laser propulsion
AFRPL	Air Force Research Propulsion Laboratory
ASLPE	Aerospace Laser-Propulsion Engine
BWA	Beam wave front analyzer
CJE	Chemical jet engine
COIL	Chemical Oxygen Iodine Laser
CW	Continuous wave
EJE	Electrojet engine
FIW	Fast ionization wave
GEO	GEOstationary orbit
HPLP	High-power laser propulsion
HWFA	Heterodyne wave front analyzer
IGW	Internal gravitational wave
ISBEP	International symposium on beamed energy propulsion
LAP	Laser ablation propulsion
LChP	Laser-chemical propulsion
LEO	Low earth orbit
LISP	Laser impulse space propulsion
LOTV	Laser orbital transfer vehicle
LPD	Laser-propulsion device
LPE	Laser-propulsion engine
LSD	Laser-sustained detonation wave
LSR	Laser-sustained radiative wave
LTD	Lightcraft Technology Demonstrator
LTE	Local thermodynamic equilibrium or Laser-thermal engine (Chap. 4)
LTP	Laser-thermal propulsion
MO	Master oscillator
MOPA	Master oscillator–power amplifier
OCD	Optical continuous discharge

PCM	Phase conjugation mirror
PMMA	Polymethylmethacrylate
POM	Polyoxymethylene
PRR	Pulse repetition rate
RP	Repetitively pulsed
RW	Radiative wave
SDI	Strategic defense initiative
SLM	Spatial light modulator
SMV	Space mini-vehicles
SSL	Solid-state laser

Chapter 1
A Brief History of Laser Propulsion

Abstract This chapter analyzes the basic achievements of laser propulsion made during the 1970s. The modern age of laser-propulsion development started at the beginning of the twenty-first century, and it concerns those laser technologies which are directed to practical applications of laser propulsion. Typical examples of these applications are efficient laser-propulsion engines and high-power laser-propulsion (HPLP) systems.

Advanced propulsion research and technology require launch and space-flight techniques that can greatly reduce mission costs. Laser propulsion is based on beamed energy delivered from an off-board laser to a flying vehicle. A variety of laser-propulsion concepts have been investigated theoretically and experimentally since the early 1970s. During the 1980s, the Strategic Defense Initiative (SDI) led to the invention of the Lightcraft concept. United by this idea, the US Air Force and NASA developed a laser technologic model required for launching small payloads into low Earth orbits (LEOs). The near-term objectives of the technology were to demonstrate essential launch capabilities in the near future. Both a large thrust and a high specific impulse I_{sp} in excess of 1000 s can be considered as advantages of laser propulsion. Moreover, the laser propulsion provides the simplicity and reliability of the engine due to the absence of moving parts and a simpler propellant injection feed system. However, this technology suffers from major limitations such as available laser power, absorption, and distortion of the laser beam that passes through Earth's atmosphere, and delivery of laser power to a laser-propulsion engine.

Keywords Laser propulsion · Laser-propulsion engine · The Strategic Defense Initiative (SDI) · The "4P rule" by Kantrowitz · Coupling coefficient · Dynamics of laser propulsion · Infinite specific impulse · The Lightcraft concept · Laser orbital transfer vehicle · Beamed Energy Propulsion

Y. A. Rezunkov, *High Power Laser Propulsion*, Springer Series on Atomic, Optical, and Plasma Physics 116, https://doi.org/10.1007/978-3-030-79693-8_1

1.1 Introduction

In 1924, K. E. Tsiolkovsky (1857–1935), a famous Russian science pioneer of the twentieth century, denoted certain disadvantages of rocket techniques for interplanetary human travel and the possibility of spacecraft launching into near-Earth orbits in his book [1]. One of the problems is the necessity to store a large propellant mass on the rocket, exceeding the payload mass by approximately three times. He deduced the following:

> *At last, there is a third, the most enticing method of obtaining speed. This is the energy transfer to a missile from the outside, from the Earth. The missile itself could not be supplied with material part (i.e., valued, in form of explosives or fuel, energy). It would be transmitted from the planet in form of a parallel beam of electromagnetic rays, with short wavelength.*

Tsiolkovsky did not develop this idea into a "beamed energy propulsion" concept, but his scientific foresight gave a hint to other researchers of the necessity to develop an alternative form of jet vehicle flight with remote power sources.

The exploration of near-Earth space started officially when the first space satellite was launched by the Soviet Union in 1957. The first lasers were created in both the USA and the USSR [2] approximately at the same time. Shortly afterward, in 1962, G. A. Askaryan and E. M. Moroz—two Soviet physicists—investigated the generation of a recoil impulse by evaporating solid-state materials with the laser-pulse effect:

Strong energy fluxes concentrated over small areas of solid surface are capable of producing an intense evaporation of material that results in a pressure decrease on the surface producing a strong recoil impulse during the material evaporation. The recoil pressure on evaporation may be several orders of magnitude greater than the beam pressure, and this was experimentally observed for the focused beam of a ruby laser [3].

They proposed using the recoil impulse to accelerate small particles and simulate the action of micrometeorites on the surfaces of bodies.

In 1972, Prof. Arthur Kantrowitz published his paper entitled "Propulsion to Orbit by Ground Based Lasers" [4, 5] and proposed the use of high-power lasers to launch small satellites into space instead of manufacturing large and energy-inefficient chemical rockets. Such satellites would literally straddle the tip of a laser beam, focused on their "propellant" area. When a high-power laser beam is focused (even loosely) on a solid matter, such matter is evaporated and ionized almost instantaneously, i.e., the release of energy is much higher than one used from burning hydrogen in rockets. So, laser-driven vehicle will be still flying on the same rocket principle, but exhaust energy and structural lightness will be incomparably superior to hydrogen-burning rockets.

A rocket driven by a laser will consist of very light focusing mirrors, a relatively small (energy efficient) amount of solid propellant, and the rest will be payload! The essence of laser propulsion was introduced by Kantrowitz, who named it the "4P rule." In 1989 [5], he proposed that four basic laser engineering technologies would be realized to implement HPLP in the following years:

1. The average laser power had to be increased by several orders of magnitude.
2. Atmospheric transmission problems needed to be explored.
3. Large-scale mirrors needed to be developed.
4. Technology for the highly efficient conversion of laser energy into the kinetic energy of a jet with speeds up to about 10^6 cm/s (specific impulse of 1000 s) and with thrust vectors independent of the laser beam axis needed to be developed.

The main subject of this chapter is a historical review of these topics applied to propulsion systems based on high-power lasers, starting from the 1970s.

1.2 Main Stages of Laser-Propulsion Developments

The principal feature of laser propulsion is that all mechanisms responsible for generating thrust are enabled by a remote power source, i.e., a laser. These mechanisms are determined as follows: (a) laser ablation of solid materials, (b) laser detonation waves induced by an instantaneous laser pulse energy release in a gaseous medium in a restricted gas volume, (c) internal energy liberation of chemically neutral polymers, etc.

The initial laser-propulsion investigations in the 1970s can be characterized as a search for prospective HPLP applications in the near future. The 1980s was the search for technical and scientific ways to produce effective laser propulsion, including the creation of optimal engine designs. The 1990s were devoted to the conceptual development of laser propulsion in near-Earth space. Since 2000, research has taken a business-like approach to laser propulsion applied to vehicle launches into near-Earth orbits and orbital flights of satellites with studied laser-propulsion engine.

The field has experienced various financing stages over the last 40 years. In particular, NASA provided active financial support to laser-propulsion projects during the 1970s–1980s [6]. Some principal scientific and engineering results of these investigations will be considered further.

The following laser-propulsion subjects were studied during the 1970s:

1. Mechanisms of generating laser-propulsion current by the absorption of laser power. It was necessary necessity to understand problems such as the efficient heating of a working substance (propellant) and conversion of laser power into thrust.
2. Optical mirrors that focus a laser beam to activate the laser breakdown of a gaseous medium. At that time, there was no understanding what type of laser would be best applied to laser propulsion, which led to the idea of superpositioning the beam concentrator and jet nozzle to form a single whole for CW and repetitively pulsed (RP) laser radiation.
3. Choice of propellant type and its thermo-physical parameters, which have a strong effect on the ability of the laser-propulsion engine to reach stable thrust production.

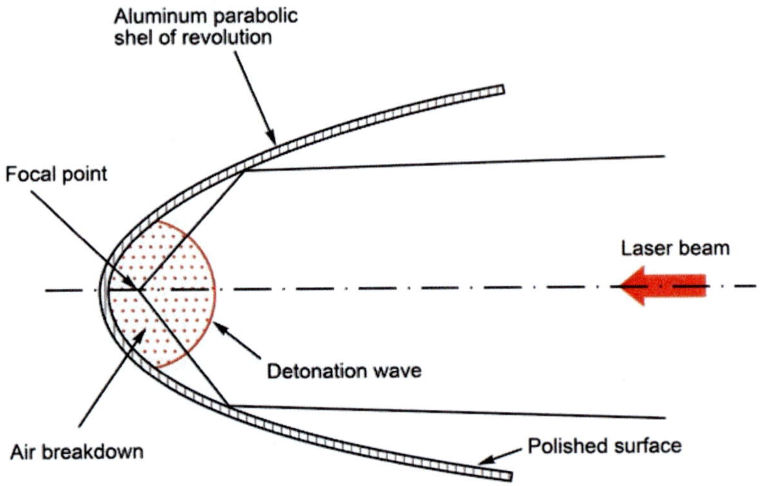

Fig. 1.1 Paraboloid of revolution for a laser-propulsion jet nozzle [8]

The main aim of these investigations was to demonstrate the reliability of laser propulsion based on high-power lasers, which could replace conventional rocket launch systems that use chemical jet engines.

The most consistent studies in this area were performed by Prof. Anthony Pirri [7], who contributed to the theory of laser propulsion produced by RP CO_2 laser power. In 1972, Pirri [8] considered the parameter called the coupling coefficient of the laser propulsion C_m, which is the ratio of produced thrust (N) to laser power (W). The coefficient characterizes the efficiency of thrust generation under the laser power effect. The propulsion mechanism studied by Pirri was the generation of strong shock waves in a gaseous propellant located at a mirror focus, which are activated by high-power laser radiation. Note that the specific impulse of the achieved exhaust jet was $I_s \sim 10^3$ s in that case.

These parameters—coupling coefficient and specific impulse—define a total efficiency η of the laser propulsion: $\eta = g{\cdot}I_s \times C_m/2$, and g is gravity.

At the reflected time, Pirri proposed the use of a laser beam concentrator in the form of a paraboloid of revolution to focus the incoming laser beam and activate a laser breakdown of onboard propellant (see Fig. 1.1). We will consider this type of laser propulsion in more detail in Chap. 2 to define the main correlations between laser and thrust characteristics.

Russian papers regarding laser-propulsion principles and theories were published a bit later, in 1976 and 1977. The main research on the theoretical and experimental investigation of laser propulsion was published under the authority of Academician A. M. Prokhorov in the Russian journals of *Quantum Electronics* and *Successes of Physical Sciences* [9, 10].

Prokhorov, along with his co-authors, developed detailed theoretical analyses of gas-dynamic processes of laser-propulsion generation via the RP laser breakdown of gaseous propellant and the laser ablation of solid materials. The results of these studies allowed the development of self-similar solutions to enable thrust production with a maximally efficient transition of laser power to thrust. These scientists demonstrated the first experiments for launching small, laser-driven objects in atmosphere.

Independently, in 1977, Pirri and Girard A. Simons [8] proposed the gas dynamics theoretical model for the laser propulsion of space vehicle flights for the first time. They also obtained self-similar solutions of the gas-dynamic equations describing the generation and propagation of shock waves in a paraboloid nozzle.

All of these theories on laser propulsion (by Prokhorov and Pirri) are considered in Chap. 2 in more detail because the received outcomes are important for understanding laser propulsion produced by the detonation mechanism of laser power added to a limited volume of gaseous propellants.

Over a period of 10 years (1972–1982) other important investigations into laser propulsion were carried out at the Air Force Research Propulsion Laboratory (AFRPL) [6]. We can make three important practical suggestions that follow from these studies in the laboratory. First, the higher the power of the laser, the larger the thrust that can be produced. Second, the investigations devoted to the achievement of "infinite specific impulse" that can be provided by outside power sources and matter are the focus. This suggestion assumes lack of a work substance on a vehicle board. Third, successful laser-propulsion technology can result in other nonconventional technology as laser power transmission through space ranges. These suggestions assume the rapid development of high-power (gigawatts at the time) lasers.

The following grouping of laser-propulsion devices (LPDs) was proposed based on (see Fig. 1.2):

- laser-supported detonation waves initiated by the laser ablation of solid matter
- pulse-produced blast waves accompanied by acoustic coupled gas flow
- laser-supported combustion-wave heating gas in a chamber

Detailed processes for laser–matter interaction applied to laser propulsion are considered in Sects. 1.2–1.5.

In the 1980s, the American program called the Strategic Defense Initiative (SDI, colloquially called "Star Wars") was proposed to protect the USA from attacks by ballistic strategic nuclear weapons (intercontinental ballistic missiles and submarine-launched ballistic missiles) [10]. The following was planned: the creation of ground-, air-, and space-based high-power lasers, including optical adaptive systems that can compensate for laser beam distortions caused by the Earth's atmosphere, the development of large-aperture telescopes for laser energy transmission through space distances, and so on. But as a consequence of the program's activity, favorable conditions appeared to develop a laser-propulsion concept based on high-power lasers.

At that time, it seemed that many of the engineering and technology problems of creating high-power lasers, including space-based lasers and laser optics, would be solved soon. It seemed possible to produce gigawatt lasers based on the design of chemical or free electron lasers. The problems of high-power radiation propagation

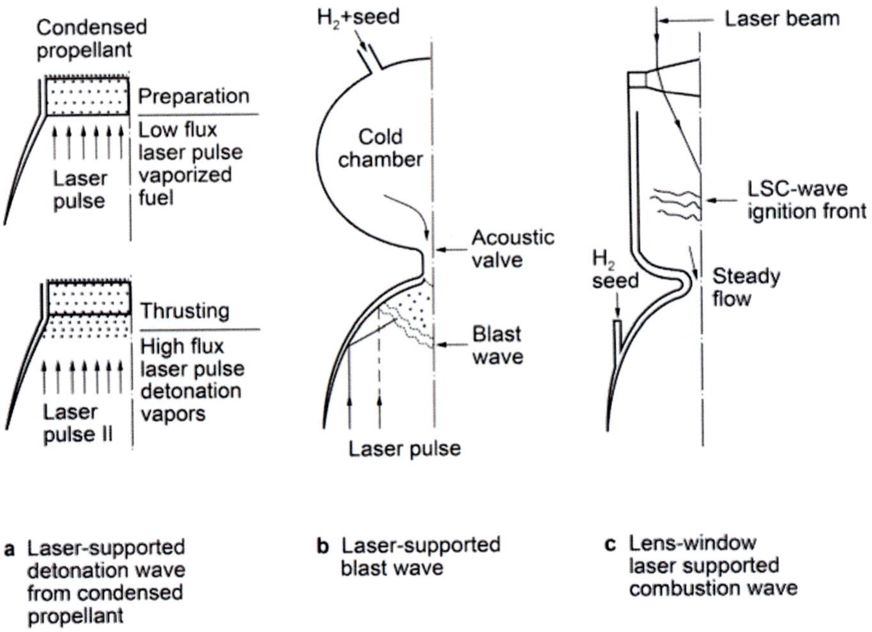

Fig. 1.2 Laser breakdown of matter applied to laser propulsion [6]

through the Earth's atmosphere were overcome by using linear and nonlinear adaptive optics. State-of-the-art optical technologies at that time were assumed to develop phased optical telescopes that consist of sub-apertures.

Some results of the HPLP investigations from that time are presented in *The Future of Flight* by Prof. Leik N. Myrabo, written jointly with Dean Ing and published in 1985 [11]. The book demonstrates the authors' point of view on further progress of beamed energy propulsion, including the application of radio-frequency and laser radiation to produce thrust. The authors also considered ways to solve the problems of HPLP, which were formulated by Kantrowitz earlier. The book contains a detailed list of papers and books issued on the considered subject to that time. Unfortunately, there were very few scientific publications on laser propulsion published in the USSR at that time.

Studies made by Myrabo [11, 12] showed that a 100-MW laser would be required for a manned flight to reach LEO. As noted in the Preface, lasers with a power of a few MW (maximum) have been developed so far; nevertheless, it was an important propulsion concept in the past. Moreover, Myrabo proposed two types of laser propulsion. The first concerns thrust production due to laser power absorption by gaseous propellant when the power is transformed into thermal energy in the gas, followed by a transformation of the energy into the kinetic energy of an exhaust jet. The second case features the intermediate transformation of laser power into the translational energy of a partially ionized gas, which is then transformed into pulsed

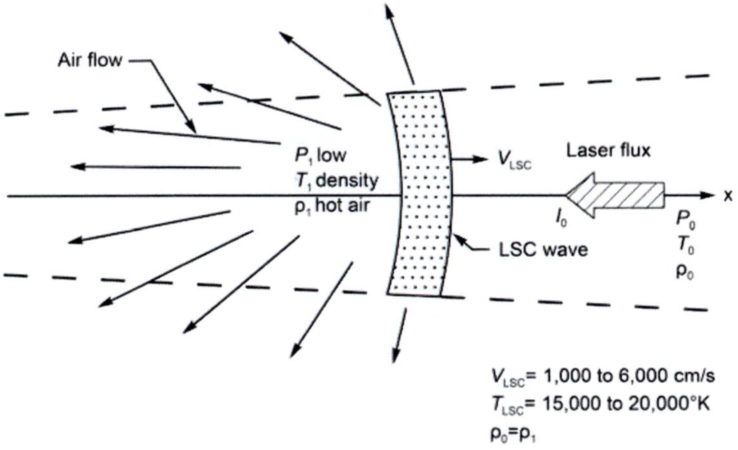

Fig. 1.3 Schematic diagram of laser-sustained, combustion-wave-mode laser beam–matter interaction

electric power by an MHD generator and finally into the jet kinetic power. All of these processes take place in a combustion chamber specially designed for each case.

The main physical premises for the laser power–matter interaction in a "combustion chamber" are as follows: (1) laser power is completely transformed into the internal energy of gas propellant (gas temperature) to obtain highly efficient laser propulsion, and (2) the gas temperature does not exceed the electron temperature of the gas molecules to provide the complete absorption of laser power. As was assumed by Myrabo, these conditions could be satisfied when the laser power interacts with a gas propellant in a laser-sustained combustion-wave mode (see Raizer [13] and Fig. 1.3).

Nevertheless, to initiate a laser detonation wave, one of the following processes has to be initiated [13]:

- laser breakdown of gas volume
- electric spark breakdown of gas
- molecular absorption of laser power
- laser breakdown of solid materials

Manned vehicle flights with laser propulsion have been considered since the 1980s, and the following engineering approaches were the foundation of their designs:

- The laser-propulsion engine has to operate in both air-breathing and rocket modes when the vehicle is launched from the Earth's atmosphere into LEOs, and it must allow braking of the vehicle when it re-enters the atmosphere during landing.
- The vehicle platform has to be designed to allow the onboard receiving optics to transmit the laser power to the laser-propulsion engine.

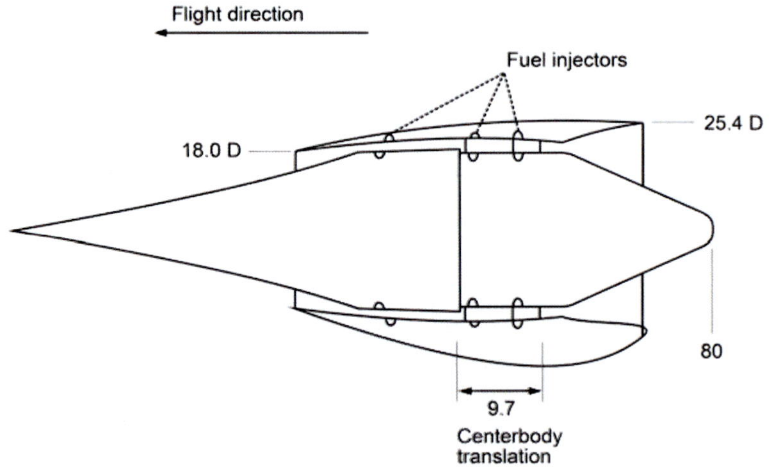

Fig. 1.4 Hypersonic air-breathing ramjet model

- Special optic-electronic equipment should adjust the laser power source with the onboard receiver optics.

These ideas were based on the development of a shuttle orbiter where the power required for vehicle orbital maneuvers would be delivered from space-based lasers. Ref. [12] assumes that 12–48 separate laser beams—each of which has 100-MW power—are used to launch a shuttle-type vehicle with laser propulsion. The laser-propulsion system with 12 beams has to provide flight to the shuttle with one man on board. The 48-beam system with a total power of 4.8 GW has to put into orbit a payload with a 23-ton mass!

One of the principal approaches to the development of reusable vehicles launched into space is the design of craft with supersonic laser propulsion. The achievement of supersonic modes of laser propulsion was assumed to be the basis for using conventional ramjet engines developed earlier by NASA [12].

A hypersonic air-breathing engine, shown in Fig. 1.4 [12], can be considered as an example of this approach. The design of the engine includes a lengthy gas channel with a central body, which allows the high-velocity transfer of incoming air flow into a combustion chamber so that the Mach cone [13] touches the inlet engine lip if an even Mach number of the flow varies sufficiently. The length of the annular duct has to satisfy the conditions of efficient supersonic detonation of injected fuel.

The version of the supersonic laser-propulsion engine developed by Myrabo is shown in Fig. 1.5 [14]. The engine device has a symmetrical structure combined with primary optics in the form of a paraboloid (plug nozzle) and with an afterbody in the form of an external expanding nozzle. The cylindrical focus of the primary mirror is located on the surface of an annular duct. The incoming laser beam is focused onto the surface, causing the breakdown of air close to the duct surface. Two versions of engine design were considered:

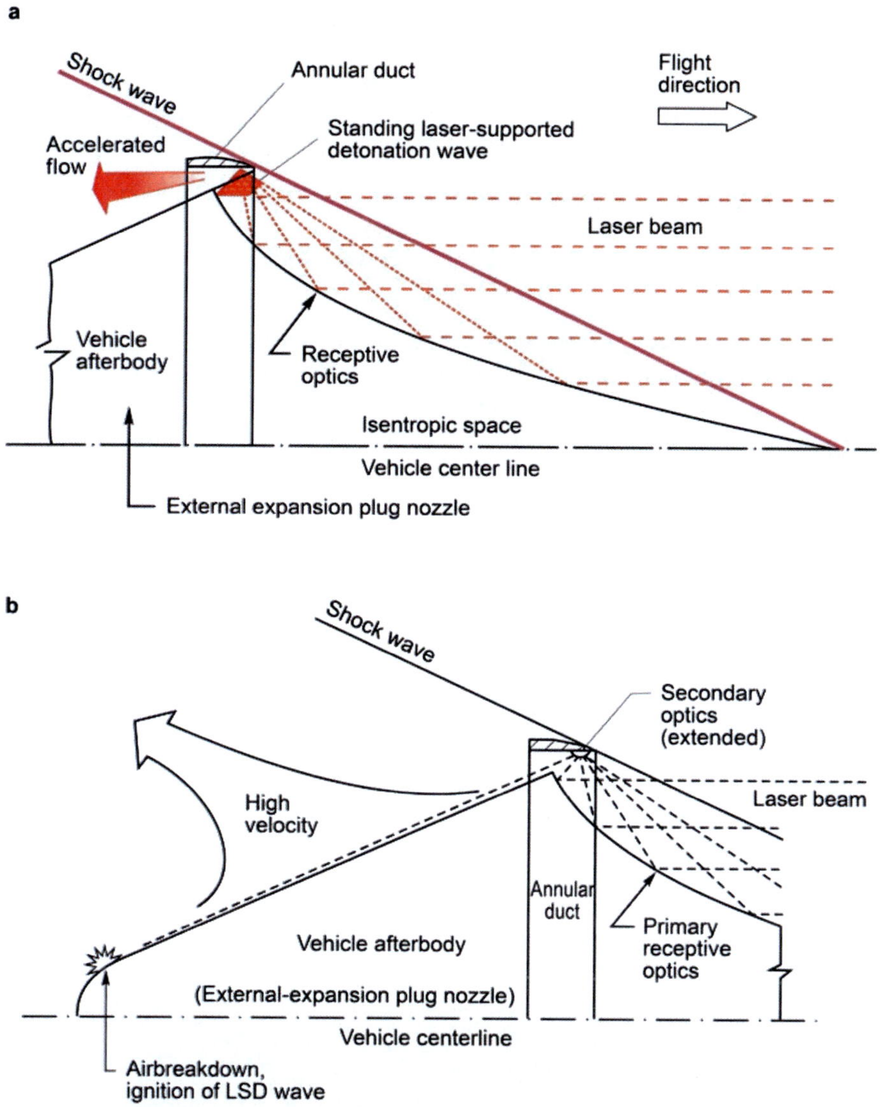

Fig. 1.5 One of the first supersonic laser-propulsion devices

- Engine devices with an inner mode of propulsion production where a quasi-stationary laser-sustained detonation (LSD) shock wave is created in the region of an air breakdown, which results in additional acceleration of incoming gas flow (Fig. 1.5(a)).
- Engines based on the laser breakdown of incoming flow close to the top of the central body where the initial shock wave could work as a "thruster" (Fig. 1.5(b)).

Both versions of supersonic laser-propulsion engines were not realized because of a few reasons. These devices were not supported by theoretical analysis of thrust production efficiency, so it took time to rebuild interest in this field. (The theory and further investigations of supersonic laser propulsion are presented in Chap. 5.)

Propulsion research from the 1980s demonstrates the following:

- HPLP systems could be built depending on the development of high-power lasers (with gigawatt power) and adaptive optics.
- The most promising laser-propulsion mechanism uses LSD waves generated by the propellant breakdown effect [15, 16].

The most outstanding vehicle design with laser propulsion is the demonstration model of Lightcraft (i.e., Lightcraft Technology Demonstrator), which was tested in atmosphere by using a 20-kW CO_2 laser at the end of the 1990s [17]. Note that the matrix beam concentrator developed at the National Lawrence Livermore Laboratory simplified the generation of LSD waves in laser-propulsion engines [18].

Some impressive results of laser propulsion were obtained also in Russia by that time. It is important to note that the laser-propulsion experiments were carried out by Dr. R. A. Liukonen using a 100-kW CO_2 laser to propel a 1-kg Lightcraft model in laboratory conditions [19]. These results are further analyzed in Chap. 2.

In the 1990s, research priorities switched from satellite launches into space to the application of laser propulsion for space missions [19, 20]. This change was caused by a few reasons. First, the "Star Wars" program had ended by that time, which resulted in less research into high-power laser development. Second, research into the efficiency of laser-propulsion launch systems at that time showed that a laser power of 1 MW was required to launch a vehicle with a 1-kg payload mass. Overly high-power laser systems must be developed to launch heavy satellites.

Within the framework of the laser-propulsion space applications, it was shown that the optimal engine must have a specific impulse of 750–2500 s to transfer a vehicle from LEO to a geostationary Earth orbit (GEO). Simultaneously, the concept of a laser orbital transfer vehicle (LOTV) was proposed to arrange regular orbital missions by using ground- or space-based lasers [11, 19, 21] (Fig. 1.6).

For example, the Laser Energy Network program (Japan [19]) was developed to enable space-mission LOTVs and develop space stations that can transmit laser power to the Earth's surface with high-power lasers [21].

There are certain limitations to laser applications for LOTV missions in space, which are caused by the laser beam divergence's sensitivity to vast space distances. A laser beam diverges due to diffraction effects, which can force the use of a large-aperture receiver telescope onboard the LOTV to collect transmitted laser power. For example, for a ground-based transmitting telescope with a 10-m aperture, a receiving telescope with a 100-m-diameter primary mirror must be installed to collect the laser power completely at a distance of 300,000 km and a wavelength $\lambda = 1.06\mu m$. The mirror will have a mass of 40 tons, made even with modern optical manufacturing techniques.

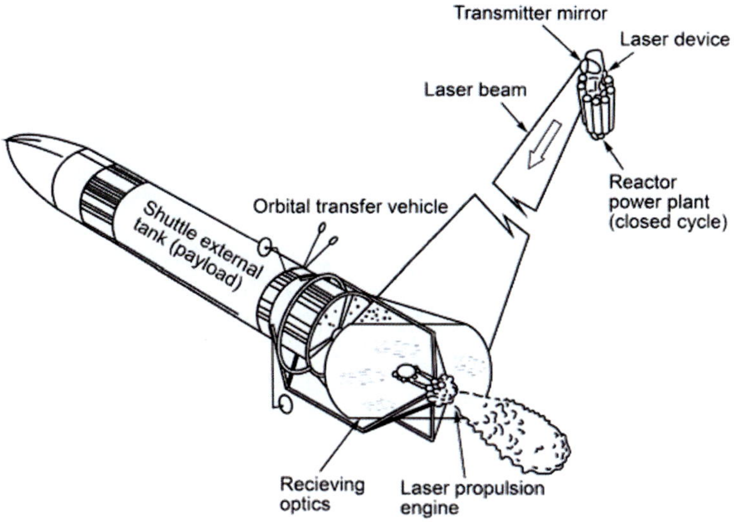

Fig. 1.6 A laser orbital transfer vehicle [11]

The results of the laser-propulsion studies performed during the 1990s defined the following limitations to the application of ground-based lasers for LOTV orbital missions, such as

1. the development of an adaptive telescope with a primary mirror that has a 12-m aperture
2. the design of a 10-MW laser system that works for a long time (approximately 10 days), continuously and persistently

The following US projects were pursued to solve these limitations [15]:

- a global space-laser energy system (SELENE)
- a large-aperture adaptive telescope, known as the Phased Array Mirror, Extendable Large Aperture (PAMELA)
- corresponding experimental equipment, known as the National Advanced Optics Mission Initiative (NAOMI)

After the close of the SDI program, a real model of the Lightcraft Technology Demonstrator (LTD) was developed by NASA [6]. The LTD is a single-stage airspace vehicle with laser propulsion that can be launched by high-power lasers (see Fig. 1.7). The LTD model as planned can carry optoelectronic equipment that provide control of the vehicle's orbital altitude and orientation in space. The LTD must be launched at an altitude of 200–2000 km. A working model of the LTD was 120 kg in mass at a maximal cross-section diameter of 1 m. An estimated laser power of 25–250 MW is required to launch it.

In 2000, the first flight experiments for launching a small LTD model were carried out at White Sands Missile Range (New Mexico, US) in an open atmosphere

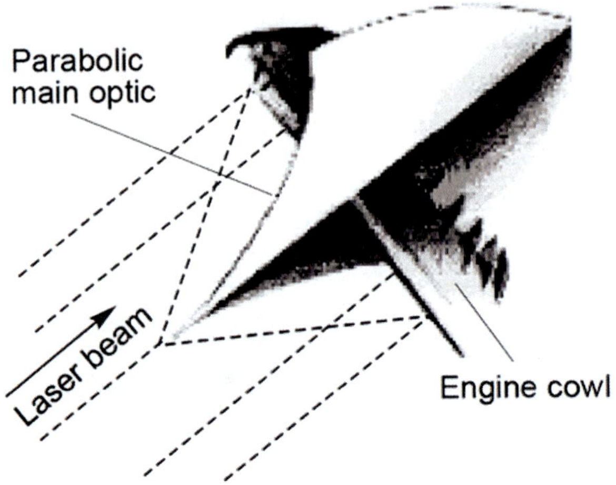

Fig. 1.7 Schematic of the Lightcraft model

[17]. The model had a 100-g mass, and it was launched at an altitude of 70 m by an RP CO_2 laser of 10-kW power.

From the 2000s until the present, there have only been two primary research trends for the development of laser-propulsion systems based on high-power lasers. These trends are devoted to (a) satellites launched into Earth orbits and (b) interorbital missions by space vehicles with laser propulsion. The laser-propulsion system (LPS) is supposed to consist of two basic components: (a) a laser source with a telescope system specially designed to transfer laser power to the space vehicle, and (b) a space vehicle with a laser-propulsion engine [21, 22]. All of these components were considered at the beginning of the twenty-first century within the framework of the International Symposium on Beamed Energy Propulsion (ISBEP). This symposium was brought to an end due to organizational problems, but the concept of laser propulsion lives on with some new ideas. For example, Yuri Milner, a Russian businessman, recently sponsored the Breakthrough Starshot project, the final aim of which is the flight of a small (nanosatellite) vehicle with a mass of 1 gm to the Alpha Centavos star system by using laser sails [23]. It is assumed that the vehicle will reach Alpha Centavos in 20 years due to super-high-power lasers (50–70 GW). The project was supported scientifically by Prof. Stephen Hawking. The Breakthrough Starshot project is based on the prospective progress of three laser engineering systems that form the main technological components of this space mission. The first one is the micro-electronics of all nano-elements in the space micro-vehicle, including the photo camera, power sources, and the navigation and communication equipment. However, the total inert mass of the vehicle cannot exceed 1 g. The second component relates to a laser sail, e.g., 10 m^2 in size and only a few grams in mass. The sail will have a thickness of only a few hundred atoms, keeping all necessary thermal properties of the material. The third component is the development of 50–70-GWlasers to accelerate the space micro-vehicle up to a velocity of one third the speed of light. Only such extreme velocity will allow the

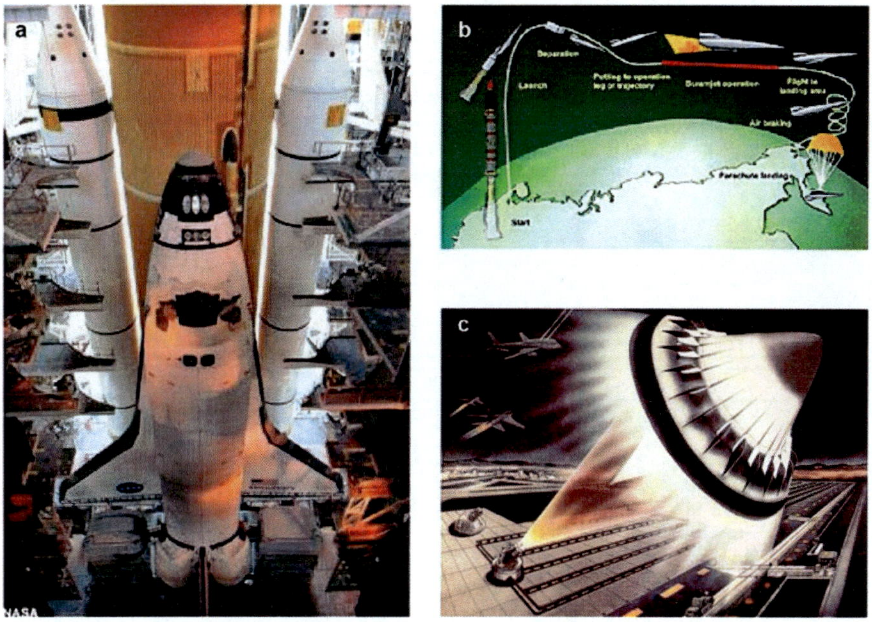

Fig. 1.8 Three types of the launch vehicles which are most technology perspective and economically sound projects: (**a**) space shuttle, (**b**) aerospace craft, and (**c**) Lightcraft technology

vehicle to reach the Alpha star quickly. The authors of the project suggest using a phase-adaptive technique to synchronize a few lasers (about 1000 lasers of lower power) to generate the required power. Some of these engineering systems are applicable to HPLP projects, too.

One of the promising paths of future space exploration is the development of reusable airspace launch systems, such as the "Space Shuttle" (US), "Buran" (USSR), and Space X (US), based on the use of reusable rocket engines (Fig. 1.8). However, the cost of launching a 1-kg payload mass into Earth orbits with these systems reaches tens of thousands of US dollars; furthermore, most of the system mass (more than 80%) consists of rocket fuel.

Progress in hypersonic aerodynamics and the corresponding aerodynamic technologies resulted in the development of single-stage supersonic aircraft such as the X-30 (US), "Hotol" (UK), and "Igla" (Russia). In accordance with the analysis made in Ref. [25], exploitation of aerospace transportation systems could decrease the launch cost of satellites to \$300. However, the authors of the systems met with some technical challenges related to the necessity to exclude the large fuel consumption at launch. This sufficiently reduced the payload mass of the aircraft.

The reason why laser-propulsion systems based on the Lightcraft vehicles are still considered for near-Earth space exploration is because this system can operate in both air-breathing and rocket modes. During the former, the atmosphere air can be used as a propellant to produce thrust, and during the latter, auxiliary propellant such

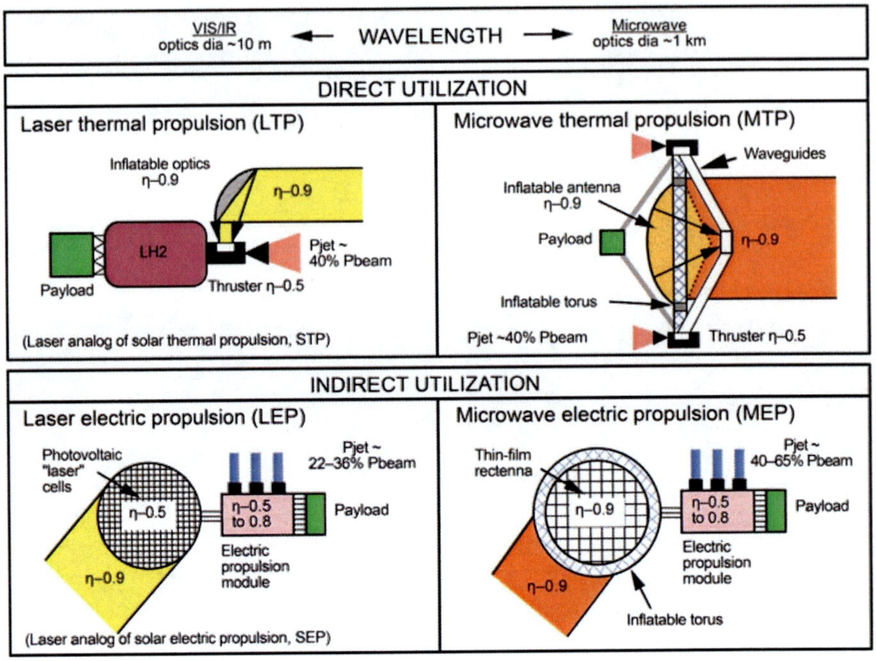

Fig. 1.9 Basic concepts of beamed energy propulsion

as liquid nitrogen stored onboard can be used [11]. One of the principal advantages of the Lightcraft is that it is a single-stage vehicle design without detached rocket stages. Moreover, the Lightcraft transportation systems can economically launch a large number of satellites, e.g., more than 1000 vehicles per year.

Note that laser propulsion is part of the more general concept of beamed energy propulsion, which is based on the use of radiation power to produce thrust (Fig. 1.9) [11]. From this point of view, a Lightcraft Technologies International project (LTI-20) developed by Prof. Leik Myrabo [24] in 2009 is an important example of a new airspace transportation system based on the thrust produced by radio-frequency radiation power. In this new form of Lightcraft evolution, Myrabo developed many aspects and engineering solutions for the most revolutionary spacecraft, including power-beaming infrastructure, emergency repair and maintenance, command systems, MHD propulsion system, and pulsed detonation engines and its theory. Taking an uncommon approach to releasing future flight technology to the general public, this comprehensive study conveys the technological possibilities for the next generation of commercial aviation. Written in superb but accessible detail, this study explains the Lightcraft will fly on microwave and laser power beamed from solar power stations in space and shows how it will ultimately replace today's commercial jet airlines.

In contrast to the LTI-20 concept, the basic subject of this book is the analysis and consideration of the physical phenomenon and technical aspects of laser propulsion that have to lead to the development of HPLP systems as a new, reusable airspace transport system. These aspects concern the following engineering systems:

- laser-propulsion engines with high technical efficiency that operate in both subsonic and supersonic air-breathing modes, as well as rocket mode in outer space
- single-stage vehicles of the Lightcraft type as a first step for the development of high-performance airspace transportation systems
- laser space mini-vehicles with an onboard optical system and laser-propulsion engines, allowing orbital mission maneuvers independent of the spatial positions of beaming lasers
- high-power lasers with adaptive laser systems, providing efficient delivery of laser power to space vehicles with laser propulsion through the Earth's atmosphere

Regarding the development of HPLP, the engineering systems are considered together with the progress of such technologies as high-power lasers, rocket engineering, and space satellites, as well as the interaction of laser radiation with various propellants, plasma physics, physics of shock waves, and so on.

1.3 Physical Processes Underlying Laser Propulsion

A number of investigations were carried out to study the processes underlying laser propulsion, and a comparison of various versions of LPDs led to conflicts due to the necessity to consider many physical and technical phenomena of the interaction of high-power laser radiation with a propellant of various physical states (gaseous, solid state, liquid, and plasma). A general approach is used here to create the processes and structures of these investigations. First, the processes defining the power efficiency of the laser propulsion are examined. The following subsections are based on the most significant laser-propulsion research.

1.3.1 General Classification of the Laser-Propulsion Phenomena

We propose the following classification of basic mechanisms responsible for thrust production under high-power laser radiation:

1. One of the first phenomena considered for thrust production is the explosive breakdown of a working substance (propellant) caused by laser pulses into gases located in the focus region of a parabolic mirror. The mirror can also be used as a jet nozzle. It is no exaggeration to say that the overwhelming majority of investigations regarding laser propulsion were made during this period. The

theoretical basis of explosive laser propulsion is stated in the papers published by Russian scientists Bunkin and Prokhorov in 1976 [5].

First, propellant such as atmospheric air is assumed to be used in a jet nozzle, and it returns to the nozzle due to the backflow after a shock wave exits the nozzle. This type of laser–pulse interaction with gas could be used to accelerate vehicles in the Earth's atmosphere, providing air-breathing propulsion.

One of the first experiments with air-breathing propulsion performed in outside atmosphere is the Lightcraft flight model realized under the guidance of Myrabo [17]. The LTD was designed to launch small satellites into LEOs as a single-stage rocket, in which both subsonic and supersonic operational modes can be attained in the atmosphere. The versatility of the Lightcraft design comes from the explosive laser-propulsion mechanism used during all stages of flight. An auxiliary liquid propellant (nitrogen) should be stored on board the Lightcraft to provide rocket-type thrust production. But in each case, there is no possibility of efficient producing thrust because of changes to the conditions of the laser radiation interaction with a propellant.

2. One of the first theoretical models of thrust production in a rocket mode of laser propulsion in vacuum conditions was proposed by Pirri [7]. He assumed that the gas propellant is stored onboard the space vehicle and that it is injected into a jet nozzle through a pinhole located at the top of a conical nozzle. The theory is based on the generation of strong shock waves, known as laser-sustained detonation waves (LDWs), and on the interaction of waves with a supersonic gas flow inside the nozzle.

   Pirri deduced semi-similar solutions for the supersonic gas-dynamic equations that describe the interaction of LDWs with a gas flow at various powers of laser radiation as well as of gas consumption that can be caused by strong detonation waves. The theory defines thrust as a function of laser-pulse energy, pulse repetition rate, and crucial nozzle section, assuming that the laser power delivered to the vehicle is transformed into thrust completely. (Chap. 2 will reveal that the last assumption is an approximation.)

3. Laser propulsion based on the thermal heating of a gas propellant and on conventional design of a jet engine that includes both a combustion chamber and a jet nozzle was studied many times, too [26]. For this laser propulsion, laser radiation is injected into the chamber through an optical window that separates the chamber inner volume from the surrounding atmosphere. Usually, gases such as Ar or H_2 are used as a propellant. To accelerate heated gas, the Laval nozzle is considered as a jet nozzle of the engine.

   The type of laser propulsion produced by CW laser power is known as laser-thermal propulsion (LTP), and it is based on the ignition of a continuous gas discharge in the engine. The first theory of the continuous discharge at various conditions of the laser radiation interaction with gases was developed by Raizer [13]. More complete investigations on LTP have been carried out by Prof. A. Sattarov at Kazan Technical University [27]. He developed a full-scale model of the LTP engine with a screw jet of gaseous propellant into a combustion chamber of the engine.

4. Laser ablation propulsion (LAP) occupies a special place in this propulsion classification. It is produced by the evaporation of a solid material surface by high-power laser radiation. Dr. Arthur Kantrovitz was the first researcher who proposed to use this type of laser propulsion to launch satellites into Earth orbits in 1972 [4].

   A clear advantage of LAP is the possibility of achieving a high specific impulse by evaporating refractory materials preferred for space-laser propulsion. The first LAP theory was developed by Prokhorov in 1976 [5]. Substantial contributions in further development of the LAP theory was made by a few Russian scientists from Physical Institute of Russian Academia of Sciences (FIAN) [28]. They considered various gas-dynamic processes running during ablating of solid materials at various gas surroundings, which define the production efficiency of thrust.

   Note that the theory of hard material ablation and the processes accompanying the ablation were developed as an applied approach to various technology applications. Laser ablation is one of the subjects of laser physics that found wide application in technology for producing new materials, composites, alloys, and so on. But, practical application of the laser ablation to thrust produce is studied much less. Nevertheless, there are a few publications in which the theoretical and experimental bases of LAP are developed.

   Proceedings by Claude Phipps [29–31] can be considered as one of the detailed investigations made in this direction, where numerous experiments of thrust production by means of the laser ablation of hard materials are analyzed, and the self-similar dependencies of the specific impulse on laser radiation intensity, laser-pulse length, and radiation wavelength are deduced.

   LAP includes three fundamental processes:

   - direct laser ablation of a solid propellant when the propulsion is produced due to the pressure of evaporated material (evaporation mechanism of propulsion production)
   - combined ablation when an auxiliary shock wave caused by the laser breakdown of evaporated material near its surface is used to produce thrust
   - laser ablation of structured materials

   The state-of-the-art theory and experiments on laser ablation propulsion are considered in Chap. 3.

5. The original technique of laser propulsion produced due to the merging of a number of weak shock waves into a low-frequency, quasi-stationary strong wave in different gases (matter vapor) was proposed in Ref. [32]. A sequence of low-intensity shock waves is induced via the laser breakdown (optical gas discharge) of a subsonic gas flow by a laser operating with a high repetition rate of pulses (up to 5 kHz). Unfortunately, the experiments carried out by the authors demonstrated high thrust instability. Detailed consideration of these experimental results is presented in Chap. 5.
6. A useful increase to the efficiency of laser propulsion can be achieved by using auxiliary energetic materials as propellants. There are a few chemical reactions

with auxiliary energy explosion that can be activated by laser power. Several experiments of laser-chemical propulsion (LChP), carried out in Ref. [33], can be supposed as proof of significant thrust production. The experiments used a 100-kW RP CO_2 laser as well as additional propellants such as polymethylmethacrylate polymers. However, the authors of these papers did not propose any theory of thrust increase in this case.

Laser-chemical propulsion will be discussed in Sect. 1.4 in more detail.

1.3.2 Basic Thrust Characteristics of Laser-Propulsion Engines

To describe the thrust properties of conventional jet engines operating with onboard fuel, a parameter known as the thrust price C_T is employed, which is a ratio of the power W generated in a combustion chamber to the thrust T produced, that is, $C_T = W/T$. However, for laser propulsion, the propellants that can be used in the laser-propulsion engine (LPE) are chemically neutral, and the notion of a thrust cost cannot be applied in this case. To describe the thrust produced by an LPE, the coupling coefficient C_m, defined as a ratio of thrust to laser power P, is noted: $C_m = T/P(N/W)$ [7].

The efficiency of propellant consumption during thrust production is characterized by a specific impulse I_{sp} that is defined as the ratio of produced thrust to rate of propellant mass per second $\dot{m}$, namely, $I_{sp} = T/\dot{m} \times g = v/g$ (s), where v is the average gas velocity of jet exhaust, and g is acceleration of gravity. The larger specific impulse is, the lower the propellant consumption.

Both of these parameters determine the LPE efficiency η:

$$\eta = C_m \cdot I_{sp} \cdot g/2 \tag{1.1}$$

Basic parameters used to describe the laser propulsion are listed in Table 1.1, including their magnitudes, which have been estimated theoretically.

Table 1.1 Thrust parameters of laser-propulsion engines. For parameter Q, P is the radiation power at the entrance of the laser-propulsion engine

Parameter	Name	Determination	Value/dimension
C_m	Coupling coefficient	$C_m = T/P$	$(10–10^3) \times 10^{-5}$ N/W
I_{sp}	Specific impulse	$I_{sp} = v/g$	$(10–10^4)$ s
Q^*	Specific heat of propellant ablation	$Q^* = P/\dot{m}$	J/g
T	Thrust		N
η	Efficiency	$\eta = C_m \times I_{sp} \times g/2$	<0.4
$\dot{m}$	Mass consumption of a propellant		g/s

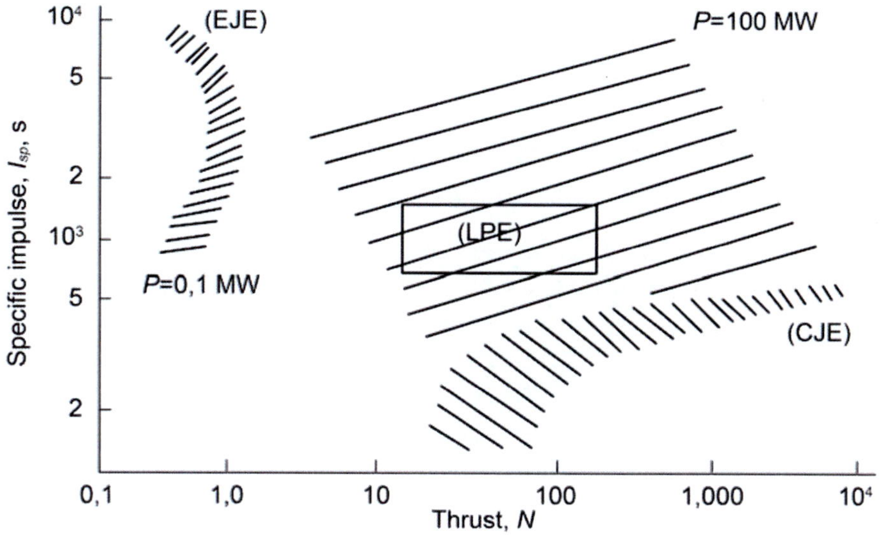

Fig. 1.10 Laser-propulsion engine parameters in comparison with other jet engines: *laser-propulsion engine* (LPE), *chemical jet engine* (CJE), and *electrojet engine* (EJE)

Note that all of these parameters are not completely independent of each other. They obey the following relations:

$$C_{\mathrm{m}} = \frac{2\eta}{I_{\mathrm{sp}} \cdot g} \quad Q^* = \frac{I_{\mathrm{sp}}^2 \cdot g^2}{2\eta} \quad T = \frac{2\eta \cdot P}{I_{\mathrm{sp}} \cdot g} \quad \dot{m} = \frac{2\eta \cdot P}{I_{\mathrm{sp}}^2 \cdot g^2} = \frac{T}{I_{\mathrm{sp}} \cdot g} \tag{1.2}$$

Figure 1.10 illustrates the thrust characteristics of the LPE as compared with other jet engines.

As one can see from Fig. 1.10, LPE is a "bridge" between electrojet engines and chemical rocket engines according to its thrust characteristics. Highly efficient LPEs that possess a specific impulse of 10^3 s and a thrust of 10^3 N are considered for use with LOTV interorbital missions (see Chap. 4).

Furthermore, one of the principal topics of this book involves the physical mechanisms that can help to develop a high-efficiency LPE. As well as for tested LPEs, several experiments with air-breathing laser propulsion were performed around the world using both RP and CW CO_2 lasers. Some results of these experiments are presented in Table 1.2.

The experiments demonstrated sufficient change of the laser-propulsion characteristics at varying of the laser power and jet engine design. For example, the data of the laser propulsion obtained in Ref. [10], where C_m is (50–60) $\times$ 10^{-5} N/W, differs from the results obtained in Refs. [17, 32], where C_m = (10–30) $\times$ 10^{-5} N/W. Approximately the same results of C_m were obtained in Ref. [34], C_m ~ (20–30)-$\times$ 10^{-5} N/W, where an e-beam-sustained CO_2 laser was used.

Table 1.2 Air-breathing laser-propulsion characteristics

Parameter	Dimension	[10]	[17]	[32]
Specific impulse C_m	N/W	50×10^{-5}	12.6×10^{-5}	30×10^{-5}
Pulse repetition rate f	Hz	100	10	100
Pulse energy E	J	0.25	10^3, or 650	4×10^3
Pulse width τ	μs	2.5	30	40
Average radiation power P	W	25	10^4	10^5
Average thrust T	N	5×10^{-3}	1.15	10
Acceleration a	m/s^2		2.3	
Specific thrust T/P	N/W	2×10^{-4}	1.15×10^{-4}	3×10^{-4}
Model mass m	kg	0.005	0.05	1.0
Jet engine design		Parabola, cone	Off-axis parabola	Bell
Beam concentrator		Parabola	Off-axis parabola	Matrix of parabolic mirrors

A considerable increase in the momentum of coupling coefficient C_m can be achieved if an auxiliary cylindrical extension is attached to the jet nozzle, as shown in Ref. [34]. In this case, the thrust increase is caused by the additional interaction of generated shock waves with the nozzle walls until the wave decays completely. A similar approach was used by Pirri in his theory of laser propulsion in space conditions [7].

A change in the geometry of the nozzle region where the laser beam interacts with a propellant can lead to a significant increase of coupling coefficient at an air-breathing mode of the laser propulsion. In that respect, a remarkable result was obtained in Ref. [33] for the engine model MLD-2, in which a beam concentrator in the form of a matrix of small parabolic mirrors was used to create a corresponding number of small shock waves that merge at a certain distance from the mirror (see Fig. 1.11). In the experiments, the coupling coefficient $C_m = 30 \times 10^{-5}$ N/W was obtained by using a 100-kW laser.

The first patent for the application of a matrix beam concentrator to laser propulsion was established by Dr. Jordin T. Kare in October 1992 [35]; however, Liukonen published his paper on MLD-2 a bit earlier, in July 1992 [33].

A promising way to achieve a large increase in the coupling coefficient C_m, for LAP is called "closed" laser ablation, whereby the interaction region of the laser radiation with a solid is confined in a volume, and an additional recoil effect is caused by this restriction [36, 37]. For example, a form of confined LAP based on multilayer (structured) optically transparent materials located on nontransparent solid substrates was studied. The confined LAP achieved a high coupling coefficient $C_m = 4.9 \times 10^{-3}$ N/W in experiments [33] with a pulsed Ne:YAG laser at a pulse duration of 85 ns and a peak power of 14 MW/cm^2. Figure 1.12 illustrates the dependences of C_m on a pulse energy density located on a multilayer target surface for single- ($C_{m(1)} = 386q^{-0.929}$) and double-pulse ($C_{m(2)} = 152q^{-0.580}$) impacts.

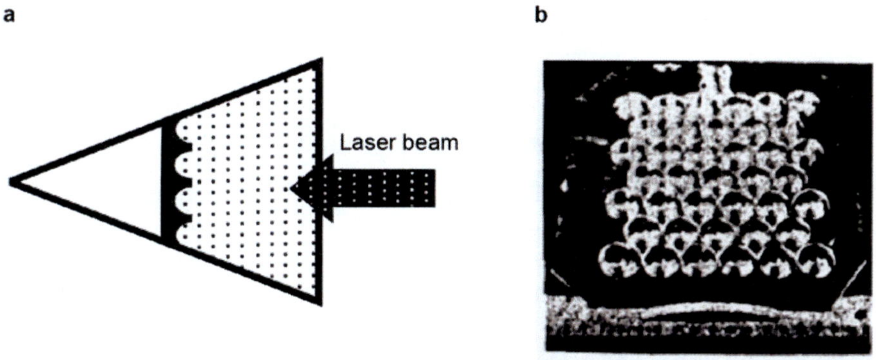

Fig. 1.11 General scheme and photo of a laser-propulsion engine with a matrix beam concentrator [33]

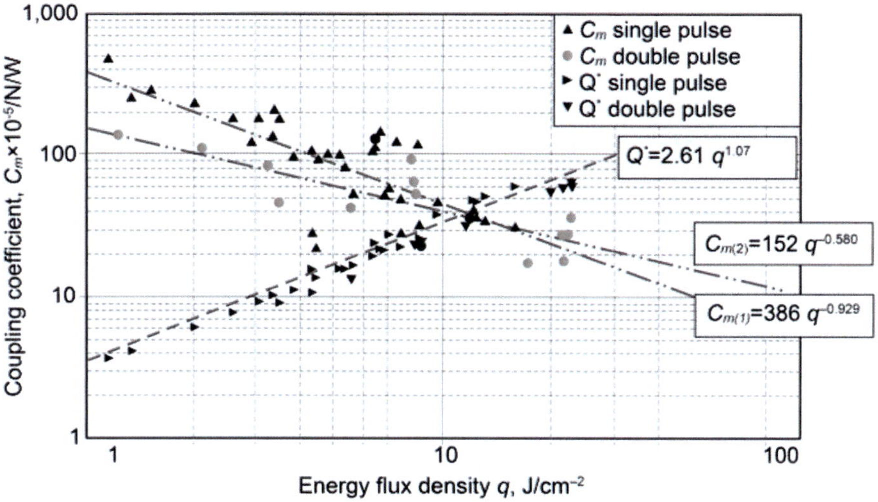

Fig. 1.12 C_m as a function of laser pulse energy density on the target surface [37]

Another way to increase the production efficiency of laser propulsion C_m the polymers of energetic materials as additional propellants, e.g., Delrin [34]. In experiments, a small amount of Delrin polymer was arranged on the tip of a thin needle that was placed at the focus of a parabolic mirror in such a way that it could be illuminated by a focused laser beam. The experiments showed a substantial increase in C_m (in factor 2) compared with the thrust produced without Delrin. The authors of the experiments proposed that the propulsion efficiency growth was the result of additional chemical reactions of Delrin vapor with atmospheric oxygen, leading to stored energy release.

Table 1.3 Types of laser propulsion

Propulsion type	Basic physical process	Principal researchers
Air-breathing laser propulsion (ABLP)	Air-breathing propulsion engines operating under RP laser radiation	Bunkin, Prokhorov, 1977
	Laser-pulsejet engine operating at space conditions	Pirri, 1977
	Subsonic laser jet engine on merging of separate shock waves	Tischenko, Appolonov, 2006
Laser ablation propulsion (LAP)	Ablation of solid and liquid materials	Kantrowitz, 1972; Phipps, 2000; Sinko, 2009
Laser-thermal propulsion (LThP)	Ignition of a continuous gas discharge in gas propellants injected into a combustion chamber	Komurasaki, 2002; Sattarov, 2010
Laser-chemical propulsion (LChP)	Combined mechanisms of laser propulsion	Liukonen, 1992; Bohn, 2002
Supersonic laser propulsion by using laser ablation jets		Rezunkov, Schmidt, 2014

Based on the brief overview of laser propulsion presented in this chapter, the techniques developed up to the present can be divided into five groups according to the underlying physical processes:

1. The characteristic feature of air-breathing laser propulsion (ABLP) is the dependence of the propulsion efficiency on the jet nozzle parameters and laser pulse features [10, 16, 17]. The coupling coefficient for ABLP engines is defined as a function of the *R*/*R*0 parameter for a local explosion theory, where *R* is the characteristic size of the engine nozzle, and *R*0 is a dynamic radius within the explosion theory. It is shown theoretically and experimentally that the maximal C_m can reach 60×10^{-5} N/W at a definite relation between *R* and *R*0.
2. Laser ablation propulsion (LAP) is based on the effects of evaporation of solid targets under high-power laser radiation, and the LAP coupling coefficient can reach (10–20) $\times 10^{-5}$ N/W [31].
3. LAP based on the usage of multilayer structured propellants provides the largest coupling coefficient up to $C_m = 10^{-2}$ N/W [37]. This technique is called confined laser ablation propulsion (CLAP).
4. The laser-chemical propulsion (LChP) technique combines the laser detonation energy and internal energy of energetic materials (propellants) to produce thrust. For example, the coupling coefficient reached the value of $C_m = 70 \times 10^{-5}$ N/W B was obtained when a Delrin polymer was placed in the engine nozzle using atmospheric air.
5. Laser-thermal propulsion (LThP) based on the ignition of a continuous gas discharge under CW laser radiation was also studied extensively.

The classification of propulsion techniques is summarized in Table 1.3. The corresponding types of developed LPEs are presented in the first column. The second column lists the physical processes underlying every laser-propulsion technique. In the third column, the main researchers are listed who proposed the theory and conducted the first experiments.

The last row in Table 1.3 includes laser-propulsion techniques that combine a few propulsion mechanisms to solve problems with laser-propulsion efficiency growth. One of the techniques is based on the interaction of an ablation jet with the supersonic flow in a nozzle. The technique would be efficient if it were applied to the supersonic mode of laser-propulsion production (see Chap. 5).

Only some principal examples of laser-propulsion techniques are presented in the table, but all of them are described in the following chapters, along with more complete references.

1.4 General Concepts of Laser Propulsion

In the 1970s–1980s, laser propulsion was considered an alternative rocket system that could reduce the cost of launching satellites into Earth orbits and create an energy-effective space vehicle launch site. In 1972, Prof. Arthur Kantrowitz proposed the use of LAP to launch satellites into space. He wrote the following:

> *One opportunity for dramatic improvement is to transmit the orbital energy from a laser on the ground to the ascending vehicle. Lasers can easily vaporize any material and it is possible to transfer energies to the vapor which is large compared to chemical energies. The evaporated material produces a jet which propels the vehicle, and the kinetic energy of the propulsive jet can be a large fraction of the energy absorbed from the laser.... The laser which is a dominant component remains on the ground.*

At the same time, a few concepts were developed to use high-power lasers located on Earth and in space to realize interorbital missions involving specially designed space vehicles, the so-called LOTV. All of these concepts must be considered as an attempt to predict the development of scientific trends of new space transportation systems.

The requirement of energy resources at these both technological fields of the laser-propulsion application is the one of the principal factors that will determine power of used lasers. Estimation of energy resources requires the parameter of energy cost C_e, which is determined as the total energy that is consumed to deliver a unit mass of the vehicle payload from the initial orbit to the final orbit.

The following subsections consider both of these concepts as applied to the laser launching of satellites into LEO and then GEO to determine the requirements for the laser-propulsion characteristics that would satisfy the minimal energy cost of each mission.

1.4.1 Launching Space Vehicles into Low Earth Orbits with Laser Propulsion

Following Kantrowitz, this subsection considers the launching of vehicles into LEOs with the LAP method. The impulse imparted to the vehicle in free space is assumed to equal the exhaust jet power; the following relation between the parameters defining LAP takes place [31]:

$$C_{\mathrm{m}} \cdot Q^{*} = g \cdot I_{\mathrm{sp}} \tag{1.3}$$

Here, $Q^{*} = E/\dot{m}$ [J·s/g] is a specific ablation energy, E is the laser pulse energy, and $\dot{m}$ is the ablated mass rate of an irradiated solid surface; $\dot{m}$ depends on the laser pulse energy, radiation wavelength, and thermo-physical and optical properties of the solid propellant (see Chap. 3). The optimization of launching a vehicle into LEO by using the laser ablative propulsion can be reduced by varying the C_{m}, Q^{*}, g, and I_{sp} parameters. The following statements may be applied to the optimization criteria:

- the energy cost of launching can be minimized by reducing the ratio of total laser radiation energy consumption E to payload mass delivered into the orbit m_{pl}, that is, $C_{\mathrm{e}} = E/m_{\mathrm{pl}}$.
- the vehicle payload mass m_{pl} can be maximized by putting it into orbit independent of the energy cost and the initial mass of the vehicle M_0.
- the launched payload mass of the vehicle can be maximized relative to the initial mass, that is, m_{pl}/M_0.

Thrust produced in vacuum is defined by the following ratio of payload mass to initial mass of the vehicle [38]:

$$\frac{m_{\mathrm{pl}}}{M_0} = \Gamma \exp\left[-\left(\frac{(\Delta V + g \cdot t_1)}{v}\right)\right] \tag{1.4}$$

where Γ is the correction coefficient magnitude, which does not exceed 1.02; ΔV is the vehicle velocity increment; t^{1} is the total vehicle flight time; and v is the average velocity of jet exhaust.

Based on this mathematical relation, the maximal payload mass delivered into low orbit and its maximal relative mass m_{pl}/M_0 can be achieved if the jet exhaust velocity goes to infinity $v \rightarrow \infty$, which means that $I_{\mathrm{sp}} \rightarrow \infty$. The energy cost can thus be expressed in the following form:

$$C_{\mathrm{e}} = v/C_{\mathrm{m}}\left[\exp\left(\frac{(\Delta V + g \cdot t_1)}{v}\right) - 1\right] \tag{1.5}$$

Figure 1.13 illustrates the dependencies of the total energy cost of launching a vehicle under vacuum conditions with a jet exhaust-specific impulse when the vehicle velocity increment is used as a parameter. In the figure, the top axis is the

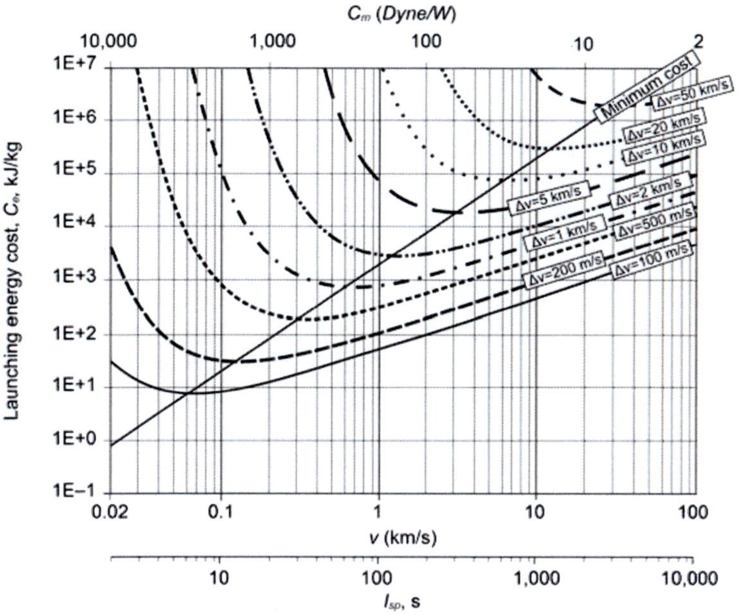

Fig. 1.13 Energy cost of launching a satellite with laser ablation propulsion [39]

momentum of the coupling coefficient C_m, and the bottom axis is a specific impulse I_{sp} of the jet exhaust. The vehicle velocity increment possesses a finite value. The figure shows that the optimal energy cost is reached when the coupling coefficient of the laser propulsion equals $C_m = (10\text{–}40) \times 10^{-5}$ N/W and the specific impulse varies as $I_{sp} = 1100\text{–}600$ s. The presented results, based on the assumption that the LAP has an efficiency $\eta = 1.0$, defines the relation between the propulsion characteristics in accordance with the ratio (1.4).

As the figure shows, the optimal launch mode of a vehicle into LEO depends on the relation between the coupling coefficient and the specific impulse of the LAP. The optimum between these two parameters is also defined by the required increment of the jet exhaust velocity. To reach LEO, this increment has to equal 7.5 km/s, which can be fulfilled if, e.g., $I_{sp} \sim 550$ s and $C_m \sim 40$ dyne/W. In this case, the energy cost will be 60 MJ/kg at a minimum! One of the last investigations into similar vehicles launched using efficient LAP showed that the minimum laser beam energy requirement per kg of payload delivered into orbit will be 80 MJ/kg [31].

1.4.2 Laser Propulsion for the Correction of LEO Satellites

One of the practical problems that could be solved by using laser propulsion is the correction of low Earth orbits of satellite [40]. Laser correction could maximize the operational duration of satellites at risk of reentering Earth's atmosphere with a high probability of returning to an operational orbit.

There are two orbital-correction scenarios. In the first case, the laser-propulsion system is switched on to produce a small thrust so that the satellite achieves a spiral orbit trajectory. The propulsion system operates continuously during the entire period that is needed for a satellite to reach into the operational orbit. The correction propulsion increases the vehicle velocity by a ΔV value so that the tangential component of its local velocity will coincide with the required orbital velocity.

In the second case, the laser propulsion is switched on during a short time period required only to impart additional velocity increment $\Delta V1$ to the vehicle and to enter with an elliptical trajectory. When the vehicle reaches operational orbit, the corrective propulsion system is switched on again to impart an accelerating or decelerating impulse corresponding to the velocity decrement $\Delta V2$, depending on its local velocity.

One option to correct the vehicle orbit depends on the particular operational feature of the laser-propulsion system. Satellite altitude correction via a spiral trajectory requires a laser operating for a long time because the vehicle has to orbit Earth a number of times, especially at the start of its motion. Moreover, a continuous supply of laser power to the vehicle requires that a number of lasers be rearranged along the vehicle flight trajectory because of the vehicle's high orbital velocity. Correction by an elliptical trajectory can be performed for a short time, and the vehicle moves by the shortest path. Moreover, only two laser systems need to be installed along the vehicle trajectory in opposite distances in the Earth atmosphere.

The laser-propulsion characteristics as applied to the satellite orbit correction can be estimated by analogous launching of vehicles into LEO. To be more precise, the following criteria apply for laser propulsion in free space:

1. minimum propellant (fuel) mass consumption while correcting a satellite orbit or maximal ratio of the satellite payload mass to satellite total mass A ($m_{\mathrm{pl}}/M_{\mathrm{st}}$), which is independent of the energy cost of the correction maneuver
2. minimum mission energy cost C_{e} of the vehicle maneuver

In Table 1.4, the correction maneuver parameters of low-orbital vehicles are presented when the vehicle moves along an elliptical trajectory from an initial orbital altitude of 200 or 280 km into a 300-km operational orbit.

Using the data listed in the table, technical requirements to the laser propulsion and laser characteristics can be defined, which will provide the satellite orbit correction at minimal energy costs and minimal propellant mass consumption. For that, let us use the well-known rocket equations in the following form [40]:

$$\Delta V_1 = -I_{\mathrm{sp}} \cdot g \cdot \ln\left(\frac{M_{\mathrm{st1}}}{M_0}\right), \text{ and } \Delta V_2 = -I_{\mathrm{sp}} \cdot g \cdot \ln\left(\frac{M_{\mathrm{st2}}}{M_{KA1}}\right) \quad (1.6)$$

Table 1.4 Satellite orbital correction that starts from two different orbits

Correction stage		Parameter	
		Velocity increment ΔV (m/s)	Maneuver duration (s)
200-km orbit	Lowering vehicle from operational orbit	n/a	158.68 days
	First correction stage	29.30	≤49.78
	Passive flight	n/a	2464.85 s (41.08 min)
	Second correction stage	30.11	≤76.07
280-km orbit	Lowering vehicle from operational orbit	n/a	65 days
	First correction stage	21.86	≤69.55
	Passive flight	n/a	1537.6 s (25.63 min)
	Second correction stage	19.23	≤76.07

where t_1, t_2 are operational duration of the laser-propulsion engine at the first and the second activation, and M_{st1} and M_{st2} are the vehicle mass after each correction cycle, determined as follows:

$$M_{st1} = M_0 - \dot{m} \cdot t_1, \text{and } M_{st2} = M_{st1} - \dot{m} \cdot t_2 = M_0 - \dot{m} \cdot (t_1 + t_2) \tag{1.7}$$

Substituting of the first equation from (1.6) into the other, and deducting one equation from other we have:

$$\exp\left(\frac{\Delta V_1 - \Delta V_2}{I_{sp} \cdot g}\right) = \frac{M_0 \cdot (M_0 - \dot{m} \cdot (t_1 + t_2))}{(M_0 - \dot{m} \cdot t_1)^2} \tag{1.8}$$

Denoting $A = \exp\left(\frac{\Delta V_1 - \Delta V_2}{I_{sp} \cdot g}\right)$, then we obtain:

$$\frac{A \cdot t_1^2}{M_0} \cdot \dot{m}^2 - (2A \cdot t_1 - (t_1 + t_2)) \cdot \dot{m} + (A - 1) \cdot M_0 = 0 \tag{1.9}$$

Or in a simpler form:

$$f(\dot{m} \tag{1.10}$$

Where the equation coefficients are expressed as:

$$a = A \cdot \frac{t_1^2}{M_0}, \quad b = 2t_1 \cdot A - (t_1 + t_2), \quad c = M_0 \cdot (A - 1) \tag{1.11}$$

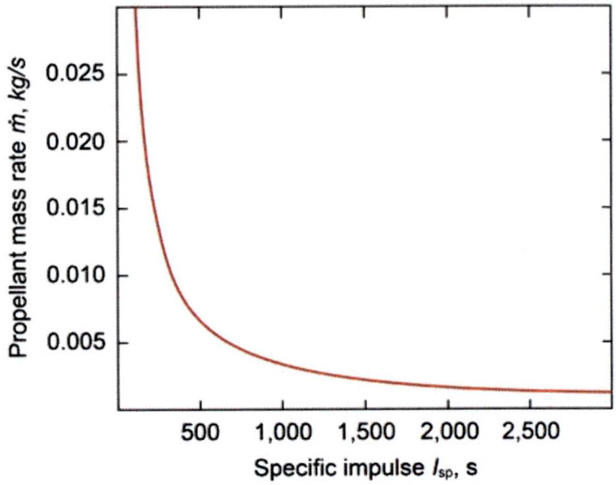

Fig. 1.14 Propellant mass rate as a function of specific impulse I_{sp}

As can be seen from the Eq. (1.10), the propellant mass consumption has an explicit minimum value at certain magnitudes of a, b, and c coefficients, which depend on such parameters as I_{sp}, t_1, t_2 in turn. The following equation may be applied to determine the mass consumption minimum:

$$\frac{d \cdot f}{d \cdot \dot{m}} = 2a \cdot \dot{m} - b = 0 \tag{1.12}$$

that corresponds to the following mass consumption:

$$\dot{m} = \frac{b}{2a} \tag{1.13}$$

at given parameters, namely, I_{sp}, t_1, t_2.

In Fig. 1.14, the dependence of propellant mass consumption rate on a specific impulse of the laser propulsion at calculated values of t_1 and t_2 is presented. It is seen that the maximum economy operational mode of the laser propulsion is provided if $I_{sp} > 10^3$ s.

The dependencies of the propellant concerning onboard storage and relative payload mass delivered into the operational orbit m_{pl}/M_{st} on the specific impulse of the laser propulsion are presented in Fig. 1.15. The calculations show that the specific impulse should be $I_{sp} > 10^3$ s if the vehicle has a mass of 100 kg.

To provide all-weather operational conditions of the satellite orbital correction, it is assumed to use airborne lasers. But in this case, such parameter as a general energy cost of the satellite orbital correction during the satellite orbital life will be a crucial one. This energy cost will define both laser power needed to transfer the satellite from one orbit into the other one and onboard propellant mass. The dependences of the energy cost on the specific impulse of the laser propulsion are presented in Fig. 1.16 for two cases such as: (a) when the laser-propulsion efficiency is 40%

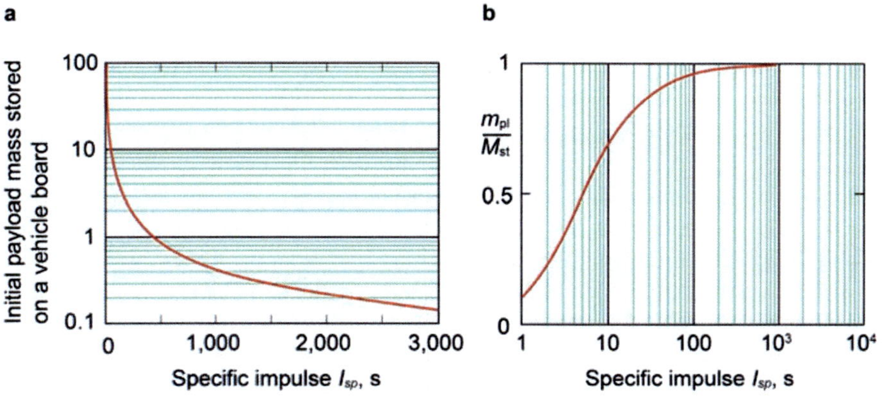

Fig. 1.15 The dependences of the (**a**) onboard propellant mass storage and (**b**) relative payload mass delivered into the operational orbit m_{pl}/M_{st} on a specific impulse of exhaust jet

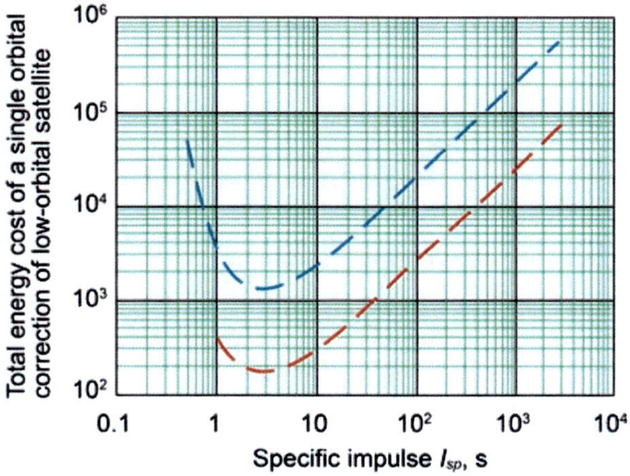

Fig. 1.16 The dependencies of the energy cost on generated specific impulse of the laser propulsion

(dashed line), and (b) the laser propulsion has maximal efficiency of 80% in the case of using laser-chemical laser propulsion (solid line). There is an apparent minimum in both curves at small specific impulse. But, large propellant mass consumption will take place if the specific impulse will be so small in these cases (see Fig. 1.15).

Thus, presented examples of possible scenarios of satellite orbit correction by using onboard laser-propulsion engine show a necessity of optimal combination of the laser propulsion and the laser radiation characteristics. It requires approximately the following $I_{sp} = 1000$ s, $C_m = 10^{-3}$ N/W at the propulsion efficiency of 70%, and the airborne laser power of a few tens of kW is required. This task is considered in Chap. 5 more detail.

1.4.3 Interorbital Missions of Space Vehicles with the Laser Propulsion

As well as for interorbital missions of the vehicle with onboard laser-propulsion engines, we can consider also two mission scenarios based on the Hohmann elliptical transfer orbit and on spiral trajectory of the vehicle flight. Let us consider the scenario of vehicle transfer from low Earth orbit (LEO) to geostationary Earth orbit (GEO) in the following way [38].

Orbital velocity of the vehicle at LEO with the orbital radius, R_{LEO}, is determined as follows:

$$V_{\mathrm{LEO}} = \sqrt{\frac{g \cdot R_{\mathrm{E}}^2}{R_{\mathrm{LEO}}}}, \text{ (or 7.8 km/s)} \tag{1.14}$$

where R_{E} is Earth radius. But at GEO, the vehicle velocity will be determined as:

$$R_{\mathrm{GEO}} = \sqrt[3]{\frac{g \cdot R_{\mathrm{E}}^2 \cdot T_{\mathrm{E}}^2}{4\pi^2}}, \text{ (or 3.08 km/s)} \tag{1.15}$$

where T_{E} is a period of Earth's rotation.

To transfer the vehicle into GEO by Hohmann trajectory, the laser-propulsion engine has to operate during a few hundred seconds and accelerate the vehicle up to perigee velocity that is equal approximately to $V_{\mathrm{per}} \approx 10.1$ km/s. For a moment, the vehicle will reach an apogee of its transfer trajectory with the velocity of $V_{\mathrm{apog}} \approx 1.6$ km/s. At this trajectory point, LPE must accelerate the vehicle up to the geostationary orbital velocity of $V_{\mathrm{GEO}} \approx 3.08$ km/s to keep vehicle motion into GEO.

The detailed calculations of the vehicle orbital mission show that the laser with a 1 MW output power must be used to launch a 1 kg payload mass into Earth orbit if the laser propulsion-specific impulse will be of 1200 s. To launch a 5-ton payload, the laser of 5 GW-power will be needed. It should be noted that the development of GW-power lasers isn't considered at present time in reality.

Nevertheless, the laser propulsion can be applied to transfer laser orbital vehicles at interorbital missions in the mode of a low thrust (100 N and less). For this purpose, various space vehicles with conventional rocket engines of low thrust are used, as it is described, for example, in Ref. [41]. Certain rocket boosters are listed in Table 1.5.

The orbital maneuvers of space vehicles with a large mass under the action of a low thrust (about~10 N) take a long time period exceeding a few tens of days. In this case, the energy costs of the orbital maneuvers grow proportionally to the flight time. To estimate efficiency of the laser propulsion as applied to this type of space vehicle missions, such parameter as the energy cost, C_{e}, of a unit mass of a payload transfer is used too. So that, the followings are determined:

Table 1.5 Characteristic parameters of the rocket systems of low thrust used for missions into GEO

Rocket booster	"Atlas-Centaurus"	"Saturn" (two rocket stages)
Initial mass, kg	3500	8000
Engine mass, kg	1500	1600
Fuel mass, kg	500	900
Specific power, kW/kg	0.04	0.18
Jet velocity, km/s	40	50
Specific impulse, s	4100	5100
Fuel mass rate, gram/s	0.07	0.20
Thrust, N	2.4	9.4
Initial acceleration, m/s^2	0.69×10^{-3}	1.17×10^{-3}
Transfer period, days	90	50
Payload mass, kg	1500	5500

Table 1.6 Main design characteristics of LOTV

Propulsion efficiency, η	0.4
Specific impulse, I_{sp}, s	500–2500
Laser power, P, kW	500
Radiation wavelength λ, μm	1.06; 0.53
Initial LOTV mass, M_0, ton	9
Transmitter telescope diameter, d_{tr}, m	1.5
Receiver telescope diameter, d_{rc}, m	5–40
LOTV initial orbit altitude, H_0, km	300

1. basic orbital characteristics of LOTV with a variable mass, namely, transfer period, maneuver velocity, payload mass, and so on
2. optimal characteristics of the laser-propulsion engine that is arranged on board of the LOTV
3. energy cost of the LOTV maneuver as diameter functions of the receiver telescope and specific impulse of the laser-propulsion engine

Because of a diffractive divergence of the laser beam generated by an airborne laser, one of the design limits of the laser-propulsion application efficiency for transferring the LOTV into high orbits is the limited optical aperture of the receiver telescope that has to collect the laser power and to direct it into the laser-propulsion engine. The limitations are caused also by the telescope primary mirror mass that grows with the mirror diameter in square proportion. For example, modern telescope mirrors have 1–5 kg/m^2 distributed mass [42], and the mirror of 60 m in a diameter would have a 12-ton mass. It is evident that decreasing the primary mirror diameter results in decreasing of collected radiation power and in decreasing of the total propulsion efficiency of the mission consequently.

As an illustration of the laser-propulsion capability, let us consider the LOTV system with parameters listed in Table 1.6. A high-power Nd:YAG laser is assumed

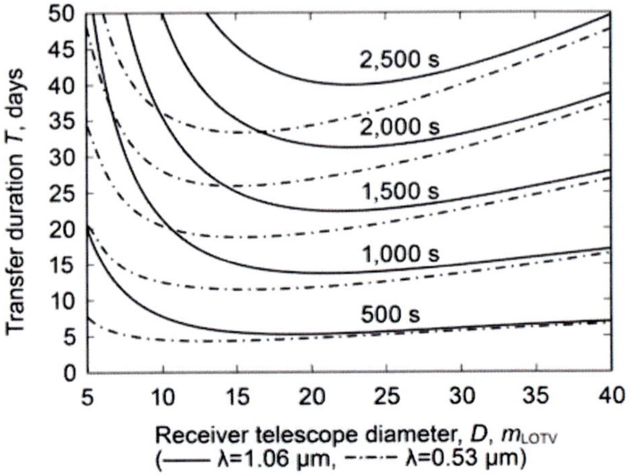

Fig. 1.17 Days quantity needed to transfer LOTV from LEO into GEO as a function of receiver mirror diameter at different specific impulse of the laser-propulsion engine

to be used, operating on one of the following two wavelengths, namely, 0.53 and 1.06μm. The choice of the laser wavelength influences on the laser beam divergence.

Because of the laser beam divergence, maximal orbit altitude where the laser power is still collected completely by the receiver telescope of LOTV is determined by following ratio $r_{\text{difr}} = \frac{d_{nm} \cdot d_{nk}}{\lambda} + R_{\text{E}}$ at the diffraction limit. Over high the orbit, the collected part power will be [43]:

$$H_{\text{difr}} = 1 - J_0^2 \left(\frac{\pi}{2} \cdot \frac{d_{\text{tr}} \cdot d_{\text{rc}}}{\lambda (r - R_{\text{E}})} \right) - J_1^2 \left(\frac{\pi}{2} \cdot \frac{d_{\text{tr}} \cdot d_{\text{rc}}}{\lambda (r - R_{\text{E}})} \right) \tag{1.16}$$

if $r \geq r_{\text{dif}}$, where J_0 and J_1 are Bessel functions. To be more accurate, we separate the total mass of the vehicle into two parts, namely, part 1 includes the vehicle operational mass consisting of the payload vehicle mass m_{pl} and stored propellant mass m_{pr}, $M_{\text{st}} = m_{\text{pl}} + m_{\text{pr}}$, and part 2 includes the receiver telescope mass, M_{mr}, that means $M_0 = M_{\text{st}} + M_{\text{mr}}$. In this, $M_{\text{mr}} = \rho_{\text{mr}} \times S_{\text{mr}}$ where ρ_{mr} is a distributed specular mass of the mirror, and S_{mr} is a total mirror square. As it follows from the definitions stated above, M_{st} parameter decreases over time during LOTV orbital mission.

Figure 1.17 illustrates the dependences of LOTV time period of transmission into GEO as a function of receiver mirror diameter for various values of the laser propulsion-specific impulse [44].

As one can see from Fig. 1.17, the LOTV transfer period depends on the telescope mirror diameter and its mass and on specific impulse of the laser-propulsion engine. And the dependencies have a minimum that changes a little bit if the specific impulse changes.

To determine optimal value of the specific impulse, let us consider the energy cost of this transfer scenario, C_{e}, as function of the same parameters. The results of corresponding calculations are presented in Figs. 1.18 and 1.19. In these cases, there are also evident minima of the dependencies C_{e} (D_{LOTV}) varying with different values of specific impulse I_{sp}.

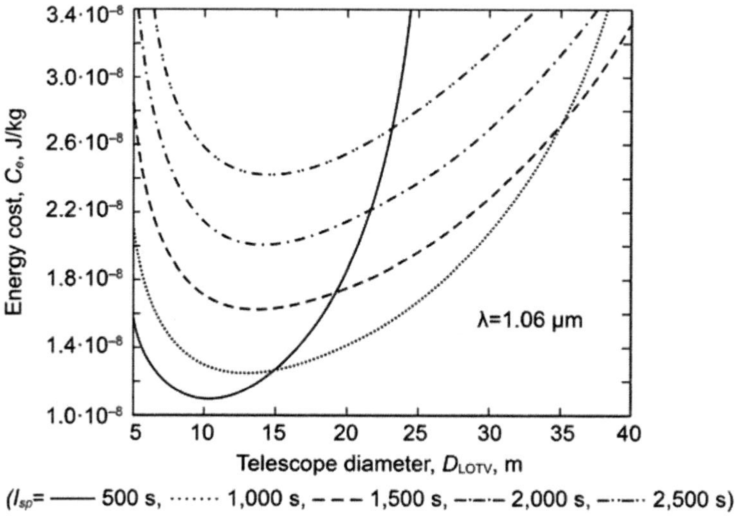

Fig. 1.18 Energy cost of the LOTV transfer mission, C_e, from LEO into GEO

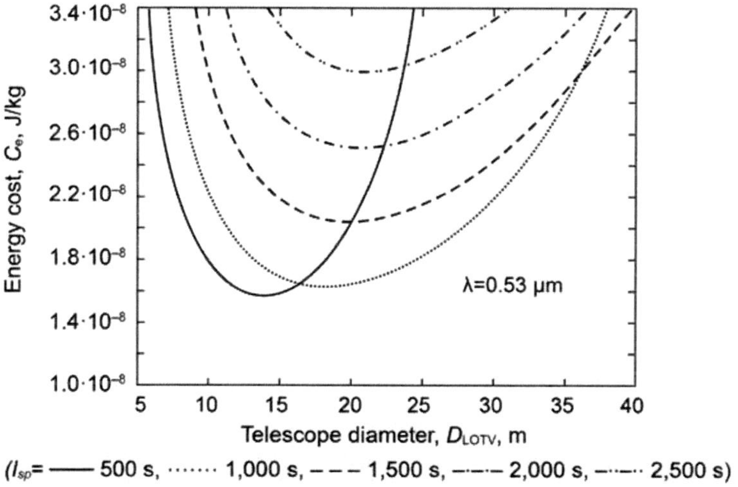

Fig. 1.19 Energy cost of the LOTV transfer mission, C_e, from LEO into GEO

For example, $C_e = 1.1 \times 10^8$ J/kg if the laser radiation is $\lambda = 0.53\mu$ and $C_e = 1.6 \times 10^8$ J/kg for $\lambda = 1.06\mu$ at $I_{sp} = 10^3$ s in both cases. For example, the energy cost of transferring vehicles by such rocket boosters as the IUS (Inertial Upper Stage) and "Centaurus" equal to $C_e = 5 \times 10^7$ J/kg and $C_e = 2 \times 10^8$ J/kg, respectively.

By using of additional requirements to the transfer scenario, we can define the LOTV device parameters more exactly. For example, one of the definitions of such requirements is that stored propellant mass does not exceed the payload mass. In this case, the optimal receiver telescope diameter will be $D_{LOTV} \sim 17$ m at a specific impulse of 10^3 s, but the minimal energy cost will be achieved if $D_{LOTV} \sim 19$ m. At $I_{sp} = 1500$ s, these parameters will be equal to $D_{LOTV} = 20$ m and $D_{LOTV} = 28$ m respectively.

1.5 Original Concepts of High-Power Laser Propulsion

A sufficiently original approach to the development of high-power laser propulsion is presented in [20], including such topics as (a) laser power transmission in the Earth's atmosphere, (b) production of the laser propulsion, and (c) lasers providing the LOTV orbital missions and its flight to the Moon. As critical parameters of the missions, propellant mass flow rate and specific impulse of LPE are used for the development of the laser-propulsion systems. Similar approach was considered at projecting of small vehicles to launch into LEO [45].

In that sense, it is very useful to consider some projects on the laser-propulsion application, which were examined in 1980s.

1.5.1 The "4P" Vehicles

One of the first proposals for the development of high-power laser propulsion was formulated as the basis of four key principles (4P), briefly formulated as payload, propellant, photons, and period [4]. The laser ablation propulsion is a basic technique of the thrust generation, which was assumed to be used in this case. So that, the double laser pulses had to ablate solid materials to initiate detonation waves in an ablated vapor. The first pulse had to ablate target surface, and the second pulse had to initiate a shock wave closely by the target surface as a result of laser-initiated breakdown of the vapor.

As well as for payload, the vehicle looks like a pyramidal cone, where bottom part is continuously faces the incident laser beam. Flat-bottom cone allows to provide a permanent thrust direction since the ablated vapor flow is always directed perpendicular to the irradiated plane.

As well as for propellants, different auxiliary substances were considered to be stored on the vehicle, namely, water, ammonium NH_3, ethane C_2H_6, carbon C, carbon dioxide CO_2, and lithium. It was indicated for the first time that the most efficient propellant for the laser propulsion would be chosen after investigating various chemical reactions running in plasma reactions induced by the high-power laser radiation.

Basic characteristics of the laser-propulsion system were estimated on the basis of 4P Conceptions, the results of which are presented in Table 1.7 [4].

We can see from the table that general efficiency of the 4P system is quite low, and it equals to 1.28% only. It should note the launching efficiency of the system can

Table 1.7 "4P" Laser-propulsion system efficiency

Laser	Laser power/power consumption	20% (for CO_2 and Free electron lasers) 25% (for solid-state lasers)
Laser power transmittance	Delivered laser power/laser power	80% at diffraction approach
η of LPE	Jet power/delivered laser power	40% is theoretical limit for combustion-wave mechanism of laser propulsion
Trajectory efficiency	Payload energy delivered to orbit/ $\int m_1 < v_1 >^2 dt/2$	20% as estimated
η of laser rocket	Payload energy delivered into orbit/ laser power	6.4% for CO_2-laser
Total η	Payload energy delivered into orbit/ electric power consumption	1.28%

be increased if the atmosphere air is used as a propellant during the vehicle flight in the atmosphere. Moreover, in the 4P principles, such problems as delivery of the high-power laser radiation through the Earth's atmosphere to a flying vehicle was not studied, that can effect on the system efficiency too. The problems of the high-power laser radiation delivery through the Earth's atmosphere and basic aspects of the development of adaptive laser system are considered in Chap. 7.

1.5.2 Lightcraft Technology Demonstrator (LTD)

Lightcraft technology demonstrator (LTD) developed for launching satellites into low Earth orbits is the most developed design of the vehicle with the laser propulsion [17, 24]. This original laser-propulsion technology was financed within the frames of the Strategy Defense Initiative program for 4 years, from 1985 to 1989 [6]. Shortly after, the LTD project was transformed into academic study program when the SDI program was closed. Then further development of the LTD run up against some problems of the laser-propulsion production at supersonic flight speeds of the vehicle in Earth's atmosphere. We will consider the problems of supersonic laser propulsion in Chap. 5.

There is one more problem of launching the Lightcraft by using ground-based lasers, which is caused by a necessity of shining the laser at the rear of the vehicle to produce a thrust. This problem is a result of the Lightcraft design. But, the launch trajectory looks complicated in this case, and it requires using special techniques (and algorithms) to propel the vehicle.

1.5.3 Laser Impulse Space Propulsion—LISP

One of the practical applications of the laser propulsion are assumed to be the Laser Impulse Space Propulsion (LISP) projects developed to remove space debris accumulated on low Earth orbits during a long period of the space satellites exploration [46–50]. LISP is based on the laser ablation propulsion effects, induced by high-peak power laser pulses when the debris objects illuminated (see Table 1.8).

In Table 1.8, the laser system protecting space stations from orbital debris is considered also on a basis of onboard lasers [47]. It is assumed that a thrust produced by the lasers will deflect the debris away from a space station. Estimates show that the laser power of the laser protection system has to be of 10 kW at the assumption that a single debris object will be appeared in the station protection region of a 10 km radius every hour. We assume that this type of the laser protection systems will be arranged on a board of space stations in a future.

For example, some experiments on recoil impulses generated under the action of Nd-laser pulses with a 0.3 ms pulse length were carried out by testing both dielectric and metal objects [51]. The purpose of these experiments was obtaining the mechanical recoil impulse dependence on the target material and on the laser beam incident angle with respect to the target's surface. Figure 1.20 illustrates the results of these experiments as functions of the coupling coefficient of the laser ablation propulsion on the laser power flux, q, at varying of the beam incident angle, φ, and changing the pressure of surrounding gas, p_0.

Recalculation of the data, presented in Fig. 1.20, made by taking into account the target square area, S, shows the following relation between the imparted impulse, J, and the laser power flux:

$$J/S = a_i(E_0/S)_i^n \tag{1.17}$$

where a_i and n_i are experimental parameters, in particular $a_\mathrm{i} = 10^{-8}–10^{-1}$ and $n_\mathrm{i} = 1–4$; E_0 is the laserpulse energy delivered to the target. Experimental data of the parameters a_i and n_i, listed in Table 1.9, allows to estimate specific recoil impulse of the solid targets at the laser pulse intensity reaching of 10^7 W/cm^2 and at different gas pressure, such as: 1 Pa and 10^5 Pa.

For example, the average coupling coefficient $C_\mathrm{m} = \mathrm{J/E} = 6 \times 10^{-5}$ N/W results in a recoil impulse of J/S = 2000 kg/ms obtained at the laser radiation intensity of 10 MW/cm^2.

Laser-propulsion techniques, as applied to remove space debris out of low Earth orbits, are discussed round the world currently. As a general approach to solve the debris problem, we can propose the projects developed by Claude Phipps and et al. [52, 53] for the further consideration. So that, all aspects of the laser impulse space propulsion are considered, including the laser ablation propulsion, high-power lasers, and debris objects.

Table 1.8 LISP components developed in the 1990s of the twentieth century

Project	ORION [46]	[47]	[48]	[49]	[50]
Description	Cleaning of near-Earth space	Cleaning of near-Earth space	Cleaning of near-Earth space	Cleaning of near-Earth space	Protection of space stations
Operation distance	1000 km	10 km	300 km	1000 km	10 km
System arrangement	Ground	Space	Ground	Ground	Space
Main laser characteristics					
Type of the laser	Nd-laser	KrF-laser	CO_2-laser	CF_3J-photo dissociate laser	Nd-laser
Laser power	20 kW	–	100 kW	–	–
Radiation wavelength	530 nm	249 nm	10.6μm	1.315μm	1.06μm
Pulse energy	20 kJ	10 kJ	1–3 kJ	up to 250 kJ	~2 kJ
Pulse repetition rate	1 Hz		100 Hz	–	
Pulse length	40 ns	50 ns	25 ms	50 ms	0.3 ms
Transmitter telescope					
Transmitter telescope diameter	6 m		18 m	–	1 m
Number of adaptive telescope elements	500		100	–	
Laser illuminator					
Average power	200 W			–	
Radiation wavelength	589 nm			1.315μm	
Pulse length	150 ns			50 ms	
Pulse repetition rate	10 kHz			10^6 Hz	
Number of lasers	4				
System parameters					
Orbital velocity increment $\|\Delta V\|$	235 m/s	30 m/s	–	100 m/s	200 m/s
Efficiency of the laser power delivery	25%		–		
Total period of the debris removal	4 years 1 cm < d < 100 cm	5–10 years			

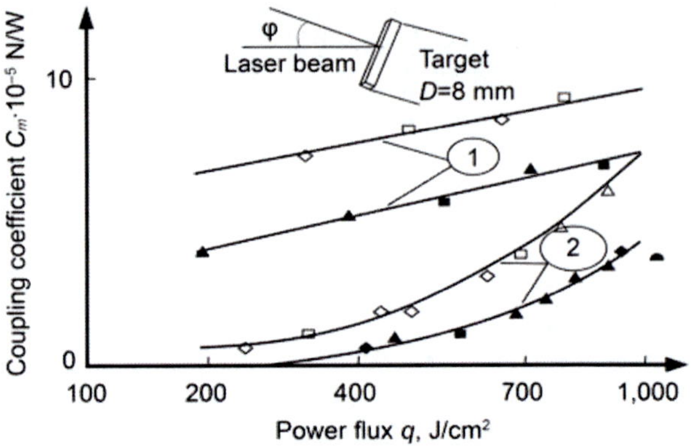

Fig. 1.20 Coupling coefficient versus laser power flux. $\varphi = 0$ (block circle, white circle), $\varphi = 30°$ (white up-pointing triangle, block up-pointing triangle), 45° (white diamond), 60° (white square, block square) and $p_0 = 1$ Pa (light symbols), $p_0 = 105$ Pa (dark symbols)

Table 1.9 Experimental parameters followed from (1.17)

Material	$p_0 = 1$ Pa		$p_0 = 105$ Pa	
	a_i	n_i	a_i	n_i
Cork	8.3×10^{-1}	1.0	1.45×10^{-1}	1.2
Graphite	1.75×10^{-1}	1.15	2.4×10^{-2}	1.4
Rubber	2.8×10^{-1}	1.17	2.9×10^{-2}	1.45
Ebonite	3.8×10^{-2}	1.5	8.3×10^{-4}	2.0
Titan	3.7×10^{-4}	2.0	4.5×10^{-5}	2.2
Steel	1.1×10^{-5}	2.55	1.5×10^{-7}	3.1
Aluminum	7.3×10^{-8}	3.25	7.8×10^{-8}	3.8

1.5.4 Principal Concept Design of the High-power Laser-Propulsion Systems

Development of HPLP systems assumes the usage of high-power ground-based or airborne lasers. In this case, the development of optic-electronic equipment specially designed for precise beaming of a flying vehicle through the Earth's atmosphere must be carried out. This task is complex even at modern progress of the laser engineering and technologies. And one vital point of the problem is a necessity of bringing a lead angle into the laser output beam to beam a vehicle that is orbited (see Chap. 7).

In our opinion, the option of preferable laser for the HPLP system will be defined by total efficiency of the laser-propulsion production under the laser radiation. The experiments on laser propulsion which have been made to the present time show the CO_2 lasers are preferable as a power source to launch vehicles into low Earth orbits. First of all, the CO_2 laser engineering systems are completely developed [54], and there are no technological problems to create a 10 MW CO_2 laser. The CO_2 lasers can also operate for a long time with a stable output. Secondly, the CO_2 laser radiation propagates in the Earth's atmosphere without high-power losses. Thirdly, numerous experiments on laser propulsion (see Chap. 2) demonstrated high efficiency of the thrust production with various designs of the laser-propulsion engines.

But at the same time, working distance of CO_2 laser is limited because of a large divergence of its radiation because of large radiation wavelength (10.6μm).

Therefore, solid-state lasers (SSL) with a shorter wavelength (1.06μm) are considered to be used more effectively. Moreover, there are a few projects demonstrating 100 kW from an SSL for orbital missions of space vehicles [55], including diode-pumped lasers and disk lasers. Development of SSL of large power is usually limited by a problem of removal thermal energy released in its active optical elements, solving of which leads to increasing of the laser device total mass and overall dimensions.

Another type of lasers which could be considered is the chemical HF and DF lasers operating in wavelength spectral range of 2.7–3.0μm and of 3.7–4.3μm at output power up to megawatt. Iodine lasers (Chemical Oxygen Iodine Laser—COIL) with a wavelength of 1.315μm can be allocated to a type of high-power lasers developed at the same time, which reached a few megawatts of power. COIL was tested in the experiments at real atmosphere conditions in 2010 [56].

Free electron lasers applied to the laser propulsion were considered also during the 1980s. Radiation wavelength of 3μm and average power up to 100 kW look attractive to develop proper HPLP system.

Perspectives of the laser-propulsion application in near-Earth space are based on two principal conditions which have to be fulfilled, namely: (1) technical ability of arrangement of three large-size optical reflectors onto GEO as a minimum to cover all low Earth orbits by the laser radiation and (2) development of long focus diffractive optics[15] decreasing a distributed mass the reflector mirrors up to 0.1–0.2 kg/m^2. These optical components propose using of the ground-based high-power lasers in the HPLP system. But in this case, the Earth's atmosphere will strongly restrict the laser power transfer to the space vehicles. First of all, the atmosphere limits a spectral range of laser wavelengths that could be used. Secondly, the atmosphere causes the laser beam divergence to be increased due to the refraction effects and turbulent phase distortions of the laser beam wavefront. To solve these problems, the development of linear and nonlinear adaptive laser systems has to be performed [57] (see Chap. 7).

General view of the HPLP system can be presented by a few optic-electronic equipment, including (see Fig. 1.21):

(a) high-power laser (source of laser power) based on "master oscillator-power amplifier" configuration
(b) laser beam control system that includes an adaptive optical system itself to compensate for the laser beam wave front distortions
(c) receiver-transmitter telescope system providing precise beaming of a vehicle with laser propulsion
(d) finally, space vehicle with the laser power receiver telescope and the laser-propulsion engine, providing maneuver of the vehicle

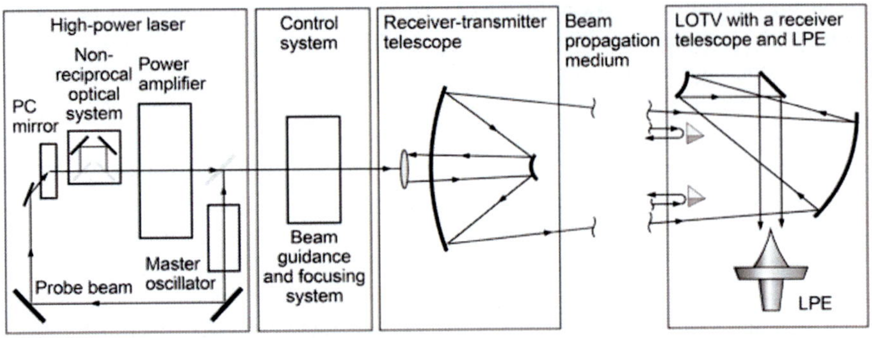

Fig. 1.21 General concept of the HPLP system

References

1. Tsiolkovsky, K.E.: Proceedings on rocket engineering. OboronGiz, Moscow (1947)
2. Durrani, M.: The Laser at 50. Phys.World. **23**, 16–62 (2010)
3. Askarian, G.A., Moroz, E.M.: Recoil impulse generated at laser-induced evaporation of solid targets—light-ablated pressure. J. Exper. Techn.Phys. (Rus.). **43**, 2725 (1962)
4. Kantrowitz, A.: Propulsion to orbit by ground-based lasers, pp. 74–76. Astron & Aeron (1972)
5. Bunkin, F.V., Prokhorov, A.M.: Laser power source application to produce a thrust. Adv. Phys. Sci. (Rus.). **119**, 425–446 (1976)
6. Myrabo, L.: Advanced beamed and field propulsion concepts final report. Principal investigator. Department of Mechanical Engineering, Rensselaer Polytechnic Institute, Troy, New York (1983)
7. Pirri, A.N., Weiss, R.F.: Laser propulsion. AIAA Pap. No 72-719. Boston, MA (1972)
8. Pirri, A.N., Simons, G.A.: The fluid mechanics of pulsed laser propulsion. AIAA J. 15, 16 (1977)
9. Barchukov, A.I., et al.: Laser air-breathing engine. Lett. J. TEPh. **23**, 273–240 (1976)
10. Ageev, V.P., et al.: Laser air-breathing jet engine. J. Quant. Electron. (Rus.). **4**, 2501–2513 (1977)
11. Myrabo, L.N., Ing, D.: The future of flight. A Bean Book, New York (1985)
12. Myrabo, L.N.: A concept for light-powered flight. AIAA Pap. 82-1214 (1982)
13. Zeldovich, Y.B., Raizer, Y.P.: Physics of shock waves and high-temperature gas-dynamic processes. Nauka, Moscow (1966)
14. Myrabo, L.N.: Laser propelled vehicle. US Patent 6,488,233, 3 Dec 2002
15. Rather, J.D.G.: Ground to space laser power beaming: mission, technologies, and economic advantages. Beamed Energy Propulsion. AIP Conf. Proc. **664**, 37–48 (2003). https://doi.org/10.1063/1.1582094
16. Jones, L.W.: A brief history of laser propulsion at the Marshall Space Flight Center. Beamed Energy Propulsion. AIP Conf. Proc. **664**, 61–78 (2003). https://doi.org/10.1063/1.1582096
17. Myrabo, L.N.: Brief history of the Lightcraft Technology Demonstrator (LTD) project. Beamed Energy Propulsion. AIP Conf. Proc. **664**, 49–60 (2003). https://doi.org/10.1063/1.1582095
18. Kare, J.T.: Laser Launch—the second wave. Beamed Energy Propulsion. AIP Conf. Proc. **664**, 22–36 (2003). https://doi.org/10.1063/1.1582093
19. Niino, M., et al.: LE-NET and multi-purpose laser propulsion system. AIAA Pap. No. **2176** (2002). https://doi.org/10.2514/6.2002-2176
20. Kare, J.T.: Vehicle and system concepts for laser orbital maneuvering and interplanetary propulsion. Beamed Energy Propulsion. AIP Conf. Proc. **664**, 662–673 (2003). https://doi.org/10.1063/1.1582151

21. Nakai, S.: Progress of power lasers and its application to space. Beamed Energy Propulsion. AIP Conf. Proc. **702**, 3–22 (2003). https://doi.org/10.1063/1.1582151
22. Rezunkov, Y.A.: Laser propulsion for LOTV space missions. Beamed Energy Propulsion. AIP Conf. Proc. **702**, 228–241 (2003). https://doi.org/10.1063/1.1721003
23. Lubin, P.: A roadmap to interstellar flight. UC Santa Barbara submitted JBIS April 2015 lubin@deepspace.ucsb.edu
24. Myrabo, L.N., Lewis, J.S.: Lightcraft. Flight handbook. (Hypersonic flight transport for an era beyond oil). Apogee Books (2009)
25. Koroteev, A.S.: New phase of space-rocket technology development. Proc. Mosc. FTI. **3**, 40–44 (2011)
26. Orbit-raising and maneuvering propulsion: research status and needs. Caveny, L. H. (ed.) Progr. Austron. & Aeron. 89 (1986)
27. Dreglin, A.F., et al.: Experimental and theoretical study of laser propulsion engine developed on the basis of CW optical discharge. News Univers.: Aeron. Techn. No 4 (2010)
28. Danilychev, V.D., Zvorykin, V.D.: Experimental study of radiative gas-dynamical processes generated under the high power laser pulses. Proc. FIAN (Rus.). **142**, 117–171 (1983)
29. Laser ablation and its applications. Phipps, C. R. (ed.). Ser.: Springer Series in Optical Sciences, p. 129 (2007)
30. Sinko, J.E., Phipps, C.R.: Modeling CO^2 laser ablation impulse of polymers in vapor and plasma regimes. Appl. Phys. Lett. **95**, 105–131 (2009). https://doi.org/10.1063/1.3234382
31. Phipps, C.R., et al.: Review: laser-ablation propulsion. J. Prop. Pow. **26**, 609–637 (2010). https://doi.org/10.2514/1.43733
32. Apollonov, V.V., Tischenko, V.N.: Laser propulsion on the basis of the effect of merching shock waves. J. Quan. Electron. (Rus.). **36**, 673–683 (2006)
33. Liukonen, R.A.: Efficiency of energy transfer into a recoil impulse for laser propulsion engine. Lett. JTP (Rus.). **18**, 81–85 (1992)
34. Schall, W.O., et al.: Lightcraft Experiments in Germany. High-Power Laser Ablation III. Proc. SPIE. **4065**, 472–481 (2000). https://doi.org/10.1117/12.407369
35. Kare, J. T.: Reflector for efficient coupling of a laser beam to air or other fluids. US Patent 5,152,135, 6 Oct. 1992
36. Fabbro, R., et al.: Physical Study of Laser-Produced Plasma in Confined Geometry. J. App. Phys. **68**, 775–784 (1990). https://doi.org/10.1063/1.346783
37. Yabe, T., et al.: Laser-driven vehicles—from inner-space to outer-space. High-Power Laser Ablation. Proc. SPIE. **4760**, 1–12 (2002). https://doi.org/10.1007/s00339-003-2125-5
38. Phipps, C.R., et al.: Transfers from Earth to LEO and LEO to Interplanetary Space using Lasers. Acta Astron. **146**, 92–102 (2018). https://doi.org/10.1016/j.actaastro.2018.02.018
39. Phipps, C.R., Reilly, J.P., Campbell, J.W.: Optimum parameters for laser launching objects into low Earth orbit. Las. Powr. Beam. **18**, 661–695 (2000). https://doi.org/10.1017/S0263034600184101
40. Bronstein, A.: Laser propulsion system for space vehicles. J. Prop. Pow. **14**, 61–268 (1998). https://doi.org/10.2514/2.5377
41. Garrison, P., Stoky, G.F.: Space rocket engines of the future. J. Prop. Pow. **4**, 520–525 (1988)
42. Next generation of Space Telescopes. Web-site: ngst.gsfc.nasa.gov
43. Born, M., Wolf, E.: Principles of optics. Nauka, Moscow (1973)
44. Rezunkov, Y.A., Pakhomov, A.V.: Perspective in-space laser propulsion demonstrator mission. Beamed Energy Propulsion. AIP Conf. Proc. **702**, 205–215 (2003). https://doi.org/10.1063/1.1721001
45. Kare, J. T.: Executive summary of the SDIO/DAPRA Workshop on Laser Propulsion. Proc. LLNL, CONF-860778 (1986)
46. Phipps, C.R., et al.: Clearing of near-Earth space debris using 20 kW, 530 nm, Earth-based, repetitively pulsed laser. Las. Part. Beam. **14**, 1–44 (1996). https://doi.org/10.1017/S0263034600009733

47. Kuznetzov, L.I., Yarygin, V.N.: Laser-rocket technique to clean near-Earth space against space debris. J. Quant. Electron. (Rus.). **21**, 600–602 (1994)
48. Belousova, I.M., et al.: On the effective transportation of laser radiation through the atmosphere. Proc. SPIE. **2771**, 252–262 (1995). https://doi.org/10.1117/12.238073
49. Basov, N.G., et al.: Laser system for observation and removal of space debris. Proc. SPIE. **3574**, 437–439 (1998). https://doi.org/10.1117/12.334467
50. Kuznetzov, L.I., Savichev, V.Y., Tikhonov, N.N.: Laser protection system for space stations against small-size debris. J. Quant. Electron. (Rus.). **25**, 372–376 (1998)
51. Kuznetzov, L.I.: Recoil impulse produced at illuminating of solid targets in the mode of developed evaporation. J. Quant. Electron. (Rus.). **20**, 1191–1195 (1993)
52. Phipps, C.R., Michaelis, M.M.: LISP: laser impulse space propulsion. Las. Par. Beam. **12**, 23–54 (1994). https://doi.org/10.1017/S0263034600007217
53. Phipps, C.R., et al.: Removing orbital debris with pulsed lasers. AIP Conf. Proc. **1464**, 1283–1300 (2012). https://doi.org/10.1063/1.4739901
54. Bulaev, V.D., et al.: Experimental laser facility based on the high-power repetitively-pulsed E-beam sustained CO_2 laser: Beamed Energy Propulsion. AIP Conf. Proc. **766**, 361–372 (2004). https://doi.org/10.1063/1.1925157
55. Sherstobitov, V.E., Leshchev, A.A., Soms, L.N.: Nonlinear optics techniques of laser beam control for laser propulsion application. Beamed Energy Propulsion. AIP Conf. Proc. **664**, 620–633 (2002). https://doi.org/10.1063/1.1582148
56. Bilimana, K.W., Horwitz, B.A., Shattucka, P.L.: Airborne laser system common path/common mode design approach. Proc. SPIE. **3706**, 196–203 (1999). https://doi.org/10.1117/12.356958
57. Higgs, C., et al.: Atmospheric compensation and tracking using active illumination. Lincoln Lab. J. **11**, 5–26 (1998)

Chapter 2
Basic Gas-Dynamic Theories of the Laser Air-Breathing and Rocket Propulsion

Abstract The interaction of laser radiation with gases and plasma is considered theoretically and experimentally as one of the main processes defining production of efficient laser propulsion in the air-breathing and rocket regimes. Laser breakdown of gases is examined as a phenomenon initiating laser power input into plasma ignition. To determine such characteristics of multi-ionized plasma as degree of ionization and its thermodynamic functions as a dependence on the initial density of gas and its temperature, semi-empirical model of the plasma is applied. The model allows specifying the thermodynamic functions of the plasma and estimating efficiency of the thrust production under the high-peak power of laser pulses.

The experimental data on laser-induced plasma are examined to determine efficiency of laser power input into the plasma on basis of the local explosion theory. At that, shock waves propagating out of the plasma region are assumed to possess the information on specific energy that was released in the breakdown region.

Keywords Air-breathing propulsion · Rocket propulsion · "Explosive" mechanism of plasma ignition · Laser breakdown of gaseous propellants · "Pulsejet" mode of thrust production · Local explosion theory · Laser mover · Beam array concentrator · Bell type vehicle · Gas-dynamics of laser propulsion · Multi-ionized plasma · Equilibrium plasma · Degree of plasma ionization · Raizer's theory · Boltzmann distribution function · Non-equilibrium plasma · Saha's equations · Inverse Bremsstrahlung effect · Non-isentropic gas flow · Parabolic nozzle · Thrust impulse

2.1 Introduction

In this chapter, basic gas-dynamic theories of laser propulsion are examined as applied to various designs of a laser-propulsion engine. The accepted approach to laser-propulsion theory is directed to achieve two purposes. The first is aimed to explain principal effects which are responsible for laser-propulsion production on the basis of theory developed at earlier stages of the laser-propulsion history in the beginning of the 1970s. The second determines positive ways of laser-propulsion engineering directed to efficient production of a thrust that corresponds to coupling

Y. A. Rezunkov, *High Power Laser Propulsion*, Springer Series on Atomic, Optical, and Plasma Physics 116, https://doi.org/10.1007/978-3-030-79693-8_2

coefficient of $C_{\mathrm{m}} \sim 10^{-3}$ N/W and production efficiency of 70%. Achieving the second purpose is related with a true understanding of gas-dynamic processes running under laser radiation interaction with a matter as well as with plasma ignited under the action of high-peak power laser pulses. It should be noted also we consider two modes of the laser-propulsion production, namely, repetitively pulsed (RP) and continuous wave (CW) modes which correspond two modes of laser radiation generation.

RP Laser-Propulsion Engine (RP LPE) operation based on "explosive" mechanism of plasma ignition is considered at first. In this case, thrust is produced by the action of high-power laser pulse on a propellant that results in (a) optical (laser) breakdown of gaseous propellant in the case of air-breathing propulsion, or (b) laser ablation of solid material surface in the case of rocket mode of thrust generation. In the case (a), the input of laser power into the gas is realized due to interaction of the laser radiation with the plasma generated. Plasma can also be generated in a vapor of ablated material, and this mechanism of laser propulsion should be referred to a combined type of thrust production. The processes of laser radiation interaction with plasma as applied to laser propulsion are considered in Sect. 2.3, basing on the theory developed by Prof. Yu. P. Raizer.

As a rule, there is no separate explosive chamber in RP LPE that operates as a special device in which the propellant's thermal energy is released. In RP LPE, gaseous propellant is injected directly into the engine nozzle that operates as a concentrator of laser beam at the same time. Usually, various types of parabolic mirrors are used as the beam concentrator. The laser-induced breakdown of the gas, a strong shock wave is initiated in a focus region of the mirror to increase a gas pressure inside the nozzle when the shock wave moves inside the nozzle. The pressure increase is accepted as additional force producing a thrust.

Laser ablation propulsion is realized because of fast evaporation of a target surface due to absorbing the laser power and bypassing a liquid stage of the evaporation. High-performance characteristics of the laser ablation propulsion are achieved if the beam concentrator is combined with a jet nozzle. The process of thrust production is complicated significantly if the pulse peak power is so strong that it initiates an ionization of evaporated materials and ignition of plasma (see Chap. 3).

So, to determine the application perspectives of laser propulsion, we have to consider some basic physical phenomena which define the production efficiency of laser propulsion. And as a first step of the consideration, the following processes are examined, namely:

- laser radiation interaction with an ignited plasma
- gas-dynamic phenomena accompanying the interaction of laser radiation with a plasma
- influence of laser-propulsion engine device on air-breathing gas-dynamics

To be more specific, we consider the listed processes as applied to launching of satellites in to low Earth orbits. Figure 2.1 shows a schematic diagram of a laser-launch of a vehicle as three consecutive stages of its flight from ground to space.

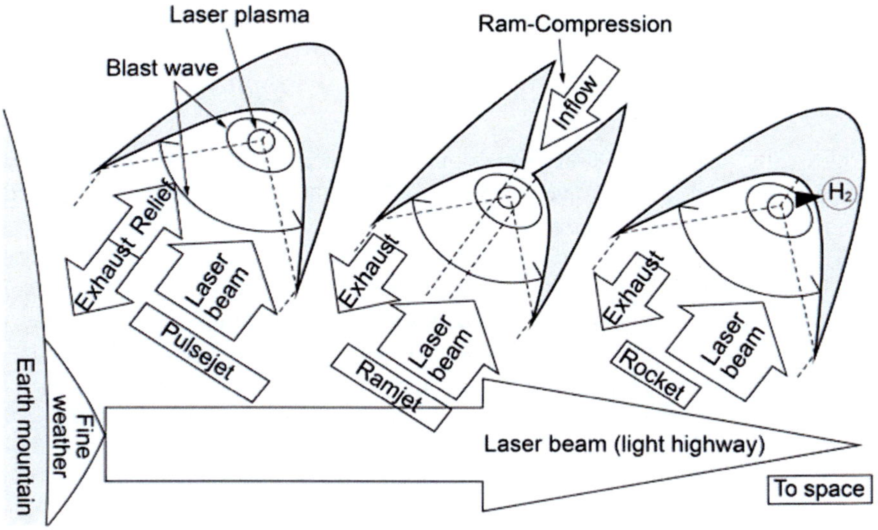

Fig. 2.1 Basic regimes of thrust production via laser propulsion of launching of vehicles to LEO [1]

First stage is determined as a "pulsejet" mode of thrust production due to pulsed input of laser power into a limited volume of a gas propellant in a focus region of beam concentrator. In the second stage named as "air-breathing" mode, incoming atmosphere air is preliminary compressed at the inlet section of the vehicle jet engine, and then the air comes to a focus region of beam concentrator where the laser power is released due to laser-induced breakdown of the air. And finally, third stage as a "rocket" one is that occurs when the vehicle leaves upper atmosphere and flies in a space. In this case, the propellant stored on the vehicle board is used to produce the thrust.

It should be also noted that the second stage involves both in the subsonic and supersonic modes of the vehicle flights when gas flow at the inlet of the engine nozzle is a supersonic one. Similar effect of formation of supersonic flow inside the nozzle occurs during the third stage of the vehicle flight.

Therefore, the consideration of both subsonic and supersonic modes of thrust production is a crucial factor of the laser propulsion to be applied in a future.

The present chapter devoted to the problems of (a) modeling of gas-dynamics in the three stages of vehicle launch to a space, (b) ignition of laser plasma in a working medium (propellant), and (c) processes of laser power release in ignited plasma, which define an efficiency of thrust production in every mode of laser-propulsion production.

2.2 Gas-dynamic Theory of Laser Propulsion

2.2.1 *Specific Properties of Pulsejet Laser Propulsion*

Explosive mechanism of laser propulsion looks especially attractive for flights of vehicles (aircrafts) in the Earth atmosphere [2, 3] because atmosphere air can be used as a propellant to produce a thrust both in the pulsejet and ramjet modes.

The first and most credible theory of the pulsejet laser propulsion is stated by Prokhorov in Ref. [2], issued in the 1970s. Prokhorov's theory is based on a local explosion theory [4], when an instantaneous input of laser power into a closed gas volume is arranged, and thrust is produced due to conversion of the heated gas energy into a generated shock wave, which accelerates gas in a nozzle.

Let us consider basis of the explosion theory and its principal outcomes. To describe the explosive processes in a gas, it is sufficient to determine thermo-physical characteristics of the heated gas and shock wave that propagates from the heated gas space. Consequently, the following is assumed:

(a) The breakdown gas space is smaller than a nozzle volume and the shock wave has propagated at a distance exceeding the breakdown region.
(b) Initial inner energy of the heated gas space can be neglected as compared with the energy released under the action of laser power.

In that case, such parameters of the gas behind the shock wave as gas density, ρ, pressure, p, and velocity, v, can be determined by following relations [5]:

$$\rho_1 = \frac{\gamma+1}{\gamma-1}\rho_0; \quad \rho_1 = \frac{2}{\gamma+1}\rho_0 \cdot D^2; \quad U_1 = \frac{2}{\gamma+1}D; \tag{2.1}$$

where $D = \frac{dr}{dt}$ is a propagation velocity of the shock wave front, index 0 denotes the gas parameters in front of the wave and subsequent 1 denotes the same parameters behind the wave; $r = \rho_0\left(\frac{E}{\rho_0}\right)^{1/5} \cdot t^{2/5}$ where r is a radius of spherical shock wave and $r \approx \left[\left(\frac{E}{\rho_0}\right) \cdot 2\Omega I\right]^{1/3} t^{2/3}$ for the wave propagating within a spatial cone with a Ω space angle of a top, E is the "explosive" energy, r^x is a radius of conjectured mass source of a propellant. Gas volume moving behind the wave is described by R radius at some moment of time, t.

Gas density ρ_1, pressure ρ_1, and velocity U_1 right behind the shock wave may be expressed as functions of non-dimensional coordinate $\eta = {}^{x}/_{r(t)}$ at some moment of time assuming correctness of the assumptions listed above. In this case, mechanical impulse of the gas volume accelerated by the wave is determined as follows [2]:

$$I = \int_0^r U_1 \cdot \rho \cdot r^2 \cdot \Omega dr = \int_0^t p_1 r^2 \Omega dt \tag{2.2}$$

where upper integration limits are defined by a time period during which the shock wave losses its power completely that corresponds to $R_{\Delta}(t_{\text{B}}) = t_{\text{B}} \times u_{\text{c}}$. Here, u_{c} is a maximum velocity of gas outflow in a nozzle exit section, and t_{B} is a time interval corresponding to the shock wave dissipation. So, the breakdown gas region can be considered as a point volume if the gas flow characteristics are analyzed at a distance exceeding the characteristic size of the volume by two to three times.

The simplification that is used in the local explosion theory is an assumption of instantaneous conversion of laser power input into a gas. Comparison of momentum coupling coefficient C_{m} obtained experimentally by varying the ratio of energy release time interval, τ, to laser pulse length τ_0, at a wide range of the ratio change, namely, from 10^{-4} up to 10^{-2}, show that the laser breakdown effect can be considered as a local instantaneous explosion. That means that if laser pulse energy is $E = 1$ kJ then the ratio $\tau/\tau_0 < 10^{-2}$ results in $\tau < 10^{-5}$ s, and if $E = 100$ J then $\tau < 3 \times 10^{-6}$ s.

The explosive theory of pulsejet propulsion allows estimation of a recoil impulse imparted to a nozzle by the initiated shock wave, as follows. The nozzle can be formed by various geometry shapes, namely, cone, paraboloid, off-axis parabolic, and so on. For the nozzle of a conical type, the theory gives the following dependences of recoil impulse on the laser pulse energy and geometric characteristics of a pressure receiver (that is a nozzle) (see Fig. 2.2) [3]:

$$
\begin{aligned}
I(R) &= 2\pi \sin^2\theta \frac{R_{\Omega}^3 \cdot \rho_0}{c_0} \int_0^{R/R_{\Omega}} x J_{\rho}^{(1)}(x)\mathrm{d}x \\
&= \pi \sin^2\theta \frac{R_{\Omega}^3 \cdot \rho_0}{c_0} \left(\frac{R}{R_{\Omega}}\right)^2 J^{(1)}\left(\frac{R}{R_{\Omega}}\right)
\end{aligned}
\tag{2.3}
$$

$$
\begin{aligned}
I(R)/E &= \frac{4\pi}{c_0}(1 + \cos\theta) \int_0^{R/R_{\Omega}} x J_{\rho}^{(1)}(x)\mathrm{d}x \\
&= \frac{2\pi}{c_0}(1 + \cos\theta) \left(\frac{R}{R_{\Omega}}\right)^2 J^{(1)}\left(\frac{R}{R_{\Omega}}\right)
\end{aligned}
\tag{2.4}
$$

where $J^{(1)}$ is Bessel function of first order with (R/R_{Ω}) as an argument, and θ is half of the cone tip angle.

It follows from these formula, the recoil impulse and coupling coefficient of the explosive propulsion will depend on two parameters, namely, R/R_{Ω} и θ. Particularly, maximum C_{m} will be in the case when apex angle of cone nozzle leads to zero at an R/R_{Ω} parameter specified. The $x^2 J^{(1)}(x)$ function in (4) has a maximum at $x \sim 0.3$, and maximal C_{m} is of 57 dynes/W in this case.

To estimate theoretically the coupling coefficient of a parabolic nozzle, one can assume the laser breakdown of a gas is initiated in the paraboloid focus and the nozzle exit angle is seen from its top at a Ω solid angle. A thrust is produced due

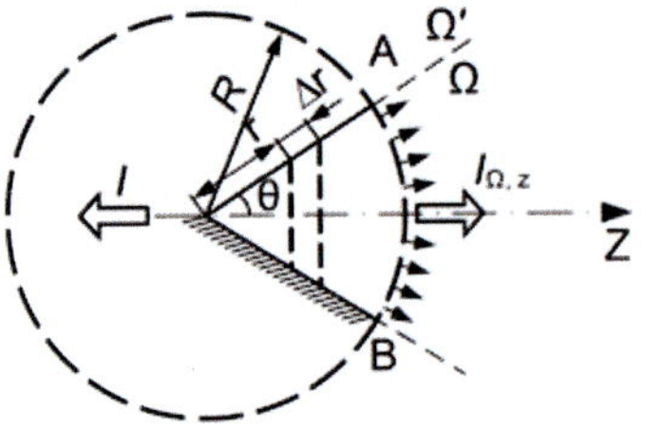

Conical nozzle

$$I(R)/E = \frac{4\pi}{c_0}(1+\cos\Theta)\int_0^{R/R_\Omega} xJ_p^{(1)}(x)dx =$$

$$\frac{2\pi}{c_0}(1+\cos\Theta)\left[\frac{R}{R_\Omega}\right]^2 J^{(1)}\left(\frac{R}{R_\Omega}\right)$$

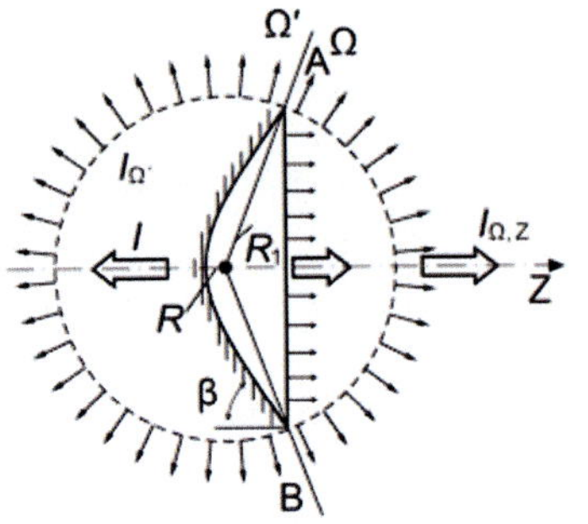

Parabolic nozzle

$$I/E = (4\pi/c_0)\left[(R_1/R_0)^2 - (R/R_0)^2\right]J^{(1)}(R_1/R_0),$$

$$R_1 = R\left[1+(4R/D)^{-2}\right]$$

Fig. 2.2 Pressure distribution models in two types of jet nozzles

propagation of generated shock wave in the paraboloid and accelerates the gas behind the wave. In this case, exhaust jet impulse will be:

$$I_{jnx} = 2\pi(1+\cos\theta)R_1^2 I^{(1)}(R_1) \tag{2.5}$$

By the analogy with a pressure receiver in the form of a cone and taking into account the fact that the paraboloid length along its axis is equal to $L = D^2/16R$, the coupling coefficient for the parabolic nozzle is expressed as:

$$I/E = (4\pi/c_0)\left[(R_1/R_0)^2 - (R/R_0)^2\right]J^{(1)}(R_1/R_0), \quad R_1 = R\left[1+(4R/D)^{-2}\right] \tag{2.6}$$

It follows from (2.6), the coupling coefficient of elongated paraboloid goes to a valve corresponding to the cone nozzle, with an apex angle approaching to zero, $\theta \to 0$.

It follows from the explosion theory that the coupling coefficient of pulsejet laser-propulsion engine operating due to a local explosion of a gaseous propellant will not exceed 60 dyne/W. Changing of the conditions of thrust production in the engine away from the optimal ones defined by this theory will result in decreasing of thrust production efficiency characterized by the coupling coefficient C_m.

Numerous experimental investigations on the pulsejet laser propulsion [3] show that C_m is a function of a complex parameter of local explosion theory, R_0 that is called like a "dynamic" radius. Maximum C_m is achieved at certain value of ratio R/R_0, where R is a characteristic geometry parameter of a nozzle (that is of a beam concentrator) (see Fig. 2.3). Maximum C_m is equal to 40–50 dyne/W at the nozzle in a form of cone or paraboloid of rotation for a wide change of laser pulse energy. Moreover, maximum C_m has a pronounced look.

In the figure,

$$R_\Omega = \left(\frac{4\pi}{\Omega} \times \frac{E}{p_0}\right)^{1/3}, \quad R_0 = \left(\frac{E}{p_0}\right)^{1/3} \tag{2.7}$$

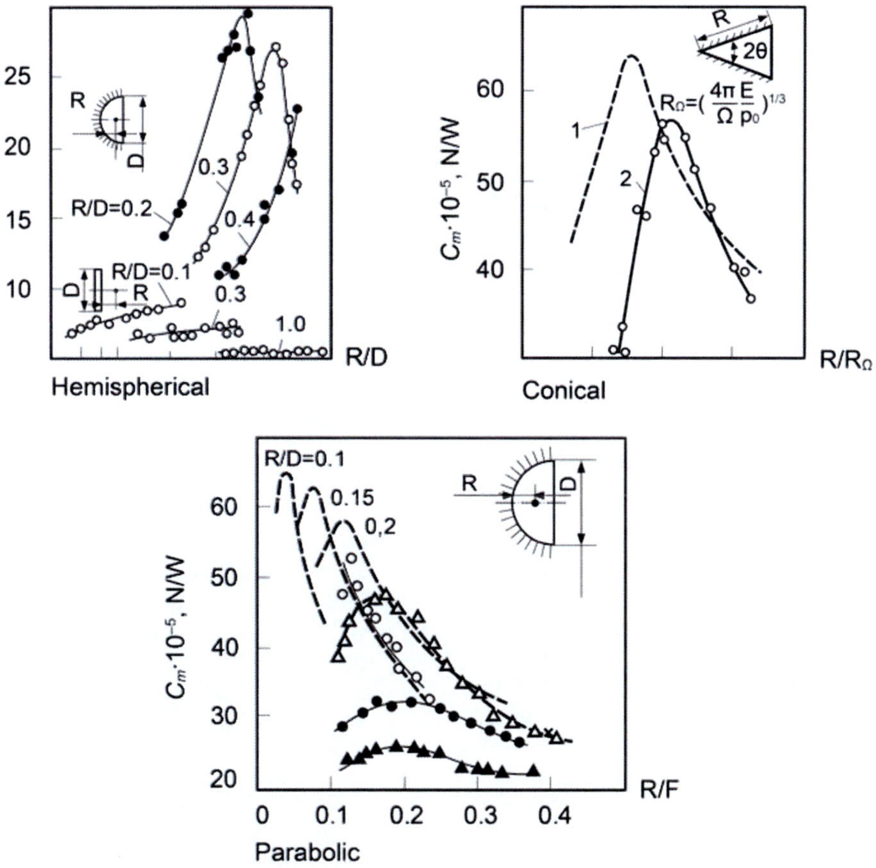

Fig. 2.3 C_m functions on dynamic radius parameter of R/R_0 or R/R_Ω. *Dashed line* is the theoretical estimations of the coupling coefficient; *points* correspond to the experimental data

where E is the absorbed pulse energy, Ω is an apex angle of a cone, ρ_0 is initial gas pressure ($\rho_0 = 0.1$–1 atm), and $(R/R_0)_{opt} = 0.3[1 + (4R/D)^{-2}]^{-1}$ for parabolic nozzle.

But, the fact that experimental coupling coefficient does not exceed 50 dynes/W for a wide range of laser energy change is evident. This restriction on C_m can be explained by two reasons. Firstly, it is a result of the experimental estimation of coupling coefficient with an error caused by uncertainty of the pulse energy released in a breakdown gaseous volume. The absorption of the laser power by plasma ignition depends on a number of effects which will be considered in the next section. The second cause is a violation of the local explosion conditions under high-peak power laser pulse, which complicates consideration of processes of thrust production.

Similar dependences of the coupling coefficient on a laser pulse energy were obtained by other researchers with the use of E-beam sustained CO_2 lasers as a power source with a long pulse (about 10μs) [6–10]. In so doing, C_m was of 20–30 dyne/W for LPE with a parabolic nozzle (see Fig. 2.4). It was discovered that a value of the coupling coefficient of maximum 30 dynes/W was achieved when the nozzle exit was made longer by a cylindrical attachment (tube) having the same diameter. So, an increase in the coupling efficiency of pulsejet propulsion can be explained by increasing the distance of a generated shock wave interaction with the

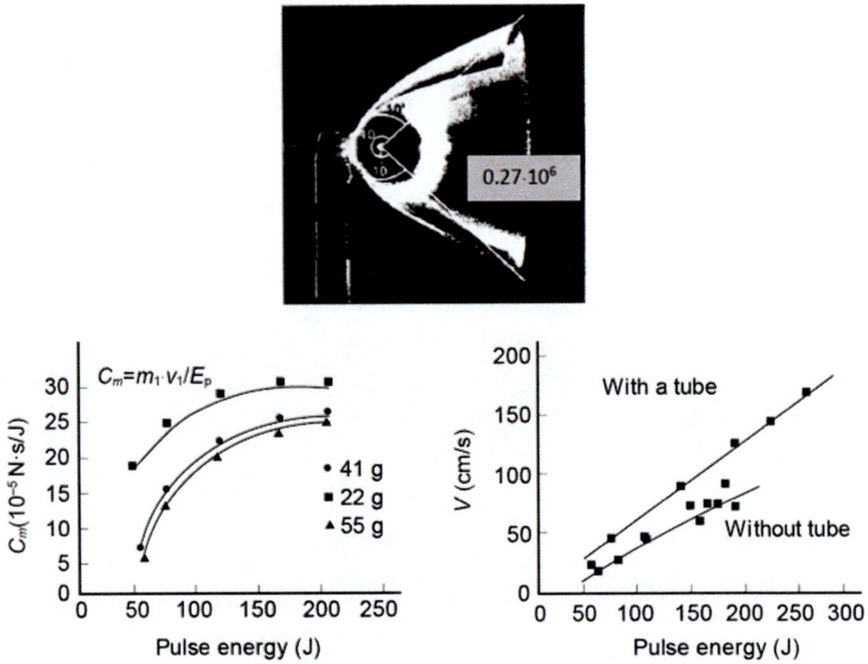

Fig. 2.4 Coupling coefficient and vehicle velocity as functions of laser pulse energy for bell type of a lightcraft [7]

nozzle walls. But these experiments demonstrated also that a total efficiency of thrust production didn't exceed 25% that is rather small.

Laser parameters	LPE properties
* pulse energy—up to 280 J	* propellant is atmosphere air
* pulse length—10μs	* total efficiency, $\eta = 25\%$
* PRR—up to 100 Hz	
* laser power—8 kW	

CO_2 lasers are usually considered as a valuable tool due to the easily accessible high output pulse energy (or average output power). It is common that industrial CO_2 lasers can generate up to 100 kW for either continuous wave or average power of repetitive pulse systems. It is recognized that the CO_2 lasers could also be applied to high-power laser propulsion. A wide range of potential capabilities of the CO_2 lasers as applied to laser propulsion is given in Ref. [11].

The experimental investigations of the laser propulsion using high-power CO_2 lasers revealed that modification of the pressure receiver design of its geometry resulted in a sufficient increase of laser-propulsion production efficiency. A remarkable example of these experiments is the LPE engine design mentioned in Refs. [12, 13] as MLD-2 (Model of Laser Mover), in which an array of small aperture parabolic mirrors was used as a beam concentrator (see Fig. 2.5). CO_2 laser with the output power of 100 kW and pulse duration of 40μs was used in the experiments.

The beam array concentrator initiated a quasi-plane shock wave at a certain distance out from the concentrator plane. A laser pulse irradiance of the concentrator reached 3–5 J/cm^2. In the experiments, maximum coupling coefficient C_{m} of 30 dynes/W was achieved under a long-duration operation of the laser.

Laser parameters	LPE properties
* pulse energy up to 4 kJ	* atmosphere air as a propellant
* pulse duration of 40μs	* LPE efficiency $\eta = 40\%$
* PRR—100 Hz (at pulse energy of 1 kJ)	

Hence, we can see the explosive mechanism of a laser power input into a gaseous propellant results in a maximum coupling coefficient of 60 dynes/W at the pulsejet laser propulsion. In this case, the thrust is produced due to accelerating a gas flow inside a nozzle by using strong shock waves generated in a focal region of parabolic nozzle or similar ones.

Let us estimate a practical applicability of this type of laser-propulsion engines as the HPLP subsystem, following Refs. [2, 3]

Maximum altitude of the air-breathing propulsion in the Earth atmosphere is limited by a dense atmosphere extended up to a 20 km altitude. Assuming a CO_2-laser with a 1 m transmitter telescope and with output beam of close to the diffractive limit to be used to launch a vehicle, we can propose the laser beam aperture may keep a constant diameter during of the vehicle flight. The last assumption allows

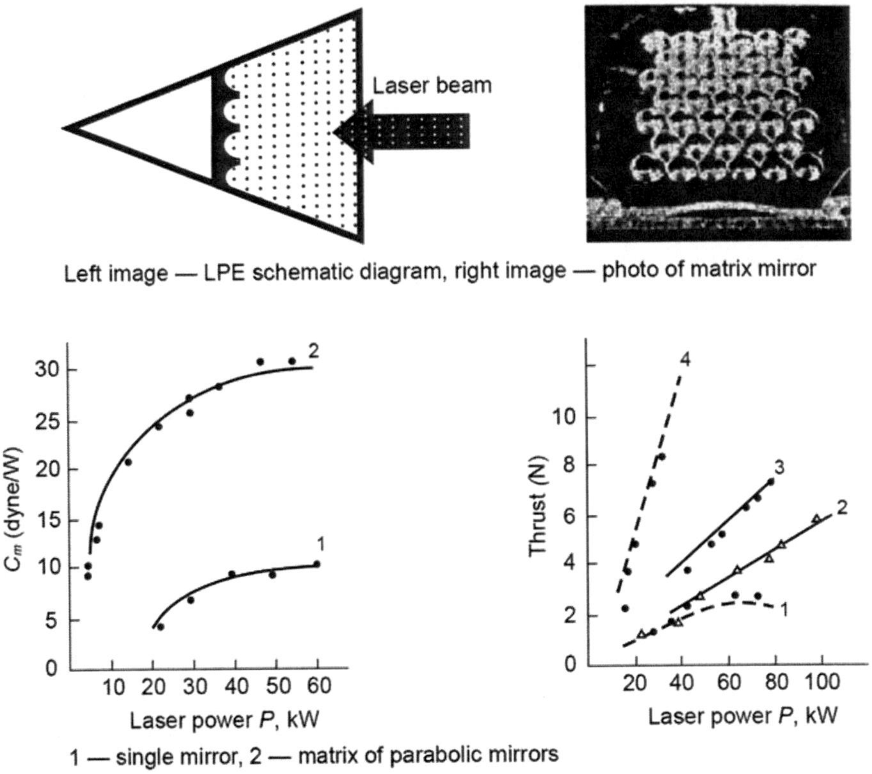

Fig. 2.5 Experimental characteristics of LPE with a matrix parabolic mirror

choosing a parabolic nozzle of 1 m in diameter as a nozzle outlet section of launched vehicle, too.

Moreover, in accordance with the Prokhorov's theory, maximum coupling coefficient of $C_m \sim 50$ dynes × s/J for paraboloid nozzle with the geometry parameter $R/D = 0.15$ at the explosive dynamic $R/R_0 \approx 0.105$ is expected. To achieve the coupling coefficient, the laser pulse energy has to have $E = 300$ kJ. Estimations show that this value of the laser pulse energy at the pulse duration of a few tens of microseconds will generate a radiation intensity that will not exceed the laser breakdown threshold of the atmosphere air that is less ~10^6 W/cm^2.

Pulsejet laser-propulsion engine will operate at a linear mode with a pulse repetition rate ν if the atmosphere air inside the engine nozzle has time to get renewed. This condition is achieved by forced replacement of the inside air mass in the nozzle after the moment when the initiated shock wave will transmit a recoil impulse to the nozzle completely. Estimations made by the theory show that this moment of time will come in $\Delta t = 1.5\ \tau_0$, where τ_0 is the laser pulse length, which assumes a pulse repetition rate has to be $V_{max} = 1/\Delta t = 160$ Hz. Considering the case when maximum replacement of the inner air is performed with the speed of sound in the atmosphere, we can have obtained that maximum pulse repetition rate will be equal to 600 Hz. Therefore, the laser-propulsion engine will produce a 50 kN thrust if the laser power of 10^8 W or 100 MW could be generated! To design such high-power lasers, it will require new technologies to be developed. Moreover, new technologies of the vehicle design will be also required to protect it against high heat flux to the jet nozzle walls of laser-propulsion engine [14, 15].

In a common case, the production efficiency of the pulsejet laser propulsion will be decreased at increasing the vehicle altitude because of decreasing the atmosphere air pressure with an altitude. This fact is illustrated in Fig. 2.6 in which the coupling coefficient of LPE engine with a parabolic nozzle as experimental functions of pressure of the surrounding gas with varying the CO_2 laser pulse energy are shown [6]. One can see that C_m changes strongly with gas pressure if the pressure is less than 300 mbar that corresponds to a 9 km atmosphere altitude above a sea level.

To understand this lowering of the coupling coefficient, the laser pulse interaction with plasma ignited has to be considered more detail (see Sect. 2.4).

There is one more problem of the pulsejet propulsion coming into being during the vehicle flights in the atmosphere. For example, the detail experiments [6, 7] on propelling of bell type vehicle by using the CO_2 laser radiation demonstrated the influence of the laser breakdown region position inside a parabolic nozzle on spatial stability of the vehicle flight trajectory. Consistent simulations of the vehicle flight based on the results of these experiments showed that the compensation of both the angular and lateral displacement of a beam focus point to stabilize the vehicle flight is not likely to be achieved by this simple device. Furthermore, even if these motions were compensated in the instantaneous plane of inclination of the vehicle, any lateral

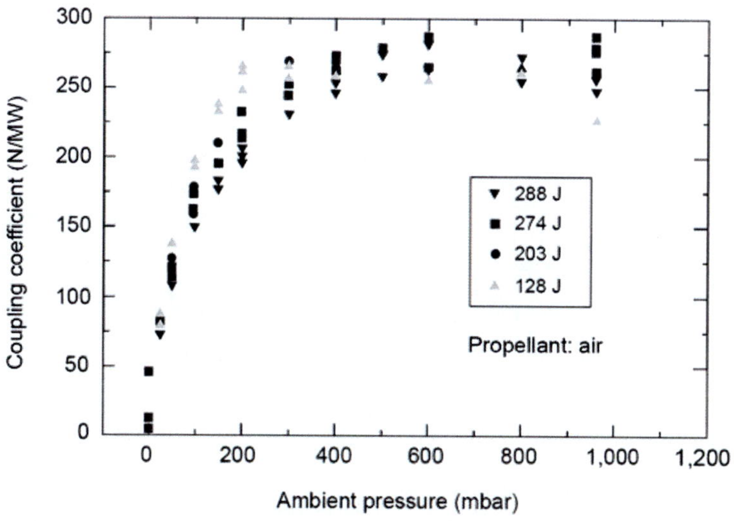

Fig. 2.6 Coupling coefficient in air as function of the gas pressure at various pulse energy

offset in the perpendicular plane would lead to lateral and angular motion there at the same time that would not be compensated.

The problem of stability of a vehicle during its flights in the atmosphere under high-power laser radiation and follow-up questions related to the problem are considered in Refs. [16, 17] and in Chap. 7 in detail.

2.2.2 Rocket Laser Propulsion at Space Conditions

The gas-dynamics of rocket laser propulsion is realized due to an injection of gaseous propellant into a jet nozzle, stored within the vehicle. And the characteristics features of repetitively pulsed (RP) mode of propulsion production is determined by the rate of the propellant consumption.

More consistent investigations on RP rocket laser propulsion were carried out by Prof. Anthony Pirri in 1972–1978 [18, 19], made with the use of RP CO_2 lasers. For the first time, Pirri proposed a simple theory of thrust production under RP laser radiation in space conditions. The theory describes the gas-dynamic phenomena in a parabolic nozzle when a gaseous propellant is injected into the nozzle through a small orifice in a vertex of the nozzle. On the basis of this theory, it is assumed that a thrust is produced due to laser power conversion into a kinetic energy of a gas when shock waves propagate inside the nozzle (Fig. 2.7). A similar approach used in that theory allows determining the thrust characteristics and requirements to the laser characteristics which have to be fulfilled to achieve a high thrust.

By the way, as it follows from the theory that the specific impulse of the RP rocket laser propulsion depends on the propellant molecular mass in a fourth root. That is why much more massive propellants can be used for space missions of vehicles under laser propulsion.

Let us consider basic statements of Pirri's theory and its variations for space missions of the vehicles with laser propulsion.

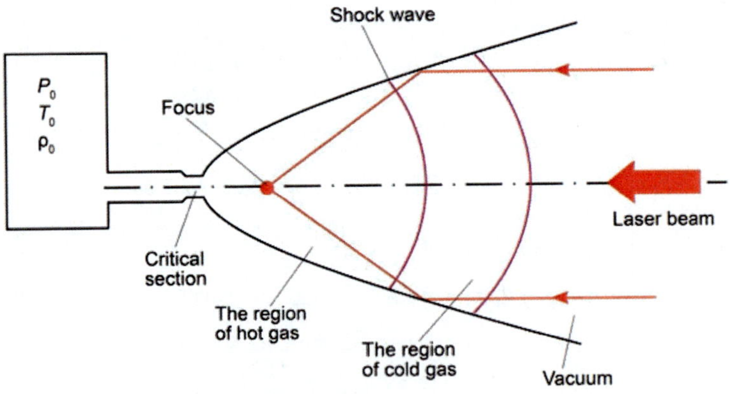

Fig. 2.7 Schematic diagram of pulsejet laser propulsion

2.2.2.1 Choice of a Propellant for Space Laser Propulsion

One of the important questions of laser propulsion under vacuum conditions is a choice of qualified propellant. As the propellant, such gases as hydrogen, helium, and air were considered in many studies [20–22].

Hydrogen. One of the principal disadvantages of using hydrogen is its low mass density in a liquid state that will require utilizing large balloons onboard storing this propellant. Chemical drawback of exploration hydrogen is also the fact that hydrogen does not dissociate completely /when it flows in a supersonic mode. That means that it does not recombine completely too. The latter results in a drastic decrease of produced specific impulse of exhaust jet.

Helium. Helium has a small molecular mass similar is hydrogen, which allows a high specific impulse of exhaust jet. But the same problem of storing the gas on a vehicle board as for hydrogen arises too. Moreover, we have to have cryogenic cooling of the gas to reach a liquid state for helium, and this fact will complicate the onboard gas storage system.

Air. Atmosphere air is considered as a propellant for laser propulsion of launching vehicles by the use of ground-based lasers as usually. But, as applied to space missions of the vehicles, one should consider some new aspects of the gas as a propellant in this case. Atmosphere air is a mixture of such gases as nitrogen, oxygen, and many others, having various thermal physical characteristics defining its storing in a liquid state. That is why the air is stored in a complicated manner under space conditions. Moreover, air is an electrically negative gas that complicates both the plasma generation processes and laser power input into this plasma.

Nitrogen. In comparison with an air, nitrogen lacks of negative properties listed above. Moreover, it can be stored at vacuum conditions relatively easy because the cryogenic engineering systems for its storage are well developed.

2.2.2.2 Determination of the Jet Nozzle Designs

As we see from Sect. 2.2.1, the nozzle geometry shape defines the efficiency of gas flow energy conversion into an exhaust jet power. Usually, axial parabolic nozzles are considered as applied to development of the pulsejet and rocket laser-propulsion engines. Other type of the nozzle geometry is an off-axis paraboloid of a rotation, proposed as the Lightcraft nozzle [22].

Following the Pirri theory [19], let us make a general analysis of a thrust production in parabolic nozzles of both types. So, we can ignore the specific way if laser beam reaches the nozzle and assume that there is an "instantaneous" release of laser pulse power in a propellant in a focus region of the nozzle. To compare both types of the nozzles, we propose to use a common means of nozzle shape description when one of the beam concentrators is enclosed into other (see Fig. 2.8). One of the principal conditions of this "inscribing" is the similarity of boundary conditions of gas flows at the outlet section of each nozzle. These conditions result in a

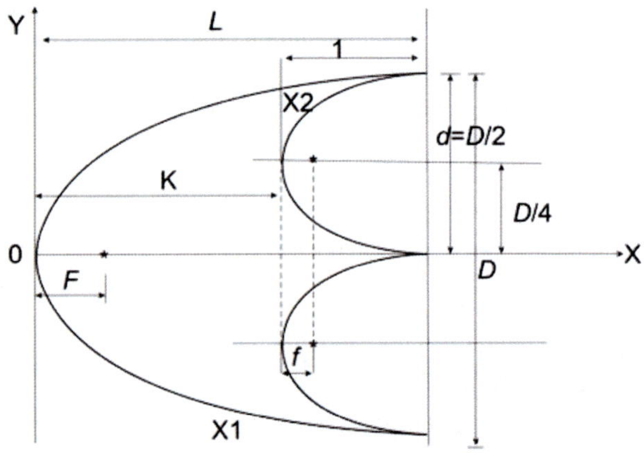

Fig. 2.8 Two types of parabolic nozzle for air-breathing laser propulsion

coincidence of not only the area of exit sections of the nozzles but also tangent lines at the nozzle edges.

Every version of the parabolic nozzles is characterized by the following parameters, namely, nozzle length along its axis (L and l for the axial and off-axis parabolas, correspondently), characteristic size of outlet section (D and d), and parabolic focus (F and f).

Axial parabolic nozzle is described by following relation:

$$x_1 = \frac{1}{4F} y^2 \tag{2.8}$$

And for the off-axis parabola correspondently:

$$x_2 = \frac{D^2}{16f} + \frac{1}{f}\left(y - \frac{D}{4}\right)^2 \tag{2.9}$$

The similarity conditions of flow over the nozzle edges are satisfied if $D/F = d/f$.

To determine a gas mass consumption for every version of nozzles, let us set some nozzle geometry parameters, on which the consumption is dependent. As for the off-axis parabolic nozzle, variation of the nozzle cross section along X-axis changes as $s = \pi\, D\, \rho$, where ρ is a nozzle radius (as it is denoted in Fig. 2.9). And gaseous propellant is injected into the nozzle through a circular slit that is located in the nozzle vertex.

Then, the propellant mass consumption of a hypothesis mass source is estimated as follows:

$$\dot{m} = \rho^* \cdot u^* \cdot \pi \cdot D \cdot L^* \cdot \theta$$

where Θ is a plane vertex angle of the off-axis parabola.

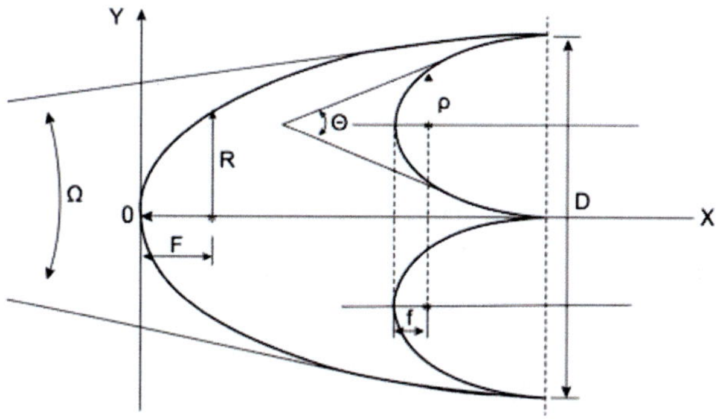

Fig. 2.9 Schematic diagrams of parabolic nozzle with a propellant mass sources

In the case of the axial parabolic nozzle, a mass source is located in the parabola vertex and gaseous propellant is injected through an orifice in it. In this case, mass consumption is defined as:

$$\dot{m} = \rho^* \cdot u^* \cdot \pi \cdot d^{*2} \cdot \Omega \tag{2.10}$$

In these formula, ρ^* and u^* are the density and flow velocity of a gas in a critical section of every hypothesis mass sources. Considering the geometric shape of every nozzle, we can deduce the following relations:

$$\rho^* u^* \frac{\pi D}{2} h^* = \rho^* u^* \frac{\pi (D^*)^2}{4} \tag{2.11}$$
$$h^* = \frac{\pi (D^*)^2 2}{4_2 \pi D} = \frac{(D^*)^2}{2D} = \rho^* \theta$$

where h^* is a slit width along a ρ coordinate. The condition of propellant mass consumptions equality through these nozzles results in following relation: $h^* = D^{*2}/4D$. This condition can be simply satisfied by varying the nozzle slit geometry parameters.

Now, let us find self-similar solutions of a gas flow in both nozzles as function of laser power characteristics for the case of thrust production under vacuum conditions, following Ref. [19]. Moreover, we consider a system of gas-dynamic equations describing non-viscous gas flows on a basis of three fundamental laws, namely:

1. mass conservation:

$$\frac{\partial \rho}{\partial t} + \operatorname{div}\left(\rho \cdot \vec{u}\right) = 0 \tag{2.12}$$

2. conservation of momentum or Newton's law:

$$\rho \cdot \frac{\partial u}{\partial t} = \rho \cdot \overline{F} + \mathrm{div}\overline{\pi} \tag{2.13}$$

3. conservation of energy:

$$\frac{\partial}{\partial t}\left(\rho \cdot \varepsilon + \frac{\rho \cdot u^2}{2}\right) = -\mathrm{div}\left(\rho \cdot \vec{u} \cdot \left(\varepsilon + \frac{u^2}{2}\right) + p \cdot \vec{u}\right) + \rho \cdot Q \tag{2.14}$$

where ε is internal energy of gas, Q is power of thermal energy sources, ρ is a gas density, u is a flow velocity, $\vec{F}$ is a vector of bulk forces, and π is a stress tensor.

To simplify solving these general equations, we assume the following:

1. Flow has a one-dimensional character that means the flow characteristics depend on only longitudinal spatial coordinate, and the parameters are averaged across every nozzle section.
2. Gas flow is under an adiabatic regime, which means there is no heat exchange between the gas flow and outside space.
3. Bulk forces are excluded from the consideration.

Similar to the consideration of pulsejet laser propulsion, we can introduce parameters of specific impulse and coupling coefficient also for the RP gas-breathing propulsion by introducing the definitions of averaged thrust T, mass consumption M, and laser power P. Then by the analogy:

$$I_{\mathrm{sp}} = \frac{\overline{T}}{\dot{\overline{M}} \cdot g} \tag{2.15}$$

where g is a gravity, and

$$C_{\mathrm{m}} = \frac{\overline{T}}{\overline{P}} = \frac{2 \cdot F^2(\theta) \cdot e_{\mathrm{R}}}{g \cdot I_{\mathrm{sp}}} \tag{2.16}$$

where $F(\theta)$ is a factor characterizing specific impulse decrease caused by changing of jet nozzle shape. Nevertheless, we can assume that $F(\theta) = 1$ without losing any generality, and θ is nozzle outlet angle.

Auxiliary parameter, e_{R}, is introduced to take into account thrust losses caused by a nonhomogeneous distribution of flow velocity in an exhaust section of every nozzle. The effect can be recorded by using average mass flow velocity:

$$\overline{u}_{\mathrm{e}} = \frac{1}{M} \cdot \int_0^M u_{\mathrm{e}} dm \tag{2.17}$$

where M is the total mass consumption of propellant. Then, the relative efficiency of thrust production at RP air-breathing propulsion as compared with conventional rocket regime can be expressed as [19]:

$$e_R = \left[\int_0^1 \left(\frac{u_e}{\overline{u}_e} \right)^2 d\left(\frac{m}{M} \right) \right]^{-1} \tag{2.18}$$

It follows from the formula that a specific impulse may be determined as the following ratio:

$$I_{sp} = \frac{(2 \cdot e_R)^{1/2}}{g} \cdot \left(\frac{\overline{P}}{\dot{\overline{M}}} \right)^{1/2} \cdot F(\theta) \tag{2.19}$$

Thus, the specific impulse of RP LPE is determined by the laser power that is transformed into specified propellant mass consumption rate. At the same time, the mass consumption is limited by the propellant mass source, $\dot{m}$, that depends on flow characteristics in an orifice of the parabola vertex.

Then, the specific impulse of the axial parabolic nozzle is:

$$I_{sp} = \frac{(2 \cdot e_R)^{1/2}}{g} \cdot \left(\frac{E}{\rho^* \cdot u^* \cdot d^{2*} \cdot (t_p - t_C)} \right)^{1/2} \cdot F(\theta) \tag{2.20}$$

where E is laser pulse energy, and $(t_p–t_c)$ is a relative pulse duration [3]. And specific impulse of the off-axis parabolic nozzle is:

$$I_{sp} = \frac{(2 \cdot e_R)^{1/2}}{g} \cdot \left(\frac{E}{\rho^* \cdot u^* \cdot h^* \cdot L \cdot (t_p - t_C)} \right)^{1/2} \cdot F(\theta) \tag{2.21}$$

These formula allow estimation of the main characteristics of a gas-breathing laser-propulsion system in RP mode at vacuum ambient conditions. As an example, let us consider the LOTV orbital maneuver (from Chap. 1) by using these formula at the condition when an onboard propellant mass is limited.

To be more specific, let us choose nitrogen as a propellant for the LOTV mission. The gas consumption during the LOTV mission will depend on a specific impulse of an exhaust jet. As it is shown in Chap. 1, optimal specific impulse has to be of $I_{sp} = 10^3$ s or so. Hence, the required laser power is determined as a variable parameter in the range from 200 kW through to 1 MW, which will guarantee getting a 20–200 N thrust at $C_m = 2 \times 10^{-4}$ N/W. It also assumes that the efficiency of laser propulsion will be of 100% that means that a decrease in the propulsion efficiency will cause only an increase of required laser power.

The flow parameters of a nitrogen in a crucial section of the mass source orifice are well known, and they must correspond to the gas flowing at the speed of sound, namely$p^* = 0.028\,\frac{\text{kg}}{\text{m}^3}, p^* = 2.083$ kPa, $T^* = 252.4$ K, $u_{\max} = 808.3\,\frac{\text{m}}{\text{s}}$, and $u^* = 321.1\,\frac{\text{m}}{\text{s}}$.

So, we can describe the LOTV orbital missions with two different onboard LPE designs as the axial parabolic and off-axis parabolic nozzles in a view of following diagrams (Figs. 2.10(a) and (b)), representing relative pulse duration $t_p = 1/f$ as functions of a diameter of crucial section of mass source for different energy of the pulse. In the figures, two inclined red lines correspond to physical limits of the gas-breathing theory at ambient vacuum conditions [23]. Among them, the bottom line is defined by a minimum time interval that is required for the mass source to restore after impact of a shock wave generated in a nozzle after laser power release. The top line is a limitation on nozzle length when gas flow in an exhaust section of a parabolic nozzle can be considered as a continual one. Vertical lines in the diagrams denote laser power required.

From Fig. 2.10(a), it is clear the specified thrust characteristics of LPE ($I_{sp} = 10$ s^3 and $T = 40$ N) can be obtained if the gas mass flow rate is of 4.2 g/s under the specified mode of laser operation. The laser has to generate pulse energy of 40 J with the pulse repetition rate of 5 kHz, which corresponds to the laser average power of 200 kW. The case of increasing the laser pulse energy at fixed laser power corresponds to moving up along vertical lines of the figure from the generated range of the LPE and laser parameters, where the specified propellant mass rate does not conform to the laser pulse energy. Moving along the horizontal lines in the figures corresponds to decreasing the efficiency of laser propulsion. That means the laser has to operate with average power of 500 kW instead of the power of 200 kW if the LPE total efficiency is equal to 40%.

Figure 2.10(b) represents similar range of the permitted time intervals between laser pulses t_p as function of a width of the cylindrical slit. As in the case of 2.10(a), there are definite relationships between the laser pulse energy E and time interval between the pulses t_p, which are imposed by gas-dynamics conditions in the nozzle. It is clear from the figure, if the width of slit equals to 1 mm (that is a technological limit in the nozzle design) then the pulse period will be 5×10^{-6} s at the laser pulse energy of 10 J and average laser power of 500 kW to produce specified thrust.

As is known, a Knudsen number [5] is a criterion defining an application of a continuum gas-dynamic theory for laser propulsion at vacuum conditions. Maximum Knudsen number makes up 0.01 when gas propellant may be thought as being a continuum. For the chosen parameters of the flow in the nozzle exhaust section, the gas flow will remain as a continuum in nature at a distance of 0.5 m down from the section.

As applied to the LOTV concepts, the laser operating with a 40 J laser pulse energy and pulse repetition rate of 5 kHz is more suitable to get a specific impulse $I_s = 10^3$ at a limited propellant mass flow rate. These parameters of high-power laser operation differ sufficiently from those ones in the LOTV concept proposed by Japanese researchers [24].

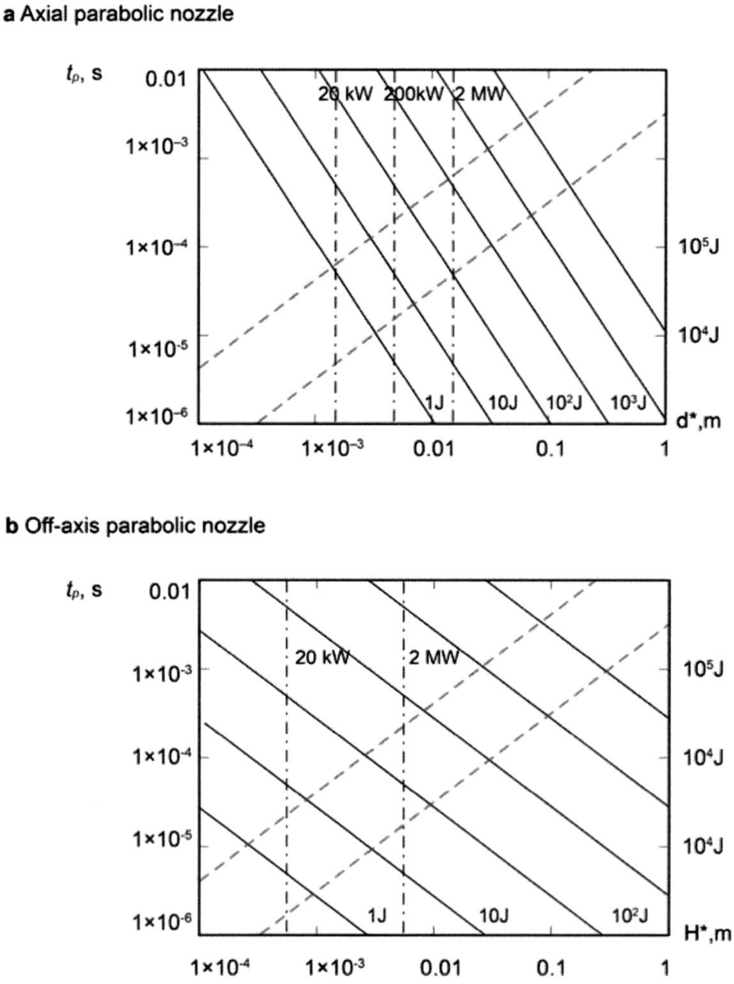

Fig. 2.10 Relative pulse duration $t_p = 1/f$ as functions of required size of required critical section at varying laser pulse energy

It should be mentioned also that considered theories of the pulsejet and rocket laser propulsion, presented above, are based on the assumption that the laser power is transformed completely into energy of a local explosion. Unfortunately, this assumption isn't quite correct because of a few reasons. Problems are caused by the physical processes running at pulsed laser radiation interaction with plasma ignition in a focal region of vehicle nozzles. Therefore, it is required to consider these processes more detail and to the define dependence of thrust production efficiency both on laser characteristics and plasma properties.

2.3 Physics of Laser Plasma Ignited in Gases as Applied to Laser Propulsion

One of the basic factors defining thrust generation efficiency of laser propulsion both in the pulsejet and ablation modes is the plasma ignited by the laser pulse during a process of laser breakdown of gases or evaporation of solid materials. The processes of laser radiation interaction with the plasma-ignited affects the laser power absorption and thrust produced as a result. The laser power release in the ignited plasma depends on many factors, including the free electron density number in the plasma, laser radiation wavelength, radiation intensity, and laser pulse duration.

As usual, a model of weakly ionized plasma is applied to describe ignition of plasma via the process of laser breakdown of a matter if the laser radiation intensity does not exceed 10^{12} W/cm^2. This plasma model allows determination of the radiation intensity threshold when the plasma is ignited as functions of gas (or vapor) components and allows description of plasma-chemical processes running in the plasma [25].

It should be noted that similar processes of the atom or molecular ionizations under laser radiation can run due to super threshold multi-photon ionization, too. Moreover, such processes as multi-stage ionization of atoms and molecules are observed in gases, but the ignition intensity threshold depends on the laser pulse length and radiation intensity. It was experimentally shown many times that twofold ionization thresholds of such gases as xenon, argon, and krypton, which possess ionization potential, I, in a 10–80 eV range, belong to 5×10^{13}–10^{14} W/cm^2, and threefold intensity thresholds is of (2–3) $\times$ 10^{14} W/cm^2. Moreover, of the results of Ref. [26], it follows the temperature of plasma ignited in an air of atmosphere pressure by the laser radiation with a 10^{12} W/cm^2 intensity under the action of Nd: YAG laser pulse can reach 50 eV. That means that the plasma cannot be considered as a weakly ionized plasma. But at the same time, the electric field formed in a gas by the laser radiation of $I = 10^{12}$–10^{14} W/cm^2 can be considered as a weak one when compared with electrical fields existing in intra-atomic space. This allows excluding various nonlinear phenomena of the laser radiation interaction with a matter from the consideration.

Of the arguments listed above, one can conclude the multi-ionized gas-plasma model should be able to predict adequately the laser power release into plasma temperature if the laser radiation with an intensity of $I = 10^{12}$–10^{14} W/cm^2 is applied as original process of laser propulsion.

Below, we consider basic models of high-temperature plasma by using examples of the plasma ignited by Nd-laser in nitrogen atmosphere, which could be applied to determine the conversion efficiency of laser power into plasma temperature in a view of laser propulsion.

2.3.1 Model of Multi-Ionized Plasma Ignited by Laser Pulses in Gases

The model of multi-ionized plasma is considered as applied to pulsejet laser propulsion, proposing that the laser radiation intensity in a gas reaches 10^{14} W/cm^2. For the example, this kind of plasma can be ignited under the action of an Nd-laser pulse with a 10 ns pulse length [30].

To determine basic characteristics of the plasma, including a degree of plasma ionization and its thermodynamic parameters as functions of initial gas density n and its temperature T_e, it is necessary to solve a system of m equations (m is a nuclear charge) for every pair of such parameters as n and T_e [25]. But, solving these equations is a lengthy and labor-intensive process. As for practical purposes, simpler approximate method can be used to determine all thermodynamic parameters of the plasma over a wide range of the plasma temperature, when gas molecules are assumed to be multi-ionized. The simplest theory of plasma ignition in gases was developed by Russian scientist Prof. Yu.P. Raizer in 1974 [25]. His method is based on solving only one equation describing a number density of ions in plasma. In spite of its simplicity, the method possesses an accuracy that is sufficient to be applied to the majority of practical problems, and the method can be easily generalized to be applied to gas mixtures. Truncation error of the Raizer's method does not exceed 25% even in the case of a weakly ionized plasma, and the error is much less for multi-ionized plasmas.

The content of the Raizer's method consists of the assumption that every atom and ion of plasma can be interpreted as an equivalent ion possessing a charge that is equal to electron number density corresponding to one initial atom, that is, $Z = n_e/n$. Here, n_e and n are electron and atom concentrations in 1 cm^3, respectively. Equivalent charge Z can possess the fractional magnitudes, and number of the charge density is equal to n, that is a density of initial atoms. Two main assumptions lay in the method basis.

First of all, ionization potential of atoms $I(z)$ and ions density $n(z)$ are considered as continuous functions of ion charge z. $I(z)$ function is created by a linear interpolation of the discrete ionization potentials I_z of ion charge, and relations combining ion number density, which charge differs from 1, have the following view:

$$n(z+1) = n(z) + \mathrm{d}n/\mathrm{d}z, \Delta z = 1 \tag{2.22}$$

Secondly, it is assumed also that Z equivalent charge is exactly equal to $z_{\max}$ a variable parameter at which $n(z)$ function has a maximum. The assumption is more accurate if the function has a peak at this point.

To apply this method to analysis of an ignited plasma, it will require using the equations linking number of ion density of different signs, temperature, ionization potential, and so on, which are ionization equilibrium equations. These equations are formulated in the following.

Basic processes of gas ionization are the ionization of atoms and molecules via collisions with electrons with atoms and molecules and due to its photo-ionization, as well as basic processes of the plasma recombination are recombination via three-particle collision and due to two-particle photo-recombination [27]. Ionization by electron impact is realized by two ways, namely: (1) direct ionization of atoms or molecules from a ground state or (2) cascade excitation of electrons in atoms with following ionization of the atoms.

In accordance with the Raizer's theory [25], the collision ionization that is considered as electron avalanche is initiated only by those electrons which possess a kinetic energy $\varepsilon_e = mv^2/2$ exceeding ionization potential of an atom or a molecular, I, and the electron average temperature, T_e, satisfy the condition of $T_e << I$ at the range of around first ionization potential. It is assumed that electrons ionize atoms at such a moment when its energy is equal to $I + T_e$. Ionization cross section σ_e depends linearly on the electron temperature if its energy is close to the atom ionization threshold, that is $\sigma_e \approx C(\varepsilon_e - I) = C \times T_e$, where C is direct ionization constant. Then, atom ionization rate by the electron impact of atom can be written as follows:

$$Z_e^i = n \cdot n_e \cdot \alpha_e = n \cdot n_e \cdot \sigma_e \cdot v_{eav}(I/T_e + 2) \exp(-I/T_e) \tag{2.23}$$

where $v_{eav} = (8T_e/\pi m_e)^{1/2}$ is its average thermal velocity of electrons, α_e is an ionization constant, and σ_e is the electron impact ionization cross section.

By the analogy, direct photo-ionization rate from the atom ground state is:

$$Z_v^i = n \cdot a_v = n \cdot (8\pi/c^2) \cdot (I/h)^2 \cdot (T_e/h) \cdot \sigma_v \cdot \exp(-I/T_e) \tag{2.24}$$

where c is velocity of light, h is the Plank's constant, and σ_ν is the atom photo-ionization cross section.

Relationships for recombination rates can be written by the analogy as:

$$Z_e^r = n_i \cdot n_e \cdot b_e = n_i \cdot n_e \cdot \sigma_e \cdot (I/T_e + 2)(g_a/g_i)(h^3/2\pi^2 \cdot m_e^2 \cdot T_e) \tag{2.25}$$

$$Z_v^r = n_i \cdot n_e \cdot b_v = n_i \cdot n_e \cdot \sigma_v(g_a/2g_i)(I^2/m_e \cdot c^2 \cdot T_e) \tag{2.26}$$

where n_i is the ion number density, b_e and b_ν are recombination coefficients, g_a and g_i are stat weights of atoms and ions, and m_e is an electron mass. From the formula listed above, ratios of impact and photo-ionizations and recombination rates are as follows:

$$\begin{aligned} Z_e^i/Z_v^r &= n_e \cdot \alpha_e/\alpha_v \\ &= n_e \cdot \left(c^2 \cdot h^3/(8m_e \cdot \pi^3)\right)^{1/2}(I/T_e + 2)/I^2 \cdot T^{1/2} \cdot \sigma_e/\sigma_v \end{aligned} \tag{2.27}$$

$$Z_e^r/Z_v^r = b_e/b_v = n_e \cdot \left(c^2 \cdot h^3/(8m_e \cdot \pi^3)\right)^{1/2}(I/T_e + 2)/I^2 \cdot T^{1/2} \cdot \sigma_e/\sigma_v$$

In the general case, the equation of plasma ionization equilibrium can be written in a view:

$$\frac{n_{z+1}}{n_z} = \frac{Z_e^i + Z_v^i}{Z_e^r + Z_v^r} \tag{2.28}$$

It is assumed that the plasma will be in a state of complete thermodynamic equilibrium if every direct process of the plasma ionization runs at the same rate as the inverse process of its recombination. Moreover, the plasma has to be intensive enough to exclude various losses of its energy. So, the parameter characterizing the plasma extension is determined as an optical depth of the plasma, and it is equal to $\tau = d \times \mu_\omega$, where d is a characteristic size of the plasma region and μ_ω is the absorption coefficient of radiation with a wavelength ω. The plasma is considered to be thick if its optical depth exceeds unity for various radiations wavelengths. In that case, a comprehensive description of the plasma can be obtained, including a finite number of thermodynamic parameters, namely, temperature, pressure, plasma components concentrations. It is known that distribution of every component of the plasma in accordance with its excited states, belongs to the Boltzmann law. The equation of plasma ionization equilibrium can be deduced by using (2.28) equation, in which corresponding rate constant from (2.21)–(2.25) are used. As a result, the equation named as Saha's equation for every ion component of the plasma is deduced [26]:

$$n_e \cdot \frac{n_{z+1}}{n_z} = A \times T^{3/2} \cdot \frac{u_{z+1}}{u_z} \cdot \exp\left[-{}^{I_{z+1}}\!/_{T}\right] \tag{2.29}$$

where $A = 2 \times (2\pi \times m_e/h^2)^{3/2} = 6 \times 10^{21}\ \text{cm}^{-3} \times eV^{-3/2}$, T has a eV measure, and subscript denotes a discrete value of ion number density as well as ionization potential and statistical sums at a given ion charge.

To determine an ionization degree of the plasma, the investigations of plasma luminescence spectrum were carried out in Ref. [28] within a spectral range of 0.50–0.52μm. This spectral range is chosen because of the luminescence components of the plasma ignited in an air or in nitrogen at the gas pressure of (0.1–1.0) atm, via the action of a laser pulse with wavelengths of 1.06 or 10.6μm, are determined by spectral lines of NII line of nitrogen.

In the case of non-equilibrium plasma, the following relations are to be fulfilled, linking three spectral lines (n, m, and l) of the plasma luminescence [29], namely:

$$\left(I_n\lambda_n^3/g_{n1}\right)\exp\left(-E_n/T_e\right)\left(\chi(E_l/T_e) - \chi(E_m/T_e)\right) = \left(I_l\lambda_l^3/g_{l1}\right)\exp\left(-E_l/T_e\right)$$
$$\left(\chi(E_n/T_e) - \chi(E_m/T_e)\right) + \left(I_m\lambda_m^3/g_{m1}\right)\exp\left(-E_m/T_e\right)\left(\chi(E_l/T_e) - \chi(E_n/T_e)\right) \tag{2.30}$$

where I_i is radiation intensity recorded, E_i is energy of upper levels of atoms, g_{i1}—statistical weights of low levels, g_{i2}—statistical weights of upper levels,

λ_i—wavelength, T_e—electron temperature, $\chi(x) = (4/3\pi^{1/2}) \int e^{-x} x^{3/2} dx$ and $\chi 1 (x) = (1/8.86) \int e^{-x} x^{3.3} dx$ are some functions corresponding to constant coulomb logarithm and to the logarithm of the temperature. Specified expression of the functions represents a kind of non-equilibrium plasma, namely, ionization or recombination ones. If the following $x < 4$ is true, then these functions differ each other sufficiently (up to four times), but if $x > 10$ then this difference is not negligible. If the plasma is in an equilibrium state, then this basic equation is fulfilled identically.

But, a number of the experiments on investigating the nitrogen plasma luminescence, which was ignited by a laser pulse, demonstrated the fact that this equation did not transformed into an identity. From that Saha's equation describing those part of the plasma from which the luminescence was detected cannot be applied into the analysis and the plasma is in a non-equilibrium state. Nevertheless, the Saha's equations may be used for some specific cases to describe the non-equilibrium plasma generated by laser.

First case is where an optically thick plasma if the temperature and pressure gradients at a distance of a mean-free path of plasma particle or photon are less than its absolute magnitudes. It is assumed that a local thermodynamic equilibrium (LTE) state is achieved in the plasma in this case [28]. The LTE state corresponds to the plasma possessing small radiation losses when conception of common temperature is applied to a whole plasma volume. Electron density number of at which the LTE state may be achieved is determined by Griem's formula [28]:

$$n_e \geq 9 \cdot 10^{17} \cdot (E_2/I)^3 \cdot (T/I)^{1/2} \tag{2.31}$$

where E_2 is energy of first excited level. It is shown [27] that an air plasma can be considered as LTE plasma only at the condition of complete ionization of the air molecules.

Second case corresponds to a partial LTE state of the plasma when Boltzmann distribution function for charged particles is valid for those energy levels which are situated upper than n^*. In this case, low energy levels are excluded from consideration because they are emptied first of all via radiative transitions. To get to the plasma equilibrium state, many collision events are required in this case. Electron density number defining the partial LTE state of plasma can be estimated from the following formula [28]:

$$n_e \geq 7 \cdot 10^{18} \cdot \left(Z^6/n_g\right) \cdot (T/I)^{1/2} \tag{2.32}$$

where n_g is a basic quantum number of low level of equilibrium state that is included into partial LTE. And we can use a total equation of Saha-Boltzmann to calculate ionization degree and thermodynamic potentials of the plasma:

$$\frac{n_z \cdot n_e}{n_{z-1}} = \frac{A \cdot g_z}{g_{z-1}} \cdot \exp\left(\frac{E_{n*} \cdot I_{z-1}}{k \cdot T_e}\right) \tag{2.33}$$

and the plasma state can be characterized by such parameters as T_e, n_e, and some functions which depend on these parameters.

Third case. If luminescence radiation losses in the plasma are strong, then it means that every collision process is not in a balance with the corresponding inverse process. In this case, such a characteristic as the plasma temperature cannot be used. Energy is no more distributed over excited levels of atoms in accordance with the Boltzmann equation, and Saha's formulae are not implemented. But, it is otherwise assumed that electrons follow the Maxwell distribution of velocities law, and the concept of electron temperature T_e can be applied. Atoms and ions can possess other distributions by its velocities which do not follow Maxwell's law. In that case, other kinds of the ionization equilibrium can exist, which differ from LTE. If collisional ionization is in a balance with a radiation recombination of atoms and ions, then the ratio of ion number density with a z charge to ion number density with a z + 1 charge may be defined by the Elvert formula [28]:

$$\frac{n_{z+1}}{n_z} = 10^8 \cdot \frac{\xi_z}{n_g} \cdot \frac{T_e}{I_z^3} \cdot \exp\left(-{}^{I_z}\!/_{T_e}\right) \tag{2.34}$$

where ζ_z is number of electrons in valence shell, and ng is a basic quantum number of an ion ground state with a z charge. Under this type of ionization equilibrium, the plasma ionization degree does not depend on number of electrons in the shell. Reaching this plasma equilibrium state depends strongly on the gas pressure and plasma temperature. This equilibrium is a characteristic of the plasma ignited at low gas pressure and high temperature of the plasma.

Let us consider the Raizer's method as applied to solving Saha's equations. In accordance with this method and using the Eq. (2.33), as well as continuous functions $I(z + 1)$ and $n(z)$, one can obtain the system of Saha's equations in the differential species form as follows:

$$n_e \cdot \left[1 + \frac{1}{n(z)} \cdot \frac{dn}{dz}\right] = n_e \cdot \left[1 + \frac{d(\ln n)}{dz}\right] = A \cdot T^{3/2} \cdot \exp\left[-{}^{I(z+1)}\!/_{T}\right] \tag{2.35}$$

Denoting ionization potential of ions Z as I and taking into account that $n_e = n \times Z$ and the derivative of $dn/dz = 0$ at a maximum, one can obtain the ratio that unites the equivalent ion charge and ionization potential that corresponds to this charge:

$$Z = \frac{A \cdot T^{3/2}}{n} \cdot \exp\left(-{}^{I}\!/_{T}\right) \tag{2.36}$$

To convert this ratio into the equation of equivalent charge as function of the gas's initial density and temperature, it is required to exclude an uncertainty in denoting ionization potential indices. It is shown in Ref. [28] that the results of the approximate and detailed calculations of the potential correspond to each other if the equivalent ionization potential obtained for Z-ions is referred to a Z + 1/2 point, namely, $I = I\cdot(Z + 1/2)$. This rule is an important one especially for the plasma in which ions charges are only a few units. Thus, simple transcendent equation for the equivalent charge looks as follows:

$$I(Z + {}^1\!/\!_2) = T \cdot \ln\left[\frac{A \cdot T^{3/2}}{Z \cdot n}\right] \tag{2.37}$$

By using the analogy technique, we can apply Raizer's method to Elvert system of equations describing the plasma ionization equilibrium (2.34). For that, it will be necessary to derive auxiliary continues functions as $\zeta(z)$ и $n_g(z)$ further to the ionization potential $I(z)$ and ion density $n(z)$ functions, which are constructed by linear interpolating of discrete amount of ions of different multiplicity. As usually, construction of these functions does not cause difficulties. Recurrent differential relations uniting ion number of density are the same as $n(z + 1) = n(z) + dn/dz$, and $\Delta z = 1$ in this case. Substituting discrete functions by continues ones and using the recurrent ratios, we can obtain the Elvert equations in a differential form:

$$1 + \frac{1}{n(z)} \cdot \frac{dn}{dz} = 1 + \frac{d\ln(n)}{dz} = 10^8 \cdot \frac{\xi(z)}{n_g(z)} \cdot \frac{T_e}{I(z)^3} \cdot \exp\left[-{}^{I(z)}\!/\!_{T_e}\right] \tag{2.38}$$

Further, let us assume that the equivalent ion charge coincides with an average number of electrons belonging to one atom that is $Z = n_e/n$, and the charge number is equal to such a z_{max} value at which the distribution function goes through a maximum. Moreover, denoting I as ion ionization potential with the equivalent charge and taking into account that $dn/dz = 0$ in the maximum of $n(z)$ distribution, we can obtain following ratio for the equivalent ionization potential:

$$10^8 \cdot \frac{\xi(Z)}{n_g(Z)} \cdot \frac{T_e}{I^3} \cdot \exp\left[-{}^{I}\!/\!_{T_e}\right] = 1 \tag{2.39}$$

Taking a logarithm of the ratio, we have:

$$I(Z + {}^1\!/\!_2) = T_e \cdot \ln\left[10^8 \cdot \frac{\xi(Z)}{n_g(Z)} \cdot \frac{T_e}{I(Z + {}^1\!/\!_2)^3}\right] \tag{2.40}$$

The solution of this equation for an equilibrium plasma does not depend on a gas initial density in this case in contrast to solution of the Eq. (2.38).

Now, let us consider application of the Raizer's method to the Elvert equations. For that, distribution of ions by charges, $n(z)$, has to be determined to estimate the function peak width. Combining the Elvert equations for every equivalent ionization potential, following system of the equations is deduced:

$$\frac{n_{z+k}}{n_z} = \exp\left(-\sum\nolimits_{i=1}^{k} \frac{I_{z+i} - I}{T_e}\right) \tag{2.41}$$

$$\frac{n_{z-k}}{n_z} = \exp\left(-\sum\nolimits_{i=1}^{k} \frac{I - I_{z-i}}{T_e}\right) \tag{2.42}$$

Using the $n(z)$ and $I(z)$ continuous functions and expanding $I(z)$ into a Taylor serious: $I(z) = I(Z) + \frac{\mathrm{d}I}{\mathrm{d}Z}$, Gauss functions for $n(z)$ is followed: $n(z) = n_{\max} \cdot \exp\left[-((z - Z)/\Delta)^2\right]$, where $\Delta = (2T_e/(\mathrm{d}I/\mathrm{d}z))^{1/2}$ is a half width of the function peak.

Figure 2.11 illustrates the dependences of a half width of the distribution function peaks $n(z)$ on plasma temperature, obtained by using Saha's equation (solid lines) and the Elvert equations (dash lines) for plasma ignited in a nitrogen at atmospheric pressure (1.0 atm) and at low gas pressure (0.1 atm). It is seen that the half width of the peaks exceeds unity (1) in a narrow temperature range of plasma that is 11–13 eV at low plasma temperature and then when the plasma temperature exceeds 60 eV. It should be marked that the condition of $\Delta < 1$ is satisfied in a wide range of the plasma temperature change for the Elvert equations.

Application of the Raizer's method to a laser-ignited plasma shows that the ion distribution function in the plasma has a narrow peak that does not practically exceed unity in a wide range of the plasma temperature changing. That means there are many ions of the binary and triple charge order if the average electron temperature exceeds the first ionization potential of the gas molecules.

Figure 2.12 illustrates how the ionization potential degree of nitrogen is changed in variation of the equilibrium plasma (solid lines) and the Elvert plasma (dash lines) temperature for the same nitrogen pressure.

As one can see from Fig. 2.13, the electron number density in the nitrogen plasma, $n_e = Z \times n_0$, does not reach a critical value (10^{21} cm^{-3}) under the Nd-laser pulse even if the temperature of the equilibrium plasma reaches 80 eV. Nevertheless, one can expect a sufficient change in the plasma index of refraction for the Nd-laser radiation with a 1.06μm wavelength in this case.

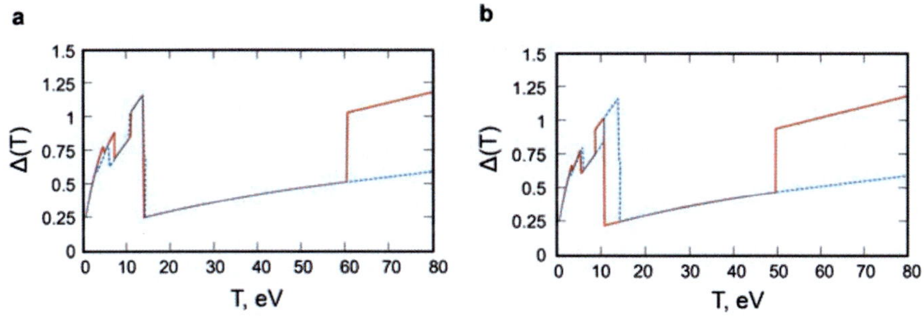

Fig. 2.11 Dependence of half width of ion charge number of density $\Delta(T)$ on plasma temperature for equilibrium plasma (solid lines) and Elvert plasma (dash lines) under (**a**) atmosphere pressure of gas and (**b**) at gas pressure of 0.1 of atmosphere one

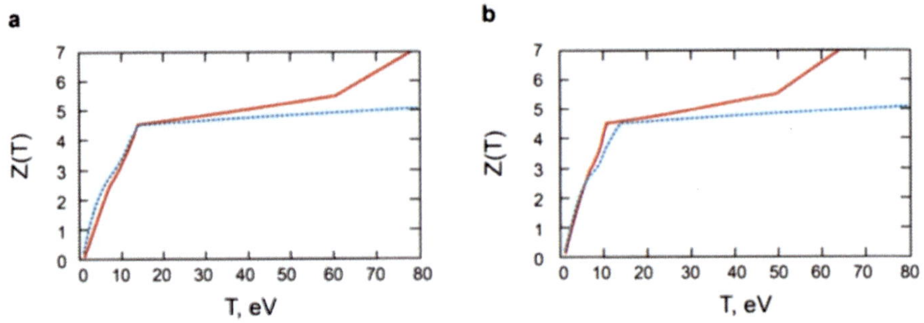

Fig. 2.12 Dependence of ionization potential degree $Z(T)$ on plasma temperature for equilibrium plasma (*solid lines*) and Elvert plasma (*dash lines*) at (**a**) atmospheric pressure and (**b**) at pressure of 0.1 of atmosphere one of nitrogen

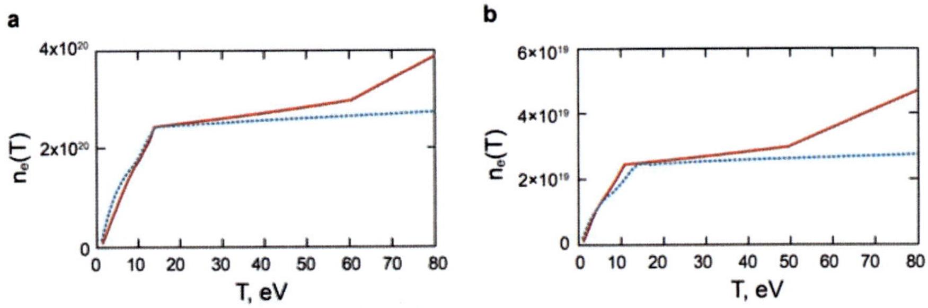

Fig. 2.13 Electron number density of equilibrium plasma (*solid lines*) and Elvert plasma (*dash lines*) at (**a**) atmosphere pressure and (**b**) at decreased pressure of nitrogen

It is known [29] that a distinctive feature of the plasma being ignited by an Nd-laser in nitrogen atmosphere is necessary to consider the laser radiation absorption not only via Inverse Bremsstrahlung effect but also via multi-photon ionization of nitrogen molecules (Fig. 2.14). The ratio of relative absorption coefficients of these processes is estimated as follows ($e^{h\nu/kT}-1$):1. The photo-ionization predominates the Inverse Bremsstrahlung effect if laser radiation quantum of energy, hν, exceeds 0.7 kT. Noting that the Nd-laser quantum of energy is of 1.13 eV.

To estimate the plasma absorption of Nd-laser power in the range of a first ionization potential of gases, corresponding to $h\nu$ > 0.7 kT, one can use the Biberman-Norman formula that includes not only Inverse Bremsstrahlung effect but also photo-ionization of excited atoms, namely [30]:

$$\mu_{\omega}=\frac{0.14\xi(\omega)p^2x_e^2\exp\left(h\Delta v/kT\right)\left(\exp\left(hv/kT-1\right)\right)}{\left(T/10^4\right)^{5/2}\left(hv_{eV}\right)^3} \qquad (2.43)$$

where $x_e = p_e/p$ is a mole fraction of electrons, $p_e = n_e$/kT is electron pressure, p is total pressure, g_1 and g_0 are statistical weights of ions and neutral atoms,

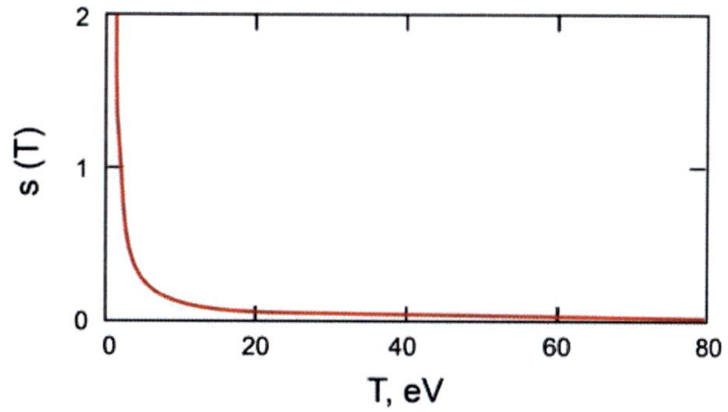

Fig. 2.14 Ratio of multi-photon ionization by Nd-laser radiation to electron collision ionization of nitrogen

respectively, and $\Delta\nu$ is a decrease of bottom boundary of plasma luminescent continuous spectrum being estimated by the Ingless-Teller formula:

$$h\Delta\nu/kT = 0.68p^{0.27}(T/10^4)^{-1.27} \tag{2.44}$$

And $\xi(\omega)$ is Biberman-Norman factor specified for every kind of gas and depending on a wavelength. For nitrogen and a 1.06μm wavelength, one can $\xi(\omega) = 0.6$.

But at high plasma temperature, the estimation of the radiation absorption coefficient is determined by the inverse Bremsstrahlung effect corrected by stimulated emission [27], namely:

$$\mu_\omega = \frac{3.1 \cdot 10^{-31} Z^3 n_0^2 g}{(T)^{3/2}(h\nu_{\text{eV}})^2}, \text{ where } g \approx 0.55 \ln \frac{2.4 \cdot 10^3 T}{Z^{4/3} n_0^{1/3}} \text{ is the Gaunt factor.} \tag{2.45}$$

Figures 2.15 and 2.16 illustrate the absorption coefficients of Nd-laser radiation power in a plasma as the plasma temperature functions. These data were obtained for the LTE plasma.

As a result of the brief analysis of plasma ignited by Nd-laser pulse (laser radiation) in a nitrogen atmosphere, one can conclude that photo-ionization process predominates a triple recombination in nitrogen, in the energy range of a first ionization potential of (1–2 eV) of nitrogen molecules, only in that case when $n_e < 10^{17}$ cm^{-3}, in this condition is true independently on the plasma state. Similar estimations of the plasma ionization processes show that its ionization degree can reach 4 (four) at the plasma temperature of 15 eV and nitrogen pressure of 1 atm. As it follows from these estimates, the laser radiation intensity should be about 10^{14} W/cm^2 and higher to reach the effects of multiple ionization of nitrogen.

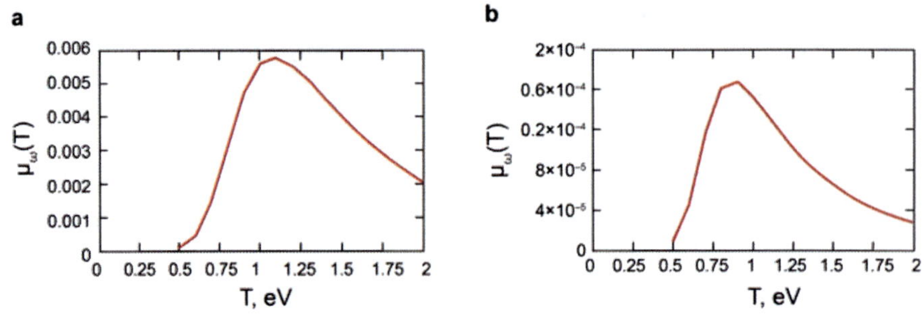

Fig. 2.15 Radiation absorption coefficient $\mu_{\omega}(T)$ for the Nd-laser-ignited plasma in nitrogen of (**a**) the atmosphere pressure and (**b**) low (of 0.1 atm) nitrogen pressure in the range of a first ionization potential of nitrogen

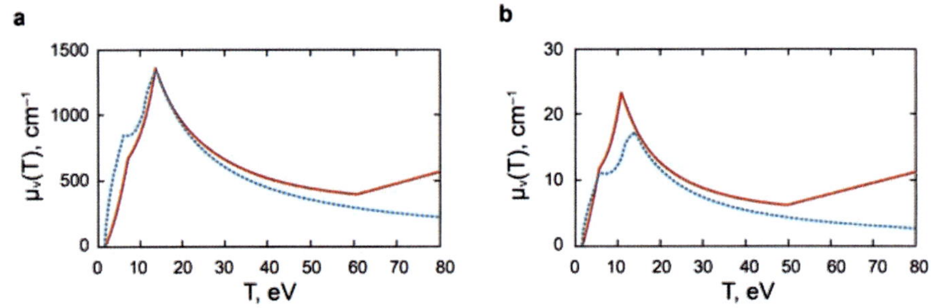

Fig. 2.16 Multi-photon ionization absorption coefficient of the Nd-laser radiation as function of the plasma temperature at (**a**) atmospheric and (**b**) low (0.1 atm) pressure of nitrogen

2.3.2 *Conversion Efficiency of Laser Power into Plasma Temperature*

Now, using the main conclusions derived in the previous section, let us consider the conversion efficiency of laser pulsed power into plasma energy (temperature), which is ignited via laser-induced breakdown of gas as compared with the experimental data obtained in Ref. [31]. As we mentioned above also, the conversion efficiency defines thrust production of the pulsejet and rocket laser propulsion. We can determine the conversion efficiency as a ratio of the pulse energy absorbed by a gas-plasma in a breakdown region, E_{ab}, to the laser pulse energy, E_{lp}, delivered to vehicle nozzle, namely, $\eta = E_{\mathrm{ab}}/E_{\mathrm{lp}}$.

Special experiments on the laser radiation interaction with gaseous plasma were made in Ref. [31] to determine the conversion efficiency. The experimental technique was based on recording of the time interval that was occupied by an ignited pressure-wave to propagate from a laser focus point to a detector being arranged at a certain distance from the focus. In these experiments, laser pulse energy, kind of gas, and its pressure were variable parameters. The opportunity to make photographs of the gas breakdown region as spatial patterns registered at the laser radiation wavelength was achieved. The effect of the radiation absorption by the ignited high density plasma allowed discovering sharp boundaries within the plasma volume.

In the figures below (see Figs. 2.17 and 2.18), shadow patterns of the plasma are shown at various pressures of the air and nitrogen. Here, dark shades are plasma ignited by the laser beam that has been focused from right to side left of every pattern.

In Fig. 2.17, it is shown that the gas volume occupied by the plasma depends on laser pulse energy. Namely, the plasma volume estimated for 100 J pulse energy is of 100 mm^3. The left part of the plasma region looks like a cone corresponding to a focused laser beam shape, but the right part of the plasma looks like a cylinder. There is no plasma shadow in the region behind a focused point of the beam. Probably, this effect illustrates laser radiation screening by the ignited plasma in front of focus point.

But, at low gas pressure, the plasma volume changes its shape, and the plasma is also ignited behind a focus point (Fig. 2.18). The plasma becomes transparent

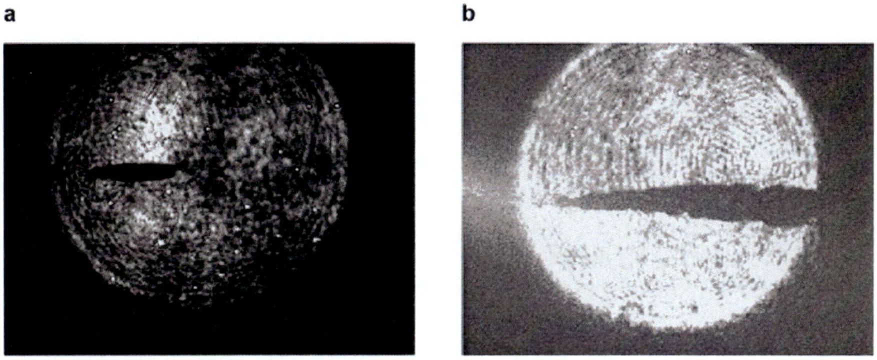

Fig. 2.17 Shadow patterns of nitrogen plasma ignited by laser pulses of (**a**) 4 J and (**b**) 100 J

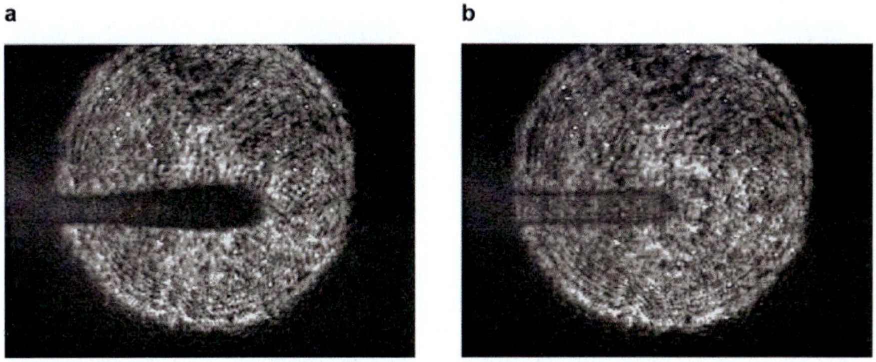

Fig. 2.18 Shadow patterns of plasma ignited at decreased nitrogen pressure (**a**) $P = 0.125$ atm and (**b**) $P = 0.05$ atm when pulse energy ~45 J

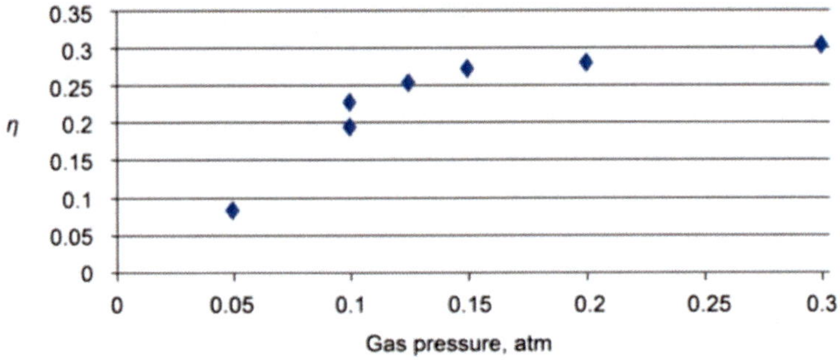

Fig. 2.19 Absorption efficiency of laser power, η, by plasma ignited in nitrogen as function of gas pressure, atm

partially for laser radiation, and the volume looks like a cylinder whose diameter exceeds the laser beam waist in the focus region. That conformities changing manner of the laser radiation interaction with the plasma. As it follows from Sect. 2.3.1, the process of multi-photon ionization of a gas exceeds the collision ionization in the case of a low gas pressure.

These processes of plasma ignition affect strongly the efficiency of laser power conversion into plasma energy (see Fig. 2.19). At low pressure (that is less than 0.2 atm), the conversion efficiency increases with the gas pressure increase, but then it remains as a constant up to atmospheric gas pressure. The conversion efficiency is also independent of the laser pulse energy if it is changed within a 20–100 J range at atmospheric gas pressure, and it keeps in a value of 35–40%. Similar data were obtained for the plasma ignited in the atmosphere air.

Another interesting fact from these experiments is radiation transmittance of the plasma. It is of 5% if the pressure ranges within 0.3–1.0 atm, and of 80% at low pressure less of 0.05 atm. And absorption coefficient of the radiation with a 1.06μm wavelength by the plasma ignited in nitrogen of atmospheric pressure is $\mu \sim 20\ \text{cm}^{-1}$.

To explain the data obtained, the model of multi-ionized plasma being considered in the previous section should be applied. Let us estimate such characteristics of the plasma as index of refraction and absorption coefficient, temperature, number density of electrons, and laser power release in the plasma.

As is known, the optical properties of the plasma are determined by its dielectric conductivity $\varepsilon(\omega)$ that depends on the ionization degree of plasma, plasma resonance frequency ωp, frequency of electron collision with hard particles of the plasma ν_{ei} as well as laser radiation spectrum. As it follows from the experiments, the following conditions are satisfied, namely, $\omega >> \nu_{ei}$, $\omega > \omega_p$. If we exclude the influence of ions on dielectric properties of the plasma, then we obtain the following ratio for the plasma dielectric conductivity that is valid in the case of a highly ionized plasma:

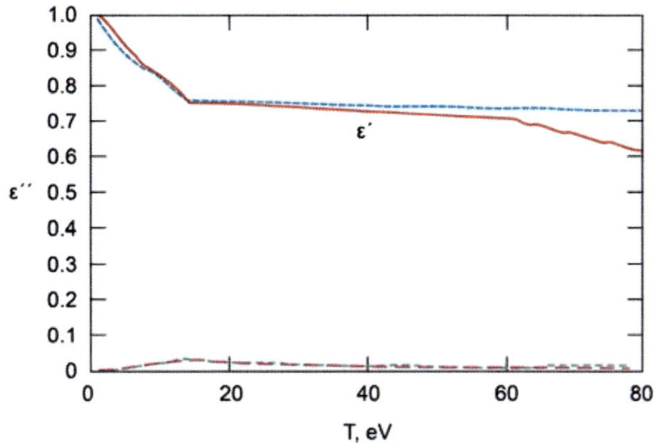

Fig. 2.20 Real and imaginary coefficients of plasma dielectric conductivity as dependencies of plasma temperature

$$\varepsilon = 1 - \frac{\omega_p^2}{\omega^2}\left(1 - i\frac{\nu_{ei}}{\omega}\right) \tag{2.46}$$

Figure 2.20 illustrates the variation of the real, ε', and imaginary, ε'', parts of the dielectric constant of plasma as functions of temperature of the plasma ignited in a nitrogen by Nd-laser pulse [30]. Real part of the dielectric conductivity is less than 0.7 and imaginary one is positive and it is much less than unit within a wide variation range of the plasma temperature.

The obtained data on the dielectric conductivity allow estimation of the absorption of laser radiation by the plasma. It is known, the transmittance of laser radiation by the plasma, $T(z)$, is determined by the plasma absorption index μ_ω and plasma thickness z, that is, $T(z) = I(z)/I(0) = \exp(-\mu_\omega z)$. In these experiments, the characteristic diameter of the plasma volume is estimated as $d = 0.2$–0.34 cm when laser pulse energy was varied from 5 through 80 J. As it is seen from the figures above, laser radiation is absorbed almost completely if the plasma temperature exceeds 1.7 eV. The plasma temperature estimated in accordance with the experimental data on plasma luminescence corresponds to 2 eV. The last result means that the plasma shadow patterns observed in the experiments were caused by the absorption of a laser radiation by the plasma.

As is seen from Fig. 2.21, the sharp boundary of the plasma volume observed in the experiments at atmosphere pressure is of 0.4 mm in a thickness, and the thickness decreases with plasma temperature increase. At low pressure of nitrogen, the plasma transmittance increases (Fig. 2.22), and it reaches $T = 0.9$ at a pressure of 0.05 atm and $T = 0.7$–0.8 at 0.1 atm.

In accordance with the accepted plasma model, the number density of electrons in plasma reaches (3–5) $\times$ 10^{20} cm^{-3} under the Nd-laser pulse. That means that the penetration depth of the laser radiation into the plasma is defined by a skin layer of the plasma, namely, $\lambda_{sl} = 1/I\ m(k)$ where k is the plasma wave number. It is seen from Fig. 2.23 that laser pulse energy is absorbed in a thin layer of the plasma, which

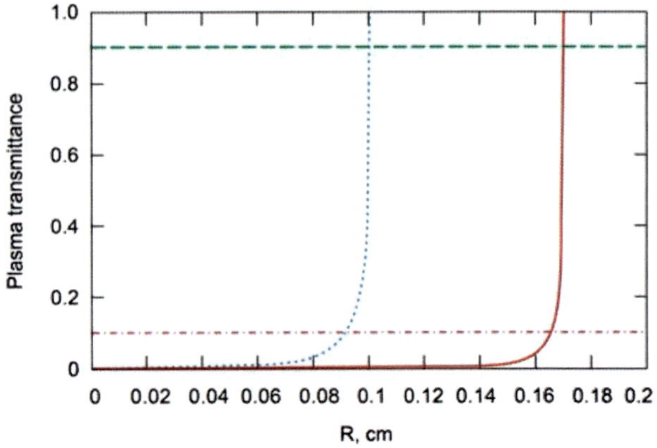

Fig. 2.21 Plasma transmittance for 1.06μm radiation at atmospheric pressure of nitrogen. *Solid line* corresponds to plasma diameter of 0.34 cm, and *dash line* to 0.2 cm, respectively

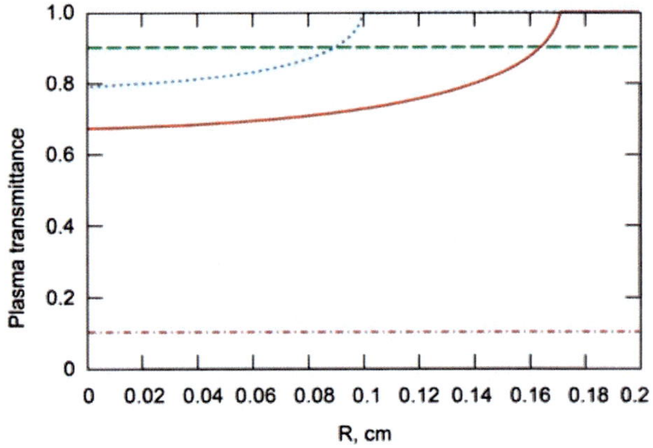

Fig. 2.22 Plasma transmittance of a 1.06μm radiation at low gas pressure

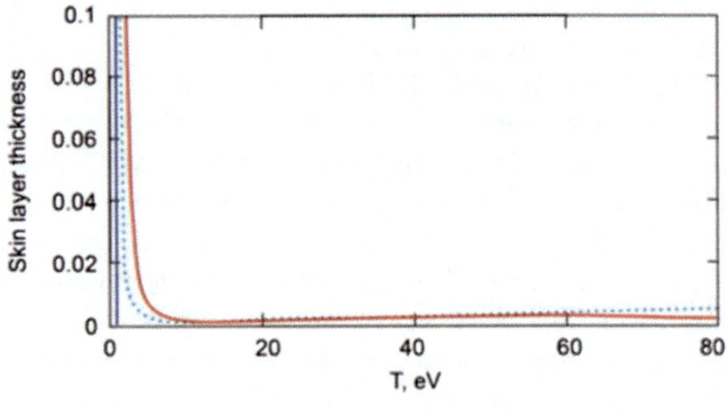

Fig. 2.23 Plasma skin layer as function of the plasma temperature. *Solid line* corresponds to equilibrium plasma, and *dash line*—to corona equilibrium plasma

is nearest to the incoming laser beam. The skin layer thickness is much less than the plasma depth along the laser beam path (1–2 cm).

As we see, the model of multi-ionized plasma allows setting all thermodynamic parameters of the plasma as functions of its temperature and initial density of a gas. And, the inner specific energy of plasma is estimated by the following formula as $\varepsilon_w = \varepsilon_{ei} + n \times Q(Z)$, where $\varepsilon_{ei} = 3/2n(1 + Z)T$ is the translational thermal energy of electrons and ions, and $n \times Q(Z)$ is the potential energy of an electron to be separated from atom or ion. The $Q(Z)$ function is formed by linear interpolation of discreet values of consecutive amount of ionization potentials I_z starting from 1 through z.

It is known [27] that the thermal capacity of a completely ionized gas is determined by the translational energy of its particles. That means the thermal capacity of multi-ionized plasma is determined by a translational energy of all particles being a part of the plasma, and the specific thermal energy of the electron and ion translational movement determine the efficiency of laser power conversion into the plasma temperature.

Characteristics of the gaseous plasma, such as specific inner energy of plasma (red line), specific translational energy of ions and electrons in the plasma (green line), and absorption coefficient of laser power (blue line) are presented in Fig. 2.24 as functions of the plasma temperature ignited in air and in nitrogen at various gas pressures. Horizontal lines of 300–500 J/cm^3 in the figure mark the specific energy of the laser power released in the experiments by Ref. [31]. It can be seen that the plasma temperature increases up to 7–10 eV if the Nd-laser pulse energy density is of 250–500 J/cm^2.

Thus, it is shown that the ionization degree of the plasma is four if its temperature reaches 10 eV, so that the electron number density is about (3–5) $\times$ 10^{20} cm^{-3} and is close to a critical density $n_e \sim 10^{21}$ cm^{-3}. Maximum absorption coefficient of laser radiation by the plasma is $\mu \sim 1400$ cm^{-1} exceeding by five orders the coefficient corresponding to the plasma temperature of first ionization potential, $\mu \sim 0.006$ cm^{-1}. All of these data were obtained at the specific laser power input of 300–500 J/cm^3 into the plasma.

A basic outcome of the considered plasma model means that the multi-ionized plasma generated by a laser pulse in a gas predicts adequately the conversion efficiency of the laser power into the plasma temperature, when compared with data obtained experimentally. In a common case, the plasma of laser-induced breakdown of gases is non-stationary and non-equilibrium thermodynamically physical phenomenon. But, the plasma may be considered to be in an equilibrium state if we record such moment of time of plasma life when all processes of the laser pulse interaction with the plasma are completed (this period takes of the order of 30 ns) but the gas-dynamic processes of its deformation have not formed yet (about 100 ns is needed). The latter statement is correct if all processes of energy transfer in the plasma are less than 30 ns duration, and the plasma temperature is sufficiently high.

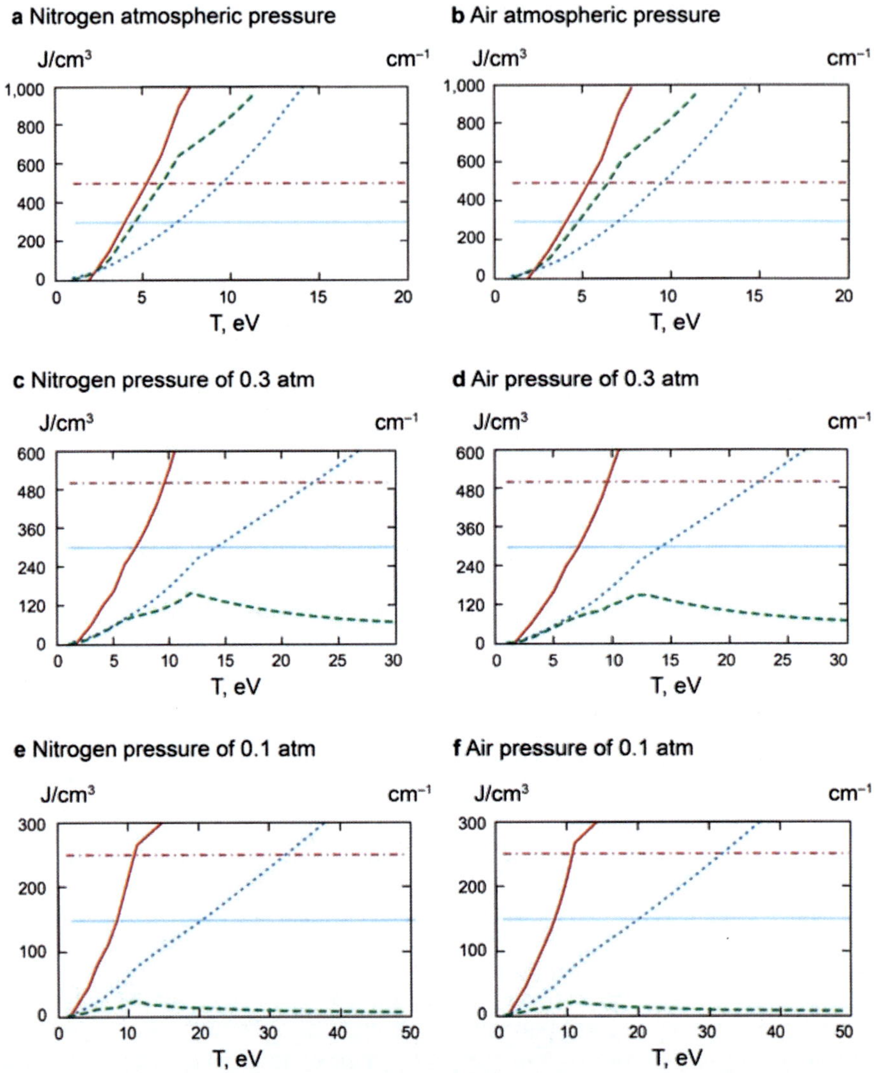

Fig. 2.24 Specific internal energy of plasma, $\varepsilon_{\omega}(T)$ (J/cm^3), specific translational energy of electrons and ions, $\varepsilon_{ei}(T)$ (J/cm^3), and laser power absorption coefficient μ, cm^{-1} as functions of the plasma temperature

2.4 Numerical Calculations of Non-stationary and Non-isentropic Gas Flows as Applied to Laser Propulsion

A number of investigations of the pulsejet laser propulsions published were performed by using numerical calculations of gas flows in the laser-propulsion engines at various conditions on the density and pressure of outside medium. The primary subject of these calculations was basic gas-dynamic processes of thrust production in the engine of a various designs. Usefulness of such type of the studies is useful and consistent because the results of the calculations discover particular properties of interaction of generated shock waves with jet nozzles of the engine. Moreover, the gas-dynamic flows in the nozzles are characterized by non-stationary and non-isotropic properties.

In the section, a few theoretical models of gas-dynamic flows in a pulsejet LPE are considered on a basis of a continuity hypothesis of flows in a nozzle under various pressure of the surrounding gas.

2.4.1 Perfect Gas Flow Models and Numerical Algorithms to Calculate Gas Flow of Pulsejet Laser Propulsion

The hypothesis of gas flow continuity in a nozzle becomes ineligible if the gas pressure inside the nozzle becomes less than a crucial value. The parameter that characterizes the crucial pressure is the Knudsen number, $Kn = \lambda/L$, that is defined as a ratio of mean free path of gas molecules, λ, at given gas pressure to characteristic size of the nozzle, L (for the example, the nozzle exits diameter). Characteristic Knudsen number is approximately equal to 0.1 if the flow continuity is valid without needing special boundary conditions of the gas flow slip. Variation of the continuity of gas flow inside a nozzle is shown in Fig. 2.25 as function of surrounding gas density and characteristic size of the nozzle.

For example, the model of flow continuity as applied to pulsejet laser propulsion will be valid if a characteristic size of the LPE nozzle is about 10^{-1} m and surrounding gas density of $\rho \geq 10^{-5}\ \rho_0$, where ρ_0 is the gas density at normal conditions. Comparing the diagram of Fig. 2.25 with a graph of the atmosphere air density variation with the atmosphere altitude increase (see Fig. 2.26), we deduce that the air density at the altitude of 80 km will define applicability of the pulsejet propulsion model.

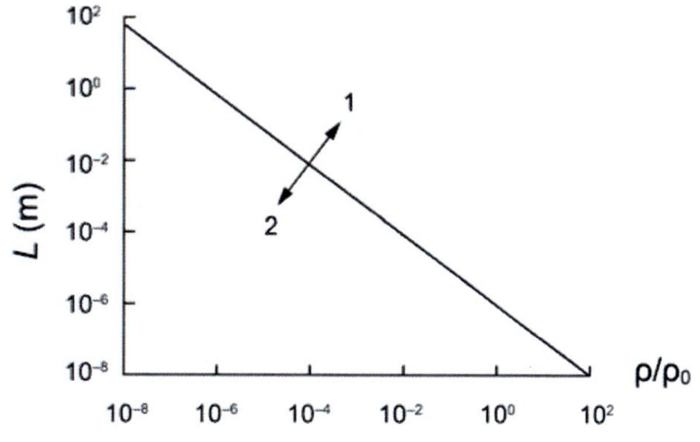

Fig. 2.25 Applicability of the continuity concept of gas flow inside LPE as function of gas density outside the engine

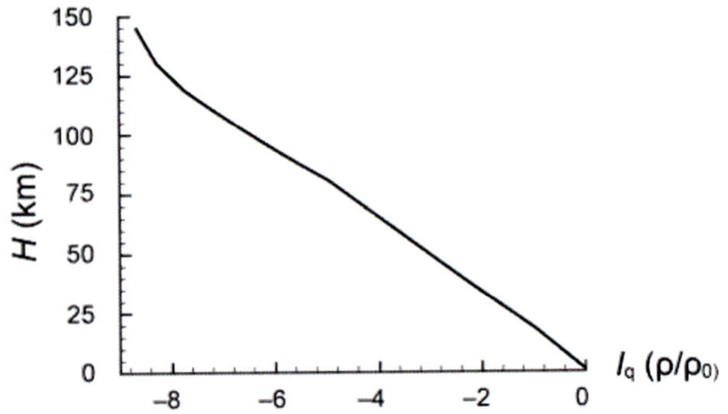

Fig. 2.26 Variation of atmosphere air density with altitude above the Earth ground

Gas-dynamic models of flows that we consider here include within them the flows of (a) a perfect gas, (b) an ideal equilibrium plasma, and (c) an ideal non-equilibrium plasma. In (a) and (b) cases, temperature of the plasma ignited due to laser breakdown of the gas as well as laser power conversion into the plasma temperature are assumed to be known. But in the case of (c), the laser power absorption by the plasma is assumed to be dependent on the radiation intensity of the focused beam as well as on molecular components of the plasma ignited. We also consider that the focus region is limited in size, and the processes of plasma ignition are limited in time. These phenomena depend also on the laser radiation wavelength, pulse energy, thermal properties of gas flow as well as on the mirror focusing parameters. In the section, we consider parabolic nozzles as a rule.

The model of perfect gas flow within a parabolic nozzle was considered in Ref. [32] as applied to pulsejet laser propulsion to determine the efficiency of thrust production as a function of the sizes of the ignited plasma. Laser energy conversion into a plasma in the parabolic mirror focus was assumed to be a well-known function of time, and the breakdown volume was considered as (a) plasma with a fixed volume during the laser pulse that corresponds to approach that of "a local explosion" examined in Sect. 2.2, and (b) the plasma volume increases proportionally to a gas volume occupied by the shock wave that is generated in the gas and propagates out of the breakdown region.

2D non-stationary equations describing evolution of the gas flow within the nozzle are written in the following form:

$$\frac{\partial U}{\partial t} + \frac{\partial F_{\mathrm{x}}(U)}{\partial x} + \frac{\partial F_{\mathrm{y}}(U)}{\partial y} = Q \tag{2.47}$$

where Q is the thermal power corresponding the laser power flux absorbed by plasma, as well as:

$$U = \begin{pmatrix} \rho \\ \rho u \\ \rho v \\ E \end{pmatrix}, \quad F_{\mathrm{x}}(U) = \begin{pmatrix} \rho u \\ \rho u^2 + p \\ \rho uv \\ (E + p)u \end{pmatrix}, \quad F_{\mathrm{y}}(U) = \begin{pmatrix} \rho v \\ \rho uv \\ \rho v^2 + p \\ (E + p)v \end{pmatrix}.$$

This system of gas-dynamic equations is enclosed by the following characteristics of perfect gas state:

$$p = p(\rho, T), \quad E = \frac{1}{(\gamma - 1)} \frac{p}{\rho} + \frac{(u^2 + v^2)}{2} \tag{2.48}$$

where $\gamma = c_P/c_V$ is adiabatic exponent.

Equations (2.47) can be solved by using the Godunov numerical scheme of a high numerical resolution on a basis of non-structured dynamic adaptive numerical grid [33]. This numerical algorithm was previously tested by solving a wide range of gas-dynamic problems including complicated shock waves in gas flows. Some examples of the numerical solutions of the Eq. (2.47) by using the model for a perfect gas (case a) as applied to pulsejet laser propulsion are presented in Fig. 2.27. Gaseous helium is considered to be injected into a parabolic nozzle through an orifice located in a vertex of the nozzle with a mass consumption rate of 5.5 g/s. The laser pulse had a pulse length of 3μs and pulse energy of 8 J. Outside gas pressure is assumed to be of 10 Pa.

Figure 2.27 illustrates intricate patterns of gas flow inside a parabolic nozzle at various moments in time after laser pulse ceases, corresponding different models of plasma. Gas flow patterns in Fig. 2.27(a, b) correspond to the model of ignited plasma with a constant volume in time. Figure 2.27(c, d) was obtained for the plasma

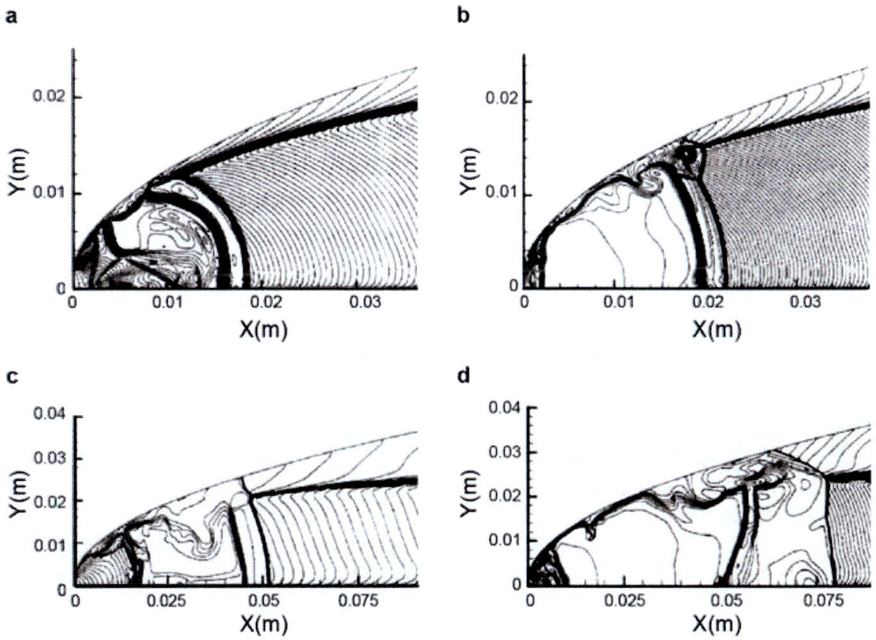

Fig. 2.27 Gas flow patterns as Mach number distribution in a parabolic nozzle after the laser pulse ceases

volume adapted to gas volume occupied by generated shock wave. Correspondently, Fig. 2.27(a, c) belong to a time delay of 3μs after the end of laser pulse, and Fig. 2.27 (b, d)—correspondently to the delay of $t = 11$μs. One can see from the figures, gas flow structures are sufficiently complicated due to the interaction of the generated shock waves with each other and with the nozzle walls for both plasma models, and the flow patterns to differ from the approximations of gas flows inside the nozzle, which are usually used in a local explosion theory.

2.4.2 Model of Equilibrium (Thermal) Plasma

It should be noted that the assumption of invariability of the adiabatic index of the plasma and gas located right behind generated shock wave is not quite correct because of the high temperature both of them. Therefore, to make correct numerical investigations of gas flows in the pulsejet laser propulsion, it is necessary to use a more accurate model of the plasma medium. Laser power release in the plasma volume, which is limited in size and is accompanied by a consequent abrupt rise of the plasma temperature, make necessary to consider the plasma medium as a multi-component medium with variable components. Moreover, the thermal and dynamic equations of the multi-component medium must include the physical-chemical reactions running in the plasma.

In this case, the model of equilibrium plasma allows simplifying the general equations which have to be used [34]. Simplification of the equations is based on the use of the concept of "averaged" medium for which all conservation equations can be written in the form of (1) instead of writing the equations for every component of the plasma. The "averaged" equations are closed by equations of plasma state at the assumption of equilibrium of physical-chemical reactions running under a laser pulse, including relations describing the medium components state by consequent thermodynamic functions. As applied to the atmosphere air that can be used as a propellant for the pulsejet laser propulsion, such a type of the equations was deduced by Kraiko in Ref. [35].

Similar numerical analysis of the averaged equations of plasma allows specifying the gas-dynamic effects running in a parabolic nozzle after the end of the laser pulse. In Fig. 2.28, the parabolic nozzle is presented schematically as the focus region of laser power conversion into a plasma. In the calculations, the laser pulse energy is assumed to be 0.012 J, the nozzle exit diameter is 20 mm, and its length is 8.3 mm.

As an initial condition, the atmosphere air at a normal pressure is considered to fill up the nozzle. Basic size of the gas-plasma volume is estimated to be equal to $d_0 = 0.8$ mm by using the data published in Ref. [32]. The gas-plasma temperature after the end of the laser pulse is estimated to be equal to $T_0 = 11{,}000$ K.

Examples of the shock waves evolution inside the nozzle are shown in Fig. 2.29, which were obtained on the basis the equilibrium plasma model. The figure illustrates pressure distributions of the plasma in the nozzle are presented at successive periods of time passed after the end of the laser pulse. The pictures of the flow

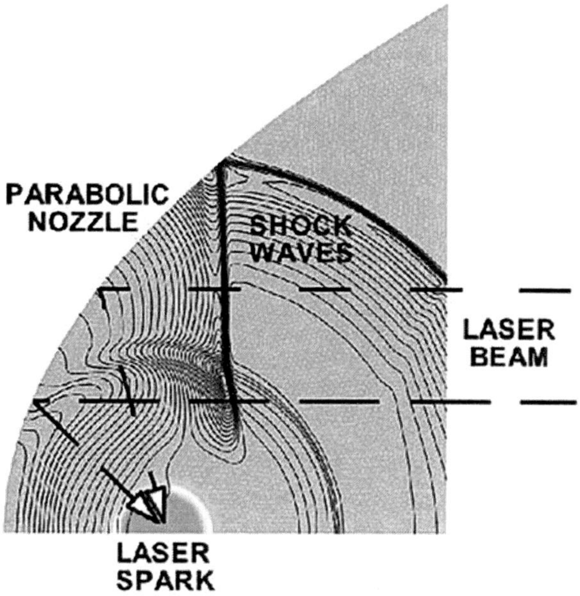

Fig. 2.28 Laser-propulsion flow structure scheme in a parabolic nozzle at some moment of time after laser pulse end

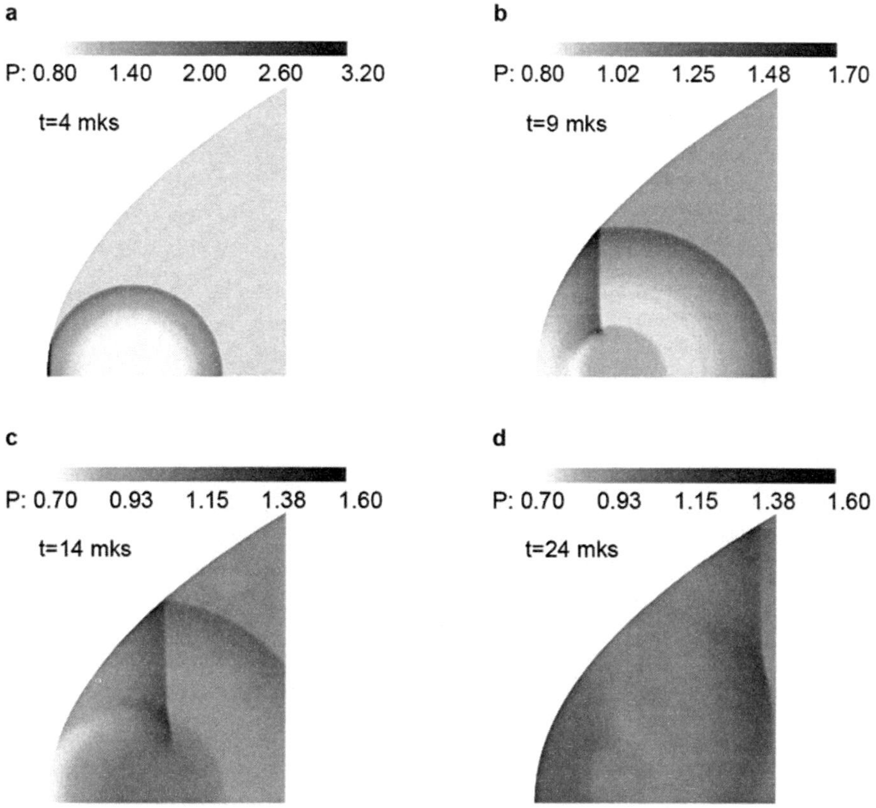

Fig. 2.29 Shock waves evolution in accordance with the equilibrium plasma model

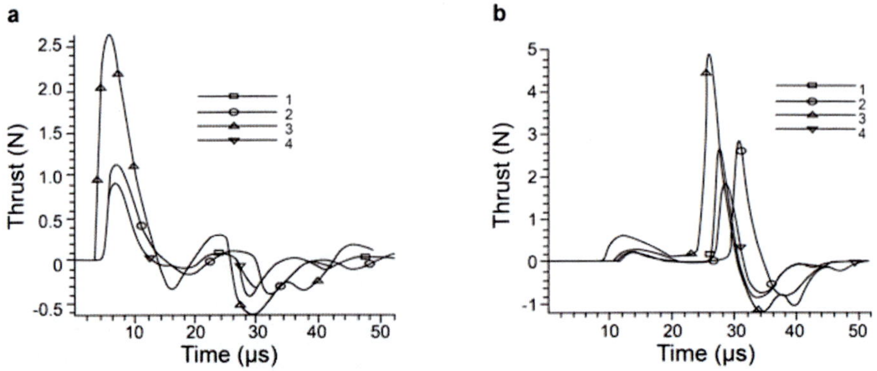

Fig. 2.30 (**a**) Pressure and (**b**) jet evolutions of thrust in time. In the figures, (1) $\gamma = 1.4$, and laser pulse energy $E = 0.0025$ J; (2) $\gamma = 1.1$, $E = 0.00983$ J; (3) equilibrium plasma model, and $E = 0.012$ J; (4) equilibrium plasma model, and $E = 0.0025$ J

structure are very complicate and characterized by complex wave structure. It is seen that a strong shock wave reflected from the nozzle wall interacts with the initial plasma volume. The pictures show also that the gas-dynamics patterns of the medium are far from its equilibrium state (that is undisturbed ones) even till 24μs after the end of the laser pulse.

The consequent time evolution of the thrust produced by the nozzle are presented in Fig. 2.30 as time-varying force that was calculated by two ways, namely: (a) integrating of gas pressure distribution along the nozzle walls (Fig. 2.30(a)) and (b) recording of the general impulse of an exhaust jet (Fig. 2.30(b)). In the figures, two different results of thrust production are presented which were calculated in accordance with two models of the gas flow, namely, the flow of perfect gas with the adiabatic index of $\gamma = 1.4$ and an equilibrium plasma with $\gamma = 1.1$. One should be mentioned that in the case of equilibrium plasma, high excitation of the inner degrees of freedom of the gas molecular is assumed.

As can be seen from Fig. 2.30(a), the thrust impulse originates at that point of time when shock wave reaches the nozzle walls, assuming this impulse is registered by a wall pressure detector. Temporal variation of the thrust possesses in nowhere monotone in character.

The instant means that thrust may be negative at that moment of time when rarefaction waves propagating from the nozzle exit reach the wall. If thrust is registered by a detector arranged at the nozzle exit, then it starts later on (Fig. 2.30(b)). The phenomenon illustrates the fact that exhaust jet is originating later at that moment of time when the generated shock wave reaches of the nozzle exit. Nevertheless, if we will integrate both time dependences of the thrust produced (a and b) then we obtain the same result on the thrust amplitude.

Unfortunately, both simple models of the laser power conversion into plasma cannot describe adequately the thrust production in the pulsejet laser propulsion because of a large rise of plasma temperature with increasing the laser pulse energy which does not correspond to the physical nature of the laser radiation interaction with the gaseous plasma [28]. That is why, the use of the model of non-equilibrium plasma and attraction of the plasma-chemical reactions running in the plasma under high-power laser radiation should be considered as well.

2.4.3 Model of Non-equilibrium Plasma as Applied to Pulsejet Laser Propulsion

The non-equilibrium plasma model includes both excitation of the internal degrees of a freedom for molecules and chemical reactions running under laser radiation. In this case, the system of the conservation equations is added to the equations of mass conservation for every plasma component as well as equations of the vibration energy conservation and electron gas energy. In the system proposed by Park [36], the gas-dynamic equations are as follows:

$$\begin{aligned} &\frac{\partial \rho_s}{\partial t} + \nabla\left(\rho_s \vec{V}\right) = \dot{\omega}_s \\ &\frac{\partial \rho e_v}{\partial t} + \nabla\left(\rho e_v \vec{V} + \rho\right) = \sum_{s=mol} \rho_s \frac{e^*_{v,s}}{\langle \tau_s \rangle} + \sum_{s=mol} \dot{\omega}_s \widetilde{D}_s \\ &\frac{\partial \rho e_e}{\partial t} + \nabla\left(\rho e_e \vec{V} + p_e\right) = \vec{V} \cdot \nabla p_e + 2\rho_e \frac{3}{2} R(T - T_e) \sum_s \frac{\nu_{es}}{M_s} - \sum_s \dot{n}_{e,s} \widetilde{I}_s \end{aligned} \quad (2.49)$$

where $\rho = \sum_s \rho_s$; $p = \sum_s p_s$; $p_s = \rho_s RT/M_s, \rho_s$ are the partial pressure and density of either of the s components of plasma, $p_e = \rho_e RT_e/M_e$, T_e are the electron gas pressure and its temperature, T is combined reciprocating and rotating temperature, $e = e_{tr} + e_v + e_e + \sum_s \rho_s h_s^0$ is total combined reciprocating and rotating energy, e_v is total vibrational energy, e_e is total energy of electron excitation, and $e_v{}^*$ is equilibrium vibrational energy. These formulas include themselves also $h_s{}^0$ the enthalpy of s-th component of plasma $\widetilde{D}_s$, as additional vibrational energy being generated at dissociation of s-th component, ν_{es} as a frequency of electron-hard particles collisions, $\widetilde{I}_s$ as ionization potential, $\dot{n}_{e,s}$ as a rate of electron increase during the ionization plasma components, $\dot{\omega}_s$ as a mass rate of s-th component generation, M_s as s-th component molecular mass, and $\langle \tau_s \rangle$ time constant of temporal vibration relaxation.

To complete this system of equations, (2.49), it is necessary to fix a set of plasma-chemical reactions running in the plasma and to determine their reactive rates as well as explicit thermodynamics ratios. As an example, let us consider nitrogen as a propellant and let us define the plasma-chemical reactions running under the action of pulsed laser radiation. So, the following nitrogen plasma components have to be considered, namely [37]:

Dissociation	$N_2 + M - N + N + M$, ($M = N, N_2, N^+, N_2^+$, e)
Associative ionization	$N + N - N_2^+ + e$
Exchange by charges	$N^+ + N_2 - N_2^+ + N$
Ionization by electron collision	$N + e - N^+ + e + e$

The consequent rate constants of these reactions are determined in Refs. [36, 37]:

$$K_{f1} = 7 \cdot 10^{21} T_a^{-1.60} \exp\left(-\frac{113,200}{T_a}\right), \quad K_{f2} = 4.4 \cdot 10^7 T_a^{1.5} \exp\left(-\frac{67500}{T_a}\right),$$

$$K_{b1} = \frac{K_{f1}}{K_{e1}}, \quad T_a = \sqrt{T_v T} \quad\quad K_{b2} = \frac{K_{f2}}{K_{e2}},$$

$$K_{f3} = 10^{12} T^{0.5} \exp\left(-\frac{12200}{T}\right), \quad K_{f4} = 2.5 \cdot 10^{34} T_e^{3.82} \exp\left(-\frac{168600}{T_e}\right),$$

$$K_{b3} = \frac{K_{f3}}{K_{e3}}, \quad\quad K_{b4} = \frac{K_{f4}}{K_{e4}},$$

Here, K_{fi}, K_{bi} are rate constants of i-th direct and inverse chemical reactions, K_{ei} is equilibrium constant for i-th reaction ($i = 1$–4). It follows from Ref. [40], the rate constant of i-th reaction may be written in the following form:

$$K_{ei}(T) = \exp\left[B_1^i + B_2^i \ln(Z) + B_3^i Z + B_4^i Z^2 + B_5^i Z^3\right] \tag{2.50}$$

where $Z = 10{,}000/T$, B^i_e is an additional constant that is determined in Ref. [8].

And specific heat capacities are determined by using approximation formulas from Ref. [38] as:

$$c_{pi} = a_i + b_i + \frac{c_i}{T^2} \tag{2.51}$$

where a_i, b_i, c_i the coefficients which are variables for every component of plasma.

The model of a non-equilibrium plasma can be used to determine the laser power conversion into plasma temperature that plays an important role in thrust production. It is known that the laser power absorption by the plasma is realized due to generation of free electrons during two processes running in the breakdown volume, namely: (1) avalanche ionization or cascade collisional ionization of gas molecules and (2) multi-photon absorption. Every process exists for certain conditions of the laser radiation interaction with the plasma, and it depends on the threshold intensity of laser radiation at which the plasma is ignited.

Therefore, one of the basic criteria of choosing a propellant for the pulsejet laser propulsion is a low intensity threshold of the gas breakdown. But, another criterion is an efficiency conversion of the laser power into the plasma energy. To determine the conditions of fulfillment of these two criteria, the processes of non-equilibrium plasma ignition under the action of laser pulse in gases have to be considered [39, 40].

2.4.4 Discussion on the Applicability of Various Models of Plasma Ignited

Let us consider the presented models of a plasma ignited under the action of a laser pulse by analyzing the experimental data on efficiency of laser power conversion into a nitrogen plasma, which are presented in Sect. 2.3.2. The efficiency was estimated by recording the time interval that was needed to shock wave run from a plasma volume to pressure sensors arranged at some distances from this plasma. In the experiments, an Nd-laser was used to ignite the plasma, pulse energy of which was 13.3 J at the pulse length of 10 ns. The results of numerical simulation of the shock wave propagation to pressure sensors are presented in Fig. 2.31 in the form of

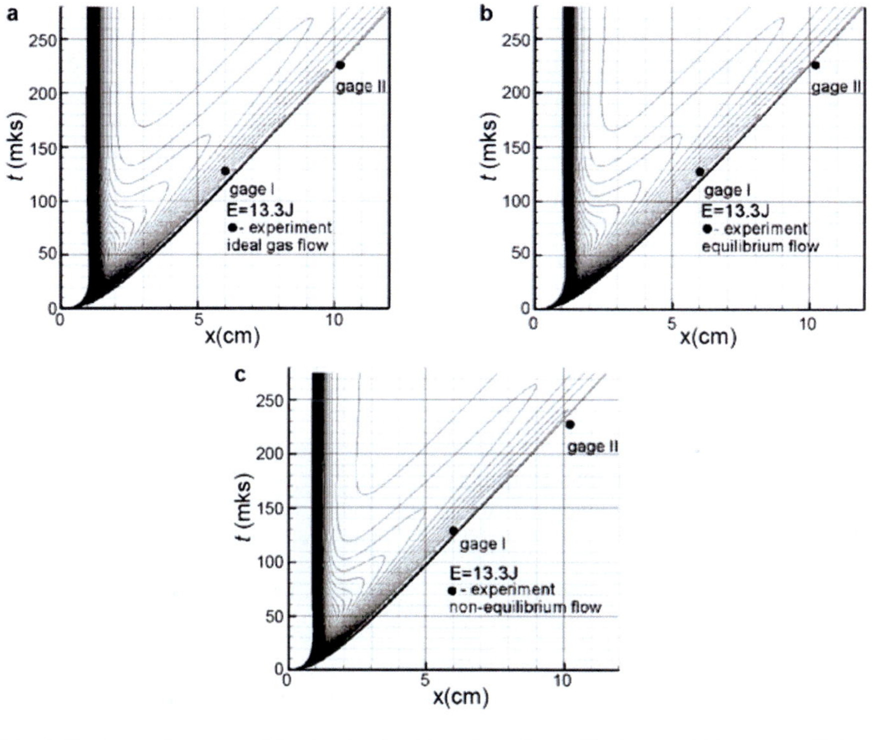

Fig. 2.31 (*x*–*t*)—diagrams of the density for (**a**) perfect, (**b**) equilibrium, and (**c**) non-equilibrium gas models. Nd-laser pulse energy 13.3 J, and pulse duration 10 ns

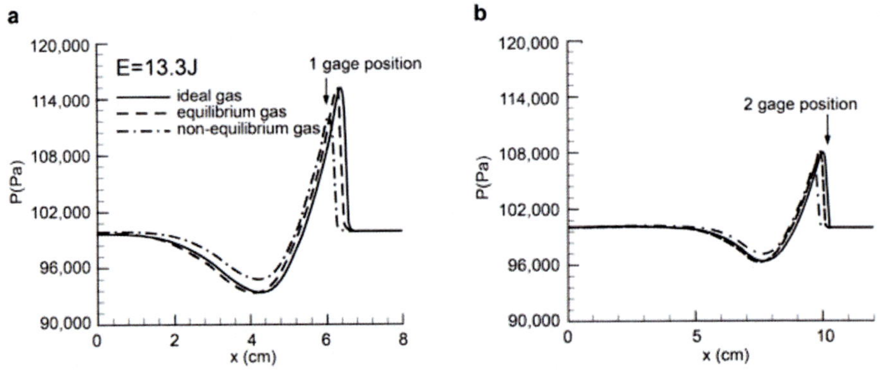

Fig. 2.32 Pressure profiles correspond to (**a**) 129μs and (**b**) 228μs laser pulse end. Arrows indicate positions of the shock wave in the experiment

x-*t* diagrams depicting variations of the plasma density in accordance with the models of (a) ideal perfect gas, (b) equilibrium plasma, and (c) non-equilibrium plasma.

These data show the nitrogen pressure profiles at the 129μs and 228μs moments of time registered after the Nd-laser pulse ends, shown in Fig. 2.32(a, b), respectively. These points of time correspond to time intervals during which the shock wave propagates to the first and second pressure sensors arranged at the distances of 6 cm and 10.2 cm from a laser beam focused point. One can see the numerical results predicted by the considered plasma models are in a good agreement with the experimental data although the models of the perfect gas and of the equilibrium plasma give a greater shock wave velocity that gives the shock wave power.

Figure 2.33 illustrates the other plasma characteristics, namely, plasma components concentration (a), density number of electrons (b), and plasma temperature (c), which are estimated at a time of 129μs. It is clearly seen that the typical time period of the plasma-chemical reactions occurring in the plasma are shorter than the relaxation time interval of the gas-dynamic processes in surrounding nitrogen.

Summering, the obtained numerical results on modeling of laser-ignited plasma (in a breakdown region) demonstrate a possibility of correctly analyzing the pulsejet laser-propulsion characteristics by using the model of non-equilibrium plasma. Moreover, the non-equilibrium plasma model plays an important role at production of laser ablative propulsion.

As for "air-breathing" laser propulsion, we consider this type of thrust production in Chap. 5. The point is that the air-breathing flow conditions are realized mainly during supersonic flight of a vehicle in the atmosphere, which requires specific consideration of both vehicle and laser radiation interactions with incoming supersonic gas flow.

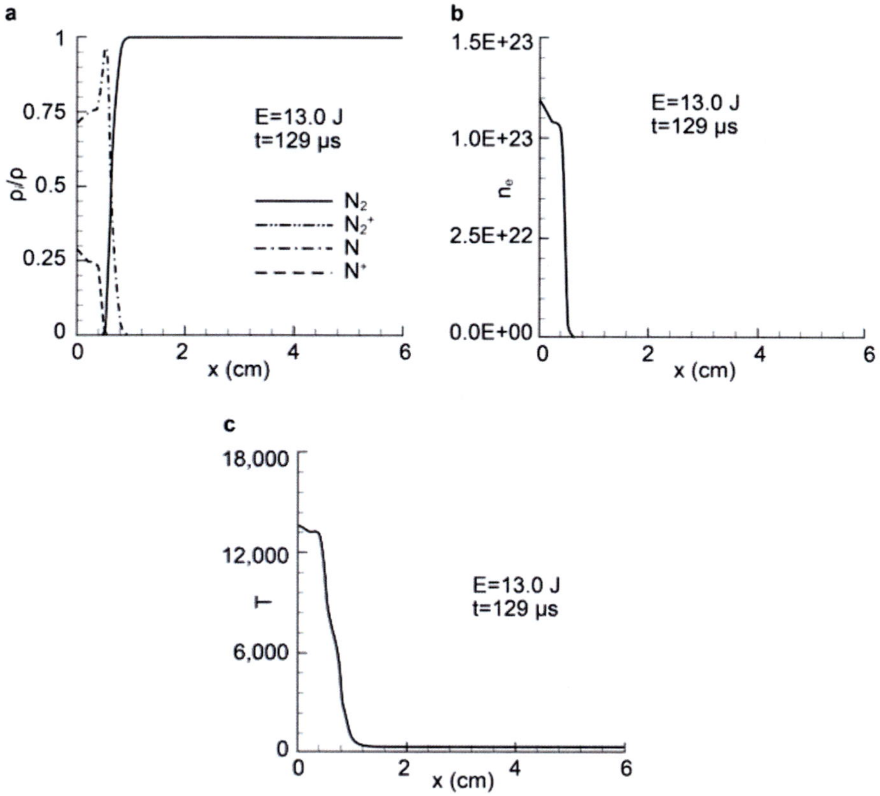

Fig. 2.33 Profiles of (**a**) the plasma component concentration, (**b**) the electron concentration, and (**c**) the temperatures

References

1. Katsurayamaa, H., et al.: A preliminary study of pulse-laser powered orbital launcher. App. Opt. **29**, 2303–2306 (1990). https://doi.org/10.1364/AO.29.002303
2. Bunkin, F.V., Prokhorov, A.M.: Laser power source application to produce a thrust. Adv. Phys. Sci. (Rus.). **119**, 425–446 (1976)
3. Ageev, V.P., et al.: Laser air-breathing jet engine. J. Quant. Electron. (Rus.). **4**, 2501–2513 (1977)
4. Sedov, L.I.: Mechanics of continues medium. Nauka, Moscow (1970)
5. Loytzjansky, L.G.: Mechanics of liquids and gases. Nauka, Moscow (1973)
6. Schall, W.O., et al.: Lightcraft Experiments in Germany. High-Power Laser Ablation III. Proc. SPIE. **4065**, 472–481 (2000). https://doi.org/10.1117/12.407370
7. Schall, W.O., et al.: Comparative lightcraft impulse measurements. Proc. SPIE. **4760**, 908–917 (2002). https://doi.org/10.1117/12.482046
8. Bohn, W.L.: Novel aspects in laser propulsion. Proc. SPIE. **7005**, 70051C (2008). https://doi.org/10.1117/12.785634

9. Scharring, S., Eckel, H.-A.: Review on laser lightcraft research at DLR stuttgart. AIP Conf. Proc. (2014)
10. Scharring, S., et al.: Downloaded laser propulsion. Standard. Iss., 19 Apr 2011 to 85.179.240.23. Redistribution subject to AIP license or copyright; http://proceedings.aip.org/about/rights_permissions
11. Bulaev, V.D., et al.: The experimental laser facility based on the high-power repetitively-pulsed E-beam sustained CO_2 laser. Beamed Energy Propulsion. AIP Conf. Proc. **766**, 361–372 (2004). https://doi.org/10.1063/1.1925157
12. Liukonen, R.A.: Efficiency of energy transfer into a recoil impulse for laser propulsion engine. Lett. JTPh (Rus.). **18**, 76–79 (1992)
13. Liukonen, R.A., Trofimenko, A.M.: Thrust force produced by laser propulsion engine. Lett. JTPh (Rus.). **18**, 81–85 (1992)
14. Apollonov, V.V., Tishchenko, V.N.: Laser air-jet engine: the action of shock waves at low laser pulse repetition rates. J. Quant. Electron. (Rus.). **37**, 798–800 (2007)
15. Apollonov, V.V., Tischenko, V.N.: Laser propulsion engine on the basis of shock waves merching effect. J. Quant. Electron. (Rus.). **36**, 673–683 (2006)
16. Takahashi, M., Ohnishi, N.: Theoretical and numerical studies of dynamic scaling of a six-degree-of-freedom laser propulsion vehicle. Hindawi Publ. Corpor. Int. J. Aerosp. Eng. **2015**, 1–12 (2015). https://doi.org/10.1155/2015/801371
17. Takahashi, M., Ohnishi, N.: Fluid-orbit coupling calculation for flight analysis of impulsively driven laser vehicle. J. Phys. Conf. Ser. **454**, 012004 (2013). https://doi.org/10.1088/1742-6596/454/1/012004
18. Pirri, A.N., Weiss, R.F.: Laser propulsion. AIAA Pap. No 72-719 (1972)
19. Pirri, A.N., Simons, G.A.: The fluid mechanics of pulsed laser propulsion. AIAA J. **15**, 835–842 (1977)
20. Kare, J.T.: Vehicle and system concepts for laser orbital maneuvering and interplanetary propulsion. Beamed Energy Propulsion. AIP Conf. Proc. **664**, 662–673 (2003). https://doi.org/10.1063/1.1582151
21. Rather, J.D.G.: Ground to space laser power beaming: mission, technologies, and economic advantages. AIP Conf. Proc. **664**, 37–48 (2003). https://doi.org/10.1063/1.1582094
22. Myrabo, L.N.: Brief history of the Lightcraft Technology Demonstrator (LTD) project. AIP Conf. Proc. **664**, 49–60 (2003). https://doi.org/10.1063/1.1582095
23. Rezunkov, Y.A.: Laser Propulsion for LOTV Space Missions. AIP Conf. Proc. **702**, 228–241 (2003). https://doi.org/10.1063/1.1721003
24. Niino, M., et al.: LE-NET and multi-purpose laser propulsion system. AIAA Pap. No. 2176 (2002). https://doi.org/10.2514/6.2002-2176
25. Raizer, Y.P.: Laser-ignited spark and development of charges. Nauka, Moscow (1974)
26. Chen, Y.-L., Lewis, J.W.L., Parigger, C.: Spatial and temporal profiles of pulsed laser-induced air plasma emissions. J. Quant. Spectr. Radiative Transf. **67**, 91–103 (2000). https://doi.org/10.1016/S0022-4073(99)00196-X
27. Luter-Devis, B., et al.: Matter under strong laser fields. J. Quant. Electron. (Rus.). **19**, 317–359 (1992)
28. Zeldovich, Ya.B., Raizer, Yu.P.: Physics of shock waves and of high-temperature gas-dynamic phenomena. Physmatgiz, Moscow (1963)
29. Method of investigating plasma (ed. Lohte-Holtgreven I.). Mir, Moscow (1971)
30. Biberman, L.M., Norman, G.E.: Continuous spectrums of atomic gases and plasma. Adv. Phys. Sci. **91**, 193–245 (1967)
31. Ageichik, A.A., et al.: Conversion efficiency of laser power into gas thermal energy as applied to laser propulsion. Op. J. **70**, 65–71 (2003). https://doi.org/10.1364/JOT.70.000274
32. Rezunkov, Y.A., et al.: Laser propulsion at ambient vacuum conditions. Rev. Las. Eng. **29**, 268–273 (2001). https://doi.org/10.2184/lsj.29.268

33. Golovachov, Yu.P., Kurakin, Yu.A., Schmidt, A.A.: Numerical investigation of non-equilibrium MGD flows in supersonic intakes. AIAA 2001-2883 (2001). https://doi.org/10.2514/6.2001-2883
34. Tarnavsky, G.L., Shpak, S.I.: Average index of adiabatic index in the problems of hypersonic streamlines of solid body by a real gas. Therm. Phys. Air-dyn. **8**, 41–58 (2001)
35. Kraiko, A.N.: Analytical formulas for thermodynamic functions of air. Eng. J. **IV**, 548–550 (1964)
36. Park, C.: Nonequilibrium hypersonic aerothermodynamics. Wiley, New York (1990)
37. Gnoffo, P.A., Gupta, R.N., Shinn, J.L.: Conservation equations and physical models for hypersonic air flows in thermal and chemical non-equilibrium. NASA Tech. Pap. 2867 (1989)
38. Balakrishnan, A.: Correlations for specific heats of air species to 50000K. AIAA-86-1277 (1986)
39. Andreev, A.A., Mak, A.A., Solovyev, N.A.: An introduction to hot laser plasma physics. NOVA Sci. Pub., Inc. (2000)
40. Andreev, A.A., Mak, A.A., Yashin, V.Y.: Generation and application of strong laser fields. J. Quant. Electron. (Rus.). **24**, 99–114 (1997)

Chapter 3
Laser Ablation of Solid Materials, Laser Ablation Propulsion

Abstract The analysis of laser pulse effects on polymers is presented as applied to laser ablation propulsion. Polymers of CHO-chemical composition are considered to use as propellants to increase the efficiency of LAP by burning of polymer vapors and producing such combustion components of this reaction as CO_2 and H_2O. Theoretical analysis of the ablation effects is considered on the base of polymers burning models.

Moreover, high-temperature oxidation of the ablated CHO-materials as well as release of combustion energy in the ablated vapor are considered as basic chemical reactions resulting from laser ablation of polymers. The first reaction is a chemical oxidation of the vapor components by oxygen entering into the polymer's composition. The second reaction is a delayed burning of the partially oxidized vapor components in the atmospheric oxygen.

In the case of the laser ablation propulsion based on the ablation of CHO-polymers, the concept of laser-propulsion efficiency is defined more correctly by the analogy with the efficiency of conventional jet engines, namely, as a ratio of exhaust jet power to the total power of the energy sources originated in a combustion chamber.

Keywords Solid targets · High-melting and low-melting materials · Vapor mass flux density · Prokhorov's gas-dynamic theory · Ablated vapor jet · Under-expanded vapor jet · Exhaust jet velocity · Solid propellant · Thrust cost · Plasma ignition · Plasma ignition time delay · Direct laser ablation propulsion · Combined laser ablation propulsion · Confined laser ablation propulsion · Multilayer structured target · "Cannon-ball" effect · High-energy polymers · Plasma-chemical reactions · K-factor for CHO-polymer

3.1 Introduction

Laser ablation propulsion (LAP) is a process of thrust generation resulting from a flash evaporation of solid (or liquid) materials under the action of laser radiation. This type of the laser propulsion was firstly proposed by Arthur Kantrowitz in 1972

Y. A. Rezunkov, *High Power Laser Propulsion*, Springer Series on Atomic, Optical, and Plasma Physics 116, https://doi.org/10.1007/978-3-030-79693-8_3

to launch vehicles into Earth orbits [1]. Until this idea was announced, the effect of a recoil impulse initiation via a laser ablation of solid targets had been discovered experimentally by Russian scientists [2].

The merits of LAP application to produce the thrust arise as follows. First of all, the ablation process can be used to produce the thrust in vacuum if a vehicle stores solid or liquid propellants on its board. To reinforce this effect, the perspective propellants such as various polymers with accumulated internal chemical energy being released under a laser pulse can be stored. Secondly, the efficiency of air-breathing propulsion tends to decrease with the increase of an altitude of the vehicle flight into the Earth atmosphere. To maintain the efficiency of the thrust production at high altitude or in space, the laser ablation propulsion could be applied. The third reason is a possibility of getting a high specific impulse of an exhaust jet through laser ablation of high-melting point materials that maintains mass consumption of the propellant. One more preference of LAP is also a property of the ablated jet be directed to the target surface perpendicularly that simplifies the thrust vector control.

It is known that the theory of the hard material ablation and the phenomena accompanying the ablation process were developed in relation to various technologies, including development and production of new materials, composites, alloys, and so on [3, 4]. But, utilization of laser ablation to produce a thrust was studied much less. Nevertheless, there have been a few scientific studies, in which the theoretical and experimental basis of the laser ablation propulsion has been developed.

Among them are the papers published by Dr. Claude Phipps [5–7] and must be mentioned first of all. In these studies, numerous experiments of the thrust production via the laser ablation of solid targets are analyzed, as well as the similar dependencies of the specific impulse and momentum of coupling coefficient of the produced thrust, on the laser radiation intensity, laser pulse length, and radiation wavelength are deduced.

There are a few regimes of the laser ablation of solid materials and laser radiation interaction with vapor of ablated material which have been discovered experimentally. The interaction depends on such characteristics of the vapor as ablated mass rate, flow velocity in a vapor jet, as well as on the laser radiation characteristics. Several detailed experimental studies in this direction which were made by the Russian Scientists Dr. Danilychev [8] and Prof. Protasov [9] are to be mentioned jointly with the investigations made by Professor Claude Phipps. Theory of laser radiation interaction with plasma ignited at laser breakdown of ablation vapor, developed by Prof. Yu. P. Raizer [10], is a common foundation of the analyses of these researches. Later, similar investigations of the laser ablation propulsion were carried out by Prof. Krokhin [11] and Dr. Zvorykin [12]. At present time, the theoretical and experimental works made by Dr. John Sinko [13, 14] must be mentioned because these papers are devoted to laser ablation of polymers and to the processes being responsible for the thrust. Some detailed investigations of the laser ablation propulsion are presented in Refs. [15, 16], where development of polymers as applied to the laser ablation propulsion is considered.

In Chap. 3, the analysis of laser pulse irradiation effects on polymers is presented as applied to laser ablation propulsion. The experiments were performed by using the pulsed Nd- and CO_2 lasers. Polymers of CHO-chemical composition are considered as propellants to increase the efficiency of LAP by burning of polymer vapors and producing such combustion components of this reaction as CO_2 and H_2O. The theoretical analysis of the ablation effects is also considered, based on some models of polymers burning.

Moreover, high-temperature oxidation of the ablated CHO-materials as well as release of combustion energy in the ablated vapor are considered as basic chemical reactions resulting from laser ablation of polymers. The first reaction is a chemical oxidation of the vapor components by oxygen entering into the polymer's composition. The second reaction is a delayed burning of the partially oxidized vapor components in the atmospheric oxygen.

In the case of the laser ablation propulsion based on the ablation of CHO-polymers, the concept of laser-propulsion efficiency is defined more correctly by the analogy with the efficiency of conventional jet engines, namely, as a ratio of exhaust jet power to the total power of the energy sources originated in a combustion chamber. As for polymers, laser radiation power released in the ablated vapor and combustion energy due to the burning of the polymers are considered to be power sources.

3.2 Physical Phenomena Underlying of Laser Ablation Propulsion

3.2.1 Basic Concept of Developed Evaporation of High-Melting and Low-Melting Materials

The developed evaporation is observed through laser radiation interaction with the solid targets provided the radiation intensity exceeds the evaporation threshold of the target material, I_{vap}. Moreover, the energy release in a thin layer of the target surface takes place very rapid due to the absorption of the radiation power within the layer before the processes of thermal conductivity and heat diffusion will divert the absorbed energy out of the layer. It is assumed that the absorbed radiation power is transformed into a vapor jet. In this case, mass flux density in the vapor, j, is determined by the ratio that follows from the energy conservation law, namely (see [4]):

$$j = \frac{(1 - R) \cdot I}{q} \tag{3.1}$$

where I is the radiation intensity on a target surface, R is a reflection coefficient of laser radiation by the target surface, and q is specific (per unit mass) evaporation heat of the target material.

If $j(T)$ function is defined by some way, then ratio in (3.1) can be used to describe the target temperature as a function of the radiation intensity I. In the case of a developed evaporation, there is $I > I_{vap}$. The threshold intensity depends on thermo-physical properties of the target material, the surrounding gas pressure, and geometrical sizes of target and laser beam aperture. With respect to the laser propulsion, main emphasis should be paid on the case when the beam aperture diameter, d, and lateral size of the target have close to each other dimension values. Moreover, the target thickness must be so large that the target is not evaporated completely during the laser-propulsion process.

Following Prokhorov's theory, let us consider heating and evaporating of the target material as a one-dimensional effect in certain geometrical conditions so that the threshold intensity doesn't depend on geometrical sizes of the ablation region. It is also assumed that the radiation intensity is uniformly distributed across the laser beam, and thermal conductivity can be neglected across lateral sides of the beam spot.

All of these basic statements allow calculation of the principal results of the developed evaporation process with respect to the laser ablation propulsion, produced under both pulsed and CW laser radiation, at vacuum and atmosphere conditions of a surrounding medium. Determination of $j(T)$ function at arbitrary pressure p_e and temperature T of surrounding medium is a sufficiently complex problem of gas-kinetic physics. But, the problem can be simplified if the radiation intensity exceeds the threshold value due to strong dependence of the gas pressure on its temperature, that is, $p_s(T) \sim \exp(-q_x T)$ where qx is the heat of the surface evaporation estimated per one vapor unit. Then, the mass flux density $j(T)$ is closely approximated by the similar formula for evaporation in a vacuum, namely:

$$j(T) = (1 - r)p_s(T)\sqrt{\frac{M}{2\pi \cdot T}} \tag{3.2}$$

where $M = A \times N_A$ is an evaporated particle mass, A is an atomic weight, N_A—Avogadro number, r is a reflection coefficient of radiation from the target ($r < 1$).

Both Eqs. (3.1) and (3.2) allow calculation of the evaporation radiation intensity, I_{vap}, as a function of thermo-physical properties of the target, namely:

$$I_{vap} = \frac{q}{1 - R} j(T_{vap}) \tag{3.3}$$

There are two different types of hard materials which could be used to produce the thrust via the laser ablation, namely, high-melting and low-melting point materials. As for high-melting point materials such as metals and graphite, the intensity of developed evaporation is determined by the following ratio:

$$I > I_{\text{vap}} + \rho \cdot p/(1 - R)\sqrt{\chi/\tau} \tag{3.4}$$

where τ is the laser pulse length. As for dielectrics, this ratio is transformed into the following relation:

$$I > I_{\text{vap}} + \rho \cdot p/(1 - R)(1/\mu\tau) \tag{3.5}$$

In (3.4) χ is a coefficient of thermal diffusivity of hard materials, and μ is a coefficient of radiation absorption by the material.

These formulas can be used to estimate the target surface temperature made of various materials. For example, the temperature of metals will change within the range of $T_{\text{boil}} < T < 1.6\ T_{\text{boil}}$ if the radiation intensity changes from 10^5 through 3×10^7 W/cm^2, where T_{boil} is the boiling point of the material. The temperature of graphite changes from 3770 K through 4150 K if the radiation intensity changes from 3×10^5 W/cm^2 to 3×10^7 W/cm^2.

3.2.2 *Simplified Gas-Dynamics Model of Laser ablation Propulsion*

Bunkin's and Prokhorov's gas-dynamic theory of the laser ablation propulsion is based on the concept of developed evaporation which describes the mass flux density of the ablated vapor jet, j, by the Eq. (3.1). In a general case, the j parameter depends on the pressure of the ambient gas and on the target surface temperature [4].

Figure 3.1 illustrates two principal schemes of the laser ablation propulsion to produce the thrust. In the first scheme (a), an under-expanded vapor jet is generated if the vapor pressure is much higher than the pressure of the surrounding gas. In scheme (b), an auxiliary nozzle is installed close to the target to transform the vapor flow smoothly into the exhaust jet in sequential mode with ambient gas pressure.

For both of these schemes of the laser ablation propulsion, the jet thrust is determined by the following relation:

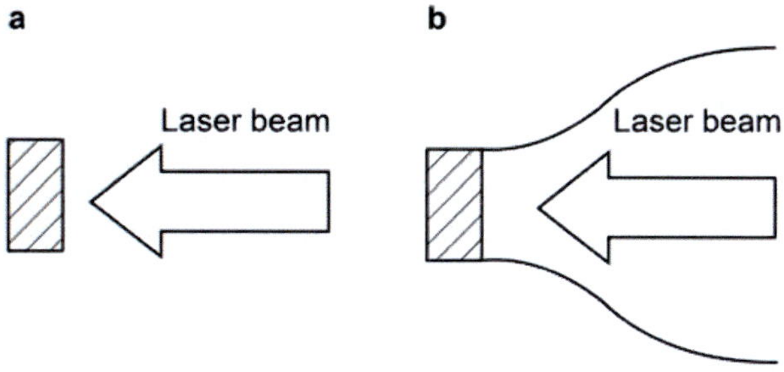

Fig. 3.1 Configuration schemes of the laser ablation propulsion

$$F = Gu_{\mathrm{e}} \tag{3.6}$$

Here, $G = S \times j = S \times (1 - R) \times I_{\mathrm{o}}\, e^{-\theta}/q$ is the target material mass consumption, S is the illuminated area of the target surface, and u_{e} is the flow velocity at the nozzle exit. In case (a), u_{e} is defined by the target surface temperature.

To increase the flow velocity of the exhaust jet in scheme (b), the laser power conversion into the flow internal energy has to be achieved due to the laser radiation interaction with the jet in a mode of the laser detonation wave. And the nozzle head should be arranged at some distance that is less than a distance of the detonation wave propagation during the laser pulse. Then the location of the nozzle head vertex will correspond to the location of a critical nozzle cross section, where the local flow velocity is equal to the speed of sound that is $c_2 = \sqrt{\gamma}\, T_2/M$. Beyond the critical cross section, the adiabatic flow of vapor starts rapidly spreading into the expanding nozzle and its velocity is determined in accordance with the Bernoulli equation:

$$u_{\mathrm{e}} = \sqrt{\frac{\gamma + 1}{\gamma - 1} c^2 \sqrt{1 - \frac{2}{\gamma + 1}} \sqrt{\left(\frac{p_{\mathrm{e}}}{p_2}\right)^{(\gamma-1)/\gamma}}} \tag{3.7}$$

where p_2 is the vapor pressure behind the laser detonation wave.

The exhaust jet velocity can reach a few kilometers per second if high-melting materials are ablated. This fact plays an important role in choosing of solid propellants in relation to the laser ablation propulsion. Another criterion of this choice is a thrust cost function that is defined as a ratio of applied laser power to produced thrust, namely:

$$\frac{P_0}{F} = \frac{e^{\theta}}{(1 - R)} \frac{q}{u_{\mathrm{e}}} \tag{3.8}$$

Here θ is an optical thickness of ablated material layer [4]. The nozzle device decreases the thrust cost function by a factor of four because $u_{\mathrm{e}} \propto \sqrt{q}$ approximately, and $P_0/F \propto \sqrt{q}$. Therefore, to get minimum thrust cost function, it is required to use low-melting propellants. But, low-melting materials do not permit achieving optimal specific thrust F_{sp} estimated as a ratio of the thrust, F, to the propellant mass consumption, G, when $F_{\mathrm{sp}} \propto \sqrt{q}$. Other definition of the specific thrust is a specific impulse that is used more frequently to study the laser propulsion. So, to get maximum specific impulse of the laser ablation propulsion, high-melting materials should be chosen. For example, $q \approx 1.2 \times 10^5$ J/g and $u_{\mathrm{e}} \approx 6$ km/s for material like graphite is used.

Let us estimate also whole efficiency of the laser ablation propulsion, which is defined by efficiency of the laser power conversion into the kinetic power of an exhaust jet. In accordance with the conventional definition of the jet power (see Ref. [4]), we have:

$$\gamma_F = \frac{G \cdot u_e^2}{2P_0} = \frac{1-R}{2e^{\theta}} \cdot \frac{u_e^2}{q} \tag{3.9}$$

Taking into account that $u_e^2 \sim q$ for high-melting materials, we can deduce that the γ_F coefficient does not depend on a specific heat of the propellant vapor, and the whole efficiency is equal only to 20–30% for laser ablation propulsion with nozzle (Fig. 3.1(b)).

It should be noted that all values of thrust characteristics of the laser ablation propulsion without a nozzle will be much less than the same characteristics of the thrust produced by the laser propulsion with a nozzle.

3.2.3 "Absorption Explosion" Model of Plasma Ignition at Laser Ablation of Solid Targets

One of the principal problems concerning the laser power conversion into a thrust at the laser ablation propulsion is a plasma ignition in the ablated vapor. The effects accompanying the plasma ignition result in various processes of the laser power conversion. In Ref. [8], the model of "absorption explosion" was proposed as the basis of analysis of a great number of experimental data on the laser ablation of solid targets, which makes these effects clearer. The key basis of this model is the criterion of the laser power conversion into the vapor flow energy that is greater than the adiabatic cooling of the vapor during its expansion into surrounding gas. Additionally, the gas cannot affect the "absorption explosion" phenomenon because there is a high pressure of a few tens of atmospheres that is reached in the ablated vapor when the pressure of surrounding gas is of 0.1–1.0 atm (see Fig. 3.2).

The "absorption explosion" model differs from the model of avalanche ionization of gases [10]. The avalanche ionization is applied usually with a bulk breakdown of gases under a laser pulse. In that case, the response time of the avalanche development, τ_i, is determined by following ratio:

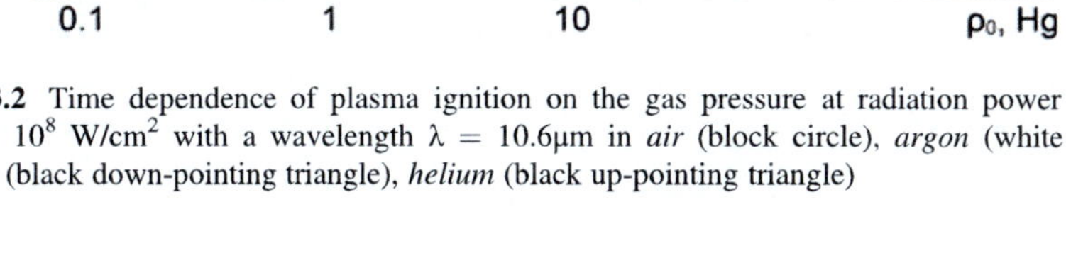

Fig. 3.2 Time dependence of plasma ignition on the gas pressure at radiation power flux of 1.4×10^8 W/cm^2 with a wavelength $\lambda = 10.6\mu$m in *air* (block circle), *argon* (white circle), *xenon* (black down-pointing triangle), *helium* (black up-pointing triangle)

$$\tau_i = 1/\nu_i \ln(N_e/N_{0e}) \tag{3.10}$$

where ν_i is the ionization frequency of atoms which depends on ionization potential of the atoms, and on gas pressure!

But, the process of plasma ignition when irradiating of a solid target must be considered as succession of heating of the target surface up to the evaporation temperature and then ionizing of the ablated vapor by the laser radiation. Therefore, the instant of plasma ignition depends on the laser irradiance flux, and the ignition belongs to threshold phenomena with respect to the flux (see Fig. 3.3).

Figure 3.3 illustrates the experimental data of the plasma ignition by CO_2 laser pulse with 150 J pulse energy. The important factor of these experiments concerns the laser pulse shape with a sharp leading edge to the pulse of 75 ns duration and with the pulse tail of 1.0μs. The laser beam is focused by a spherical mirror with a 1.5 m focal length onto a graphite target arranged in a chamber.

It was shown experimentally that the plasma was ignited only during the action of a leading edge peak of the laser pulse. If ionization process in the vapor wasn't developed during the peak, then the plasma wasn't ignited at the pulse tail despite large amount of power contained in the tail of the pulse. This fact shows the radiation intensity has an effect on the moment of plasma ignition via the electron avalanche initiation.

These experimental results show that $(E/S)/\tau^{1/2}$ = constant, where $(E/S) = \int_0^\tau \times q(t) \times dt$ is the energy density transmitted to target surface, and τ is the laser pulse length. If the radiant flux, $q(t)$, grows linearly with time as $q(t) = \alpha \times t$, then the target surface temperature can be expressed by the following relation using the thermal conductivity equation [8]:

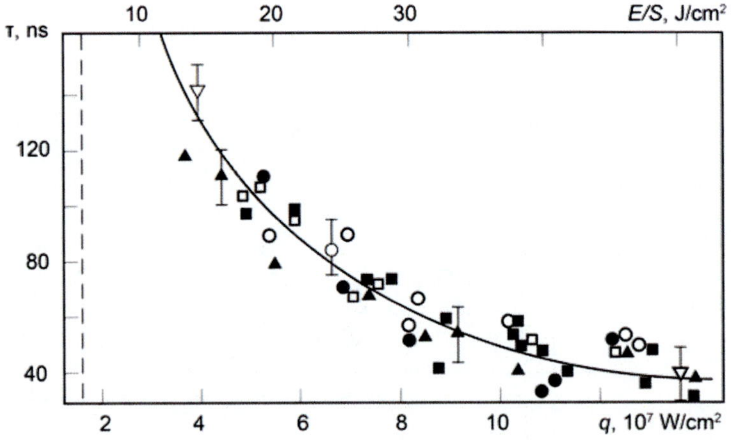

Fig. 3.3 Plasma ignition time delay with respect to the laser pulse end as function of irradiance, q, of solid target in presence of *air*, *helium*, and *argon*

$$T = \frac{8(1-R)}{3\sqrt{\pi\rho c\chi}} \cdot \frac{(E/S)t}{t^{1/2}} \tag{3.11}$$

where ρ is the density of the target material, c is the specific thermal capacity, χ is the coefficient of thermal conductivity, and R is the coefficient of radiation reflection of the target. So, the condition of plasma ignition, that is $E/S_\tau/\tau^{1/2}$ = constant, represents the fact that target is heated up to the temperature of rapid evaporation of the target surface, which is accompanied by strong ionization of vapor. This fact is also confirmed by estimation of a graphite target temperature: if $\rho = 1.7$ g/cm^3, c = 0.7 J/(g × grad), $\chi = 1.14$ W/cm × K, and $R = 0.2$ [12], then the target temperature reaches $T \sim 3700$ °C for time that is half the time required to ignite a plasma. The same time results of the plasma ignition at irradiance of metal and dielectric targets were obtained in Ref. [9], where the CO_2 laser at 10.6μm was applied (see Fig. 3.4).

So, it appears the laser ablation of solid targets is accompanied by various processes initiated by the laser radiation. As for the laser ablation propulsion, there is a great interest to investigate two processes which are (a) the developed evaporation transforms into a plasma in the vapor and (b) interaction of the laser radiation with the plasma as functions of the target irradiance [11, 12]. One of the characteristics of the vapor transformation is the radiation intensity threshold, I_0^*, exceeding of which results in increasing of the target irradiance and in changing of the evaporation dynamics. At the same time, interaction of the laser radiation with an ignited plasma will define the efficiency of the laser power conversion into the inner energy of the plasma and of a thrust production consequently. In many respects, the conversion efficiency will depend on laser radiation characteristics. Let us consider

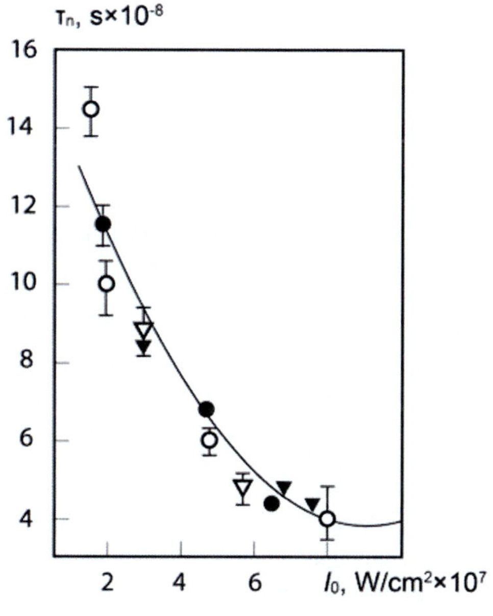

Fig. 3.4 Dependence of plasma ignition time delay with respect to a laser pulse end on the target irradiance for various target materials: *white circle*—$(C_2F_4)_n$ at $p_0 \approx 10^3$ Pa; *block circle*—$(C_2F_4)_n$ at $p_0 \approx 2.5 \times 10^3$ Pa; *black up-pointing triangle*—Al at $p_0 \approx 4 \times 10^3$ Pa; *black down-pointing triangle*—Al at $p_0 \approx 3 \times 10^4$ Pa

the processes of a laser radiation interaction with an ignited plasma following theory developed by Prof. Raizer [10].

3.2.4 Gas-Dynamic Models of the Laser Radiation Interaction with Ionized Gas (Gaseous Plasma)

Laser-sustained detonation wave. The detonation effect of the plasma development takes place if the gas located directly behind the shock wave is ionized by this wave at certain conditions. This strong shock wave is identified as the Laser-Sustained Detonation (LSD) wave.

The "gas-plasma" medium follows the LSD wave with a velocity that is equal to a local speed of sound in the medium. It is evident that the "laser power-LSD wave" process can only exist by maintaining the energy balance between the wave velocity and the laser power.

It is known that the process of gas compression behind a strong shock wave is limited by a magnitude of $(\gamma-1)/(\gamma+1)$ where γ is an adiabatic index. The condition of equality of the local gas velocity to the local speed of sound in the gas is satisfied at a Juge point of Hugoniot adiabatic, that results in the ratio: $v = a = \sqrt{\gamma \frac{p}{\rho}}$. Then, the gas compression and the LSD wave velocity are determined by the laser power flux density in accordance with the following formulas:

$$
\begin{aligned}
p &= \left\{ \frac{2\sqrt{1-\eta}S_0\rho_0^{1/2}}{[(\gamma+1)(\gamma-1)]\eta - 1} \right\}^{2/3} \\
\rho_0 D\varepsilon(T,\eta) &= S_0\beta \\
\beta &= \frac{1}{1-(\gamma-1)(1-\eta)/2\eta}
\end{aligned}
\tag{3.12}
$$

assuming full absorption of the laser power by plasma ignited behind the wave. Particularly, $V_{LSD} = [2\,(\gamma^2-1)\,q/\rho_0]^{1/3} = 8 \times 10^5$ cm/s at $q = 10^8$ W/cm^2, if specific gas energy is $\varepsilon_G = 10^{12}$ erg/g, that corresponds to the plasma temperature of $T = 24{,}000$ K behind the wave. Such a high plasma temperature conforms with the degree of plasma ionization, that is $\alpha = 0.77$, and free-electron concentration in the plasma must be equal to $N_{eG} = 3.8 \times 10^{19}$ cm^{-3} if the initial pressure of a "cold" gas is the same as atmospheric pressure.

Laser-Sustained Radiative wave. If the laser radiation flux q exceeds 10^8 W/cm^2 (for the CO_2 laser as an example), then the transition from the LSD wave to a radiative process of absorption wave propagation is occurred, which results in spreading of the shock wave front via creation of ionized gas volume in front of this wave due to heating of the gas by the UV radiation generated by the wave. The latter means that the radiative process of the laser power absorption becomes more efficient than the gas-dynamic effect.

In this case, the balance equation of power is:

$$\rho_0 V \varepsilon_* = \sigma T^4 \xi, \ \ \rho_0 V(\varepsilon - \varepsilon_*) = q \tag{3.13}$$

where σ is a Stefan-Boltzmann constant, $\xi(H/l)$—is the plasma emissivity, H—is the radiative layer thickness, l—is a radiation free path length of molecules in the plasma, and ε—is the plasma internal energy. The velocity of the plasma propagation front will be equal to:

$V \sim q^{0.73}/\rho_o{}^{0.91}$, $T \sim q^{0.18}/\rho_o{}^{0.02}$—for the atmospheric air. $\varepsilon = \frac{RT}{\mu}\frac{1}{(\gamma-1)}$ for an ideal gas.

Laser-sustained subsonic radiative wave (LSR Wave). If the ionization degree of the shock wave front is so low to absorb the laser power efficiently, then the plasma behind the wave moves due to the irradiative process but with a velocity that is less than a local speed of sound.

This mode of interaction of the laser-induced shock wave with the laser radiation is described by the following parameters:

$$V = \left[\frac{2(\gamma - 1)}{\gamma(\gamma + 1)}\right]^{1/3} \left(\frac{q}{\rho_0}\right)^{1/3}, \ p = \left[\left(\frac{\gamma - 1}{\gamma}\right)^2 \frac{\gamma_\gamma + 1}{2}\right]^{1/3} \rho_0^{1/3} q^{2/3} \tag{3.14}$$

where q is the laser radiation flux, γ—is the adiabatic index of a "cold" gas, γ—is the adiabatic index of plasma, and:

$$\frac{V_{\text{LSD}}}{V_{\text{LSR}}} = \left[\frac{2(\gamma^2 - 1)(\gamma + 1)}{2(\gamma - 1)\gamma}\right]^{1/3} \tag{3.15}$$

And this ratio is equal to 1.63 if $\gamma = 1.4$.

All types of the laser-sustained waves are graphically illustrated in Fig. 3.5 [17].

For example, the LSD wave in atmospheric air is observed if the CO_2 laser power flux changes from 3×10^7 W/cm^2 to 3×10^8 W/cm^2. Similar results were obtained by other researchers[8]. But, the LSD wave in such monoatomic gases as argon or helium was observed only within a narrow range of the radiation power flux, namely $q = (5–6) \times 10^7$ W/cm^2.

What is more important every type of laser-sustained waves generated close to a solid target surface affects strongly on a recoil impulse imparted to the target. This fact is confirmed by various experimental data, the result of one of which is presented in Fig. 3.6 as coupling coefficient function of the surrounding gas pressure produced under the action of the CO_2 laser pulse [8].

We explain the results presented in Fig. 3.6 as follows. If the gas pressure surrounding the target is rather low, namely, $p_0 < 1$ mm Hg that means the shock wave is transparent to the laser radiation and the gas is hardly ionized. As a consequence, the produced recoil impulse on the target surface is determined only by ablated vapor jet power. If the gas pressure changes within the range of 1–80 mm

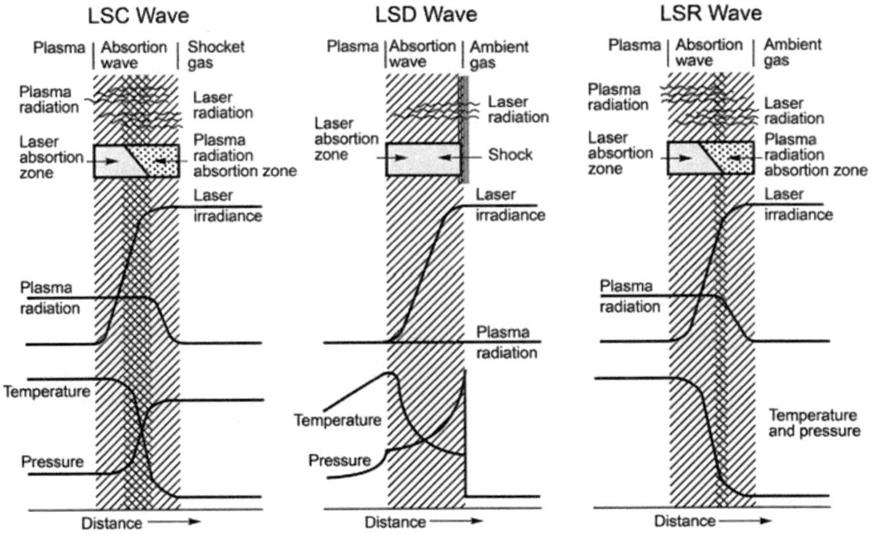

Fig. 3.5 Basic types of the laser-sustained waves generated close to a surface of solid target

Fig. 3.6 Coupling coefficient of a recoil impulse for solid target as a function of gas pressure

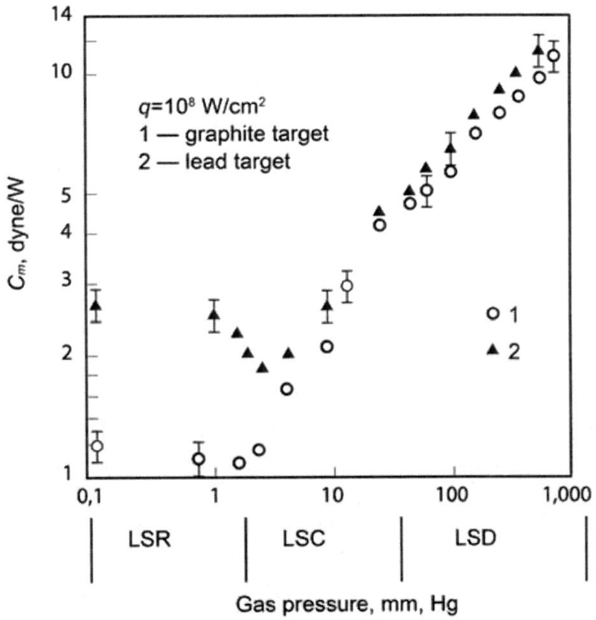

Hg, the laser radiation interaction with the generated shock wave is characterized by an LSC wave. The plasma ignited in the vapor screens the target's surface from the laser radiation that leads to decrease of vapor pressure near to the target surface. This effect results in decreasing of the coupling coefficient, as a consequence. But, the coupling coefficient increases again when increasing the gas pressure from 10 to 80 mm Hg due to auxiliary evaporation of the target surface under the luminescence radiation emitted by the plasma. Finally, if the gas pressure is more than 80 mm Hg, the laser-sustained detonation wave is produced and further rise of pressure near to the target's surface is performed.

Additionally, each type of these laser-sustained waves will be defined by the type of the target material, laser radiation spectral characteristics and its power, and so on. Consequently, we obtain different nature of the laser ablation propulsion in every case.

3.3 Effects of Solid Target Structure on Laser Ablation Propulsion

Laser ablation propulsion occupies a special research field as applied to the development of laser-propulsion engines. The laser ablation propulsion is based on three principal phenomena, including:

- direct laser ablation of a plane target when a thrust is produced due to the recoil impulse imparted to the target by a vapor jet (evaporation mechanism of the propulsion) [5]
- combined laser ablation of solid targets that is followed a laser-supported detonation wave generated at the laser breakdown of the vapor [9, 18]
- confined laser ablation of multilayer structured targets [19, 20, 23]

Now, let us consider the basic subjects concerning each laser ablation phenomena and its influence on the efficiency of the thrust production. One of the criteria of the efficiency is the momentum coupling coefficient, C_m, being identified both theoretically and experimentally.

3.3.1 *Direct Laser Ablation Propulsion*

It has been shown experimentally that the momentum coupling coefficient, C_m, of the direct laser ablation propulsion depends on both the ablated material and laser radiation characteristics. The C_m increases proportionally to the laser radiation intensity, and it reaches maximum value if the laser radiation intensity becomes equal to the intensity threshold of the laser breakdown of the ablated vapors. There are different estimates of maximum C_m in a theory of the direct laser ablation propulsion. According to Ref. [5], the upper limit of C_m for a specified material is achieved if the laser pulse is rather long of ~1 ms and is weakly dependent on the laser radiation wavelength, namely:

$$C_{\mathrm{m\,max}} = \frac{A_2}{a} \cdot \left(\frac{\gamma T_{\mathrm{b}}}{m_{\mathrm{i}}}\right)^{1/2} \tag{3.16}$$

Here, A_2 is a coefficient of absorption of the laser power by the target material during its evaporation, a is a specific heat of evaporation (J/g), i is an effective adiabatic index of vapor, and m_{i} is a mass of atoms or molecules in the vapor. Consequently, from the formula (3.16), it is necessary to choose a material with maximal ratio of $\gamma \times T_{\mathrm{b}}/a^2 \times m_{\mathrm{i}}$ to get the maximum C_{m} and to provide the maximum absorption of the laser power by the target. Here, T_{b} is a boiling temperature of the target material. Estimations of C_{m} made by the given formula closely coincide with the experimental data on C_{m}, and for example, $C_{\mathrm{mmax}} = 6 \times 10^{-5}$ N/W for an aluminum target, and $C_{\mathrm{mmax}} = 10 \times 10^{-5}$ N/W for a cooper target.

But, the plasma ignited by a long laser pulse in the vapor results in a decrease of C_{m} parameter [21]. In this case, the approximation for the momentum coupling coefficient has a good accuracy if the laser radiation intensity exceeds the breakdown threshold of the evaporated material:

$$C_{\mathrm{m}} = b\left(I\lambda\tau^{1/2}\right)^n \tag{3.17}$$

Here, I is the radiation intensity (W/cm^2), λ is the radiation wavelength (cm), τ is the pulse duration (s), b and n are empirical parameters, for example, $b = 5.6$ for aluminum alloys, $b = 6.5$ for CH-materials, but $n = -0.300$ for both materials. The expression (3.17) is a consecutive consideration of the coupling coefficient as a function of laser characteristics within a wide range, namely, laser pulse intensity from 3 MW/cm^2 to 70 TW/cm^2, which pulse duration from 500 ps to 1.5 ms, and the radiation wavelength from 0.248 to 10.6μm. It is significant that this assessment enables to generalize selection of different laser types as applied to the laser ablation propulsion, but there are some engineering restrictions as stated below. For example, the coupling coefficient, C_{m}, obtained experimentally at the direct laser ablation propulsion with the use of some metals and nature materials are presented in Table 3.1.

Here, F is a laser power flux. As it is seen, maximum C_{m} makes up a few dynes per Watt. These data show that the direct laser ablation is not an effective way of thrust production in relation to practices of the high-power laser propulsion.

Table 3.1 Experimental C_{m} of the direct laser ablation propulsion [21–23]

Target material	C_{m} 10^{-5} N/W	λ, μm	F J/cm^2	Pulse duration, ns
Aluminum 2024	4.08	1.06	3.53	25 ns
Carbon phenolic	2.91	1.06	19.8	50 and 25 ns
PMMA (polymethylmethacrylate)	1.07 10.0	1.06 10.6	21.4 20.0	63 and 65 ns 20 and 40μs
Ice	2.93	1.06	20.9	47 and 24 ns
Limestone	1.4	1.06	24.2	51 and 22 ns

3.3.2 Combined Laser Ablation Propulsion

One of the possible ways of increasing of C_m is a production of additional thrust via generation of laser-sustained shock waves in the ablated vapor jet due to the laser radiation interaction with the plasma ignited in the vapor. The laser plasma expansion is accompanied by a laser-induced shock wave which increases a pressure in plasma up to 10^3–10^6 atm, and the plasma temperature reaches tens or hundreds of electron volts. In a vacuum, this effect results in an increase of C_m by 1–2 orders as compared with the direct laser ablation propulsion. Additionally, a long lifetime of the shock wave run supports the pressure impulse on the target surface for two to three times longer than the laser pulse duration.

Theoretically, the coupling coefficient of the combined laser ablation propulsion can be estimated by the following formula [24, 25]:

$$C_m = (10)^{5/3}\chi\left(\rho \cdot 10^9/I\right)^{1/3} \cdot 10\,\mathrm{N/W} \tag{3.18}$$

where $\chi = ((\gamma + 1)/2\gamma)^{2\gamma/(\gamma-1)}$, and ρ is a density of the ambient gas. As it is seen from this formula, C_m is determined by the gas medium properties and the laser radiation intensity, and it does not depend on type of the target material and the radiation wavelength.

But, practical application of the combined laser ablation propulsion possesses a number of essential restrictions. Compression and power release over the shock wave front result in heating of the vapor in the ablation regions so that the laser radiation does not penetrate the target surface. Therefore, the laser pulse intensity must exceed 10^{11}–10^{12} W/cm^2.

Nevertheless, as it is shown in the log-pressure-vs-log-irradiance plot for typical targets (Fig. 3.7) character of the plot closely follows the power law for LSC and LSD waves generated under the action of a laser pulse. This fact suggests a simple

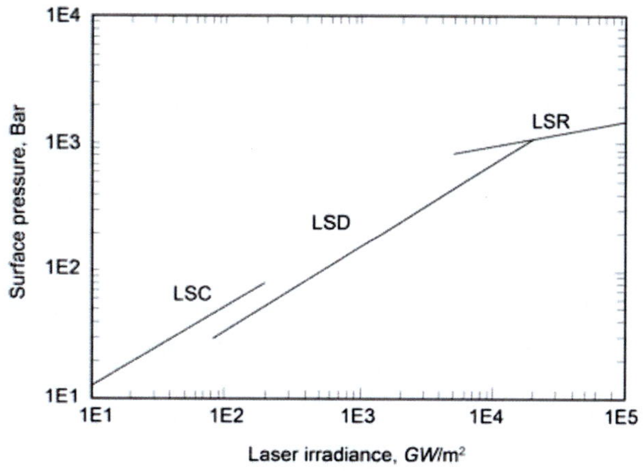

Fig. 3.7 Pressure vs laser irradiance on the target for experiments three types of absorption waves in which are induced (see Ref. [21])

model incorporating both vapor and plasma physics is needed to be considered in the detailed physics of the laser radiation interaction with ignition of plasma.

The approach used in modeling depends entirely on the available data for the physical properties of the target. Hence, we have to know the threshold radiation flux to generate a plasma, target material density, and the laser-absorption coefficient in an ablation region as it is determined in investigations made in Ref. [21]. The authors provided a more precise model that meets our accuracy requirement in modeling thrust impulse at the laser ablation propulsion.

3.3.3 Confined Laser Ablation of Multilayer Structured Targets

Confined laser ablation is the process occurring resulting from the interaction of the laser radiation with targets in the case when the interaction volume is bounded [23–25]. The volume limitation is observed in multilayer (structured) targets. Application of multilayer targets results in increasing of C_m in comparison with the direct laser ablation at similar radiation intensity. It is just the method that achieves unprecedentedly high values $C_m \sim 5 \times 10^{-3}$ N/W[25]. This effect is also called a "cannon-ball" effect.

There is general principle of designing of multilayer solid targets. A substrate material of the target (sometimes with an absorption layer) is in very close contact with a coating that is transparent at a given laser radiation wavelength. Under action of a laser pulse action on the target, the radiation passes through the transparent coating and penetrates into the substrate where it being absorbed at skin-layer depth in the substrate. If the radiation intensity exceeds the evaporate threshold intensity, then material evaporates through the gap between the substrate and coating, which serves as a shield preventing escape of the vapor. Moreover, if the radiation intensity exceeds the intensity threshold of breakdown of the vapor, then a plasma is generated in this gap, too. These processes are accompanied by two shock waves, which are generated and propagate in the both substrate and coating.

As it follows from the conservation law of momentum fluxes applied to these two shock waves, which propagates in opposite directions ($\rho_1 U_1^2 = \rho_2 U_2^2$), the ratio of the kinetic energy fluxes is determined by a ratio of material densities of the coatings ρ_1 and substrate ρ_2 materials, that is $W_1/W_2 = (\rho_2/\rho_1)^{1/2}$. Besides, the pressure impulse duration exceeds the laser pulse duration by a factor of several times as in the case of combined laser ablation propulsion. And, C_m weakly depends on the type of substrate material.

Physical phenomena of the laser radiation interaction with multilayer structured targets were studied in Ref. [20] for the first time. In addition, self-similar model of the interaction of a 1 GW laser pulse with the structured target was developed. In accordance with the model, the following formula for the pressure in the substrate-to-coating gap is determined as:

$$P_{\max}\left(10^8\,\mathrm{Pa}\right) = 0.1 \cdot \left(\frac{\alpha}{2\alpha+3}\right)^{1/2} Z^{1/2}\left(\mathrm{g/cm^2\,s}\right) \cdot I_0^{1/2}\left(10^9\,\mathrm{W/cm^2}\right) \tag{3.19}$$

Here, α is a ratio of heat energy to total internal energy, and $\alpha = 0.1$–0.15 as usual, I_0 is the laser radiation intensity, and Z is determined by the ratio:

$$\frac{2}{Z} = \frac{1}{Z_1} + \frac{1}{Z_2}$$

where Z_1 and Z_2 are acoustical impedances of the substrate and coatings, respectively, $Z_i = \rho_I D_i$, where D_i is the shock wave velocity propagating in solids, and $Z_i \approx \rho_I C_{0i}$ depends on the local speed of sound, C_{0i}.

Then, C_m can be defined by the following formula:

$$C_m = 0.10 \cdot \left(\frac{\alpha Z}{\alpha+3} \cdot \frac{1}{I_0}\right)^{1/2} \mathrm{dyne} \times \mathrm{s/J} \tag{3.20}$$

For example, $C_m = 26$ dyne × s/J if $\alpha = 0.15$, $Z = 1.4 \times 10^6$ g/cm^2 × s, and $I_0 = 1$ GW/cm^2.

The influence of coatings on C_m was thoroughly investigated in Ref. [19], including theoretical analysis and experiments of the maximum impulses transmitted to the structured target under the action of the Nd-laser pulse with the duration of 40 ns and intensity of 0.74×10^9 W/cm^2. The following materials were considered as coatings, such as plexiglass (Perspex), silicone rubber, K9 glass, quartz glass, and lead glass. The aviation aluminum alloy, type of 2024T62, was used as a substrate, on which these coatings were deposited. To increase the laser power absorption, additional absorbing black layer was arranged between the substrate and coatings.

In Tables 3.2 and 3.3, the acoustic impedance Z_1 of the coating materials being under the investigation (column 2), recoil impulse duration (column 3), P_{max} vapor pressure measured (column 4), P_{max} calculated pressure (column 5), I_m pressure impulses measured (column 6), and C_m obtained in the experiments are presented. Acoustic impedance for the aluminum substrate is $Z_1 = 1.4 \times 10^6$ g/cm × s.

Table 3.2 shows that the maximum pressure and experimentally registered pressure rise are in a good agreement with each other. Application of a transparent coating with large impedance provides a larger P_{max}. The recoil impulse duration and C_m increase with the impedance although the relation between them has not a linear character. The coating thickness affects the maximum C_m.

The problems of the radiation resistance of the coatings as well as the application of this type of laser ablation to produce a thrust with a repetitively pulsed mode of laser operation requires additional considerations. The coating layer must not be damaged under the laser radiation in the repetitively pulsed mode during the target irradiation. In the case of the coatings damage, it is necessary to move the laser beam on the coatings surface.

Table 3.2 The pressure behind a shock wave and C_m of multilayer targets with a supplemented black absorption layer

Coating material	$I_0 \times 10^9$ W/cm^2	Z_1, g/cm × s	Pressure impulse duration, ns	P_{max} calculated $\times 10^8$ Pa	P_{max} experiment $\times 10^8$ Pa	I_m experiment Pa × s	C_m 10^{-5} N/W
Perspex	0.74	0.32	53	11.0	11.3	58.8	19.9
Silicone rubber	0.74	0.47	54	12.9	13.8	72.9	24.6
K9 glass	0.68	1.14	160	17.6	15.9	265.6	97.6
Quartz glass	0.76	1.31	131	18.3	17.2	217.5	71.5
Lead glass	0.90	1.54	126	19.2	22.8	240.2	66.7

Table 3.3 The pressure behind a shock wave and C_m in multilayer targets without any black absorption layer

Coating material	$I_0, \times 10^9$ W/ cm_2	Z_1, g/ cm × s	Pressure impulse duration, ns	P_{max} calcul. $\times 10^8$ Pa	P_{max} exper. $\times 10^8$ Pa	I_m exper. Pa × s	C_m 10^{-5} N/W
Perspex	0.84	0.32	75	3.8	4.3	28.5	8.4
Silicone rubber	0.80	0.47	62	6.1	6.4	37.8	11.8
K9 glass	0.72	1.14	99	15.9	15.9	133.7	46.4
Quartz glass	0.72	1.31	81	13.9	13.9	112.6	39.1
Lead glass	0.75	1.54	89	13.8	14.2	122.8	40.9

In the case of a long-term laser-propulsion operation on the base of confined targets, liquid propellants can be utilized [20]. In particular, water could be injected from a reservoir arranged between two plates through holes of small diameter on to the ablator surface, which is retained on the surface due to the surface tension force. In this case, the hole diameter, R, is estimated from the time, t, of the water film restoration on the surfaces which is to be less than 1/f, where f is a pulse-repetition rate of the laser radiation, that is $t = (\rho R^3/3\sigma)^{1/2}$ where σ is the surface tension of a liquid film.

Thus one can conclude that a most promising way of getting a high value of the momentum coupling coefficient, C_m, reaching several hundred dynes per Watts is by the use of multilayer targets. Laser ablation propulsion produced by the multilayer targets combines such advantages as achievement of maximum values of C_m at a given laser average power, operation in a vacuum, and combination of ablation and detonation wave effects increasing a recoil impulse. But in this case, the laser ablation propulsion possesses a low specific impulse in an exhaust jet that is only of several tens of seconds resulting in a high consumption of the liquid propellant.

The fact should be mentioned that the main part of the experiments of laser propulsion in the multilayer targets was made by using Nd-lasers with a 1.06μm wavelength. Therefore, the question of the wavelength influence on this type of the laser ablation propulsion requires additional study.

3.4 Laser Ablation Propulsion Based on Ablation of High-Energy Polymers

Now, it is time to consider the laser ablation propulsion based on an ablation of energetic materials which are candidates for use as propellants. The point is that these materials release their internal energy under the action of the laser radiation [7, 16], which can be used to produce additional thrust and to increase the laser-propulsion efficiency as well.

One of the first experiments with this type of the laser ablation propulsion was carried out by Dr. Liukonen in the beginning of the 1990s of the last century by using a high-power CO_2 laser [26]. In the experiments, the propellant called “SLAVIT” arranged inside a conical nozzle was used. One of the useful characteristics of this propellant was a low value of evaporation threshold when irradiated with the CO_2 laser radiation. The experiments demonstrated a considerable increase in C_m at the repetitively pulsed mode of the laser operation, namely, from 20 dynes/W (without SLAVIT) through 60 dynes/W (with SLAVIT). Unfortunately, the author of these investigations didn’t disclose the SLAVIT chemical composition.

Similar experiments were carried out in Ref. [27] but with the use of another high-energy propellant produced as Delrin. The Delrin polymer belongs to the CHO-polymer group. A special feature of the experiments is exploration of a parabolic nozzle that was placed in an evacuated chamber filled with various gases at variable values of pressure that allowed investigating the dependencies of the momentum coupling coefficient under various operational conditions. In the experiments, a pulsed CO_2 laser with output pulse energy of 250 J was used. Delrin was arranged on a tip of thin metal needle placed inside the nozzle top, so that the needle tip was in a focus region of the laser pulse. The experiments showed a significant increase in C_m that was a factor of two when compared with the results of similar experiments made without Delrin. The experiments demonstrated also the fact that the momentum coupling coefficient registered in a medium of nitrogen as a buffer gas in the chamber was less than C_m registered in the atmospheric air.

3.4.1 Basic Plasma-chemical Reactions Proceeding in the CHO-Polymer Vapor Under Laser Radiation

We must make some principal remarks followed from these experiments. First of them followed from Derlin experiments is the fact that the momentum coupling coefficient increases significantly even if a nitrogen is used as a buffer gas in the jet nozzle. The second remark consists in a lack of publications of any model describing chemical reactions which occur under the action of the laser pulse to achieve the additional energy release.

Let us consider the plasma-chemical reactions initiated by a laser during the process of CHO-polymer evaporation and interaction of the laser radiation with the ablated vapor, following Ref. [28]. To be more precise, let us consider the polymeric and polycrystal polymers of a CHO-chemical composition, which possess a negative oxygen balance in their molecules. It is also assumed that the basic thermal characteristics of the polymers are determined, including specific heat of combustion, detonation energy, and energy of delayed burning of its detonation products.

To consider the interaction of the laser radiation with high-energetic polymers, we apply the physical-chemical model of the detonation process initiated by a laser pulse. It is known that the laser power induces evaporating and thermal decomposition of CHO-polymers. One of the processes accompanying the decomposition is pyrolysis that starts at a temperature of 500–700° C even under conditions of oxygen deficiency. In the air, the reaction of high-temperature oxidation (combustion) of the pyrolysis products can be observed due to the following reductive scheme:

$$C_aH_bO_c + O_2 \rightarrow CO_2 + H_2O + Q \tag{3.21}$$

where $C_aH_bO_c$ is a general chemical formula of the CHO-polymers, Q is a specific heat of combustion, which is released at processing of two successive chemical reactions, namely, detonation and delayed burning of the detonation products; $Q = Q_{det} + NQ_{db}$ where N is a coefficient of afterburning of detonation products.

The detonation means spreading of chemical oxidation of combustible components in a plasma with a velocity exceeding the speed of sound in the plasma. Hence, the detonation energy Q_{det} is released into the plasma due to the following reaction:

$$C_aH_bO_c \rightarrow CO_2 + H_2O + H_2 + C + Q_{det} \tag{3.22}$$

To determine Q_{det}, we use the detonation products being formed for all types of explosive materials on a basis of thermodynamics principles presented in Ref. [29]. Applying these principles to CHO polymers, we can obtain the following chemical chain of detonation products: CO_2-H_2O-O_2-H_2. According to the accepted assumption, such molecules as CO_2, H_2O, can be considered as the detonation products but not CO molecules. Moreover, generation of CO_2 prevails over H_2O.

The next chemical reaction is a delayed burning of incompletely oxidized detonation products in atmospheric oxygen, due to which the delayed burning energy Q_{db} is released. This type of the chemical reaction can be realized, for example, in a rocket engine if a carbon-to-hydrogen fuel is used, namely:

$$C + H_2 + O_2 \rightarrow CO_2 + H_2O + Q_{db} \tag{3.23}$$

Taking into account these assumptions, additional energy release due to the detonation and delayed burning reactions can be calculated by following Hess law:

$$Q_{det} = -\left(\Delta H_{pr} - \Delta H^{o}_{f298}\right) \tag{3.24}$$

$$Q_{db} = -\Delta H_{pr} \tag{3.25}$$

where ΔH_{pr} is the formation enthalpy of reaction products; $\Delta H^{\circ}{}_{f\,298}$ is the formation enthalpy of polymers.

The energy balance of CHO-polymers vapors and combustion of its components due to the oxidation by atmospheric oxygen, corresponding to standard conditions of laboratory experiments at the laser ablation of materials, can be expressed in the following form:

$$(M+m)v^2/2 = \beta[\alpha E + mQ] \quad (3.26)$$

where m is the mass of evaporated polymer, M is the mass of the air filling up the jet nozzle volume before laser pulse, v is the average velocity of an exhaust jet, α is the coefficient of the laser power conversion into the thermal energy of a propellant, and β is the conversion coefficient of the accumulated thermal energy into the kinetic energy of vapor jet.

In this case, and in accordance with a general definition of the momentum coupling coefficient C_m, we have:

$$C_{m1} = J/E = Mv/E = \sqrt{2M\beta\alpha/E} \quad (3.27)$$

$$C_{m2} = J/E = (M+m)v/E = \sqrt{2(M+m)\beta[\alpha E + mQ]/E^2} \quad (3.28)$$

where, C_{m1} is the momentum coupling coefficient of a conventional laser air-breathing propulsion using the atmospheric air as a propellant, and C_{m2} is the momentum coupling coefficient obtained with CHO-polymers as an additional propellant. To estimate an additional increase of the momentum coupling coefficient, we introduce K factor that is determined from:

$$K = C_{m2}/C_{m1} = \sqrt{(1+m/M)(1+mQ/\alpha E)} \quad (3.29)$$

Here, $Q^* = E/m^{21}$ can be interpreted as a parameter of specific ablation energy of high-energy polymers. Then, following the formula (3.29), we consequently obtain:

$$K = C_{m2}/C_{m1} = \sqrt{(1+m/M)(1+Q/\alpha Q^*)} \quad (3.30)$$

Thus, the K-factor allows to estimate the effect of plasma-chemical reactions proceeding in the CHO-type polymer vapor on the efficiency of a thrust production in comparison with the conventional laser propulsion produced at similar ambient conditions.

Figure 3.8 presents images of plasma jets generated under the action of the CO_2 laser pulse on different CHO-polymers, which were taken by a camera with an open shutter. It is seen that the luminescence of the plasma differs for every case. For polyoxymethylene, the laser power density on the target is approximately equal to a threshold of the polymer evaporation, and the plasma does not screen the laser radiation. For polyvinylchloride, the plasma luminescence occupies a larger area than for the case of polyoxymethylene. And finally, we observe a greater plasma area generated at the laser ablation of polycarbonate. The coupling coefficient measured in every case are equal to 24×10^{-5} N/W, 14×10^{-5} N/W, and 5×10^{-5} N/W, respectively.

To examine the influence of CHO-polymers on the laser-propulsion efficiency, the ablating of polymers such as polyoxymethylene (Delrin®) $[-CH_2O-]_n$, polyvinylchloride $[-CH_2CHCl-]_n$ polystyrene $[-C_8H_8-]_n$, and polycarbonate

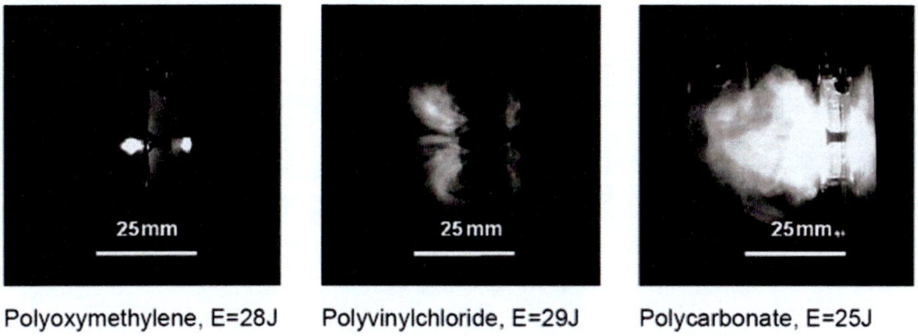

Fig. 3.8 Photographs of plasma vapors registered at ablating of the CHO-polymers

[-$C_{16}H_{14}O_3$-]$_n$ under CO_2 and Nd-laser pulses were studied experimentally in Ref. [28]. Polyvinylchloride is chosen because it demonstrates a high increase in thrust production. Moreover, first two polymers above belong to a linear aliphatic polymer, thermal decomposition of which (break of chain) is initiated at a low temperature of ~200 °C. These polymers possess a high chemical activity and belong to low-temperature constructional materials [29]. The next two polymers have benzene rings in their chemical composition. They absorb the CO_2 laser power and redistribute it over the benzene ring links, resulting in an increase of the polymer decomposition temperature up to ~350 °C. Polymers with an aromatic main chain (polycarbonate, epoxy resins) belong to the second generation of high-temperature constructional polymers with a weak chemical activity.

Preliminary thermodynamic analysis of combustion reactions accompanying the laser ablation of polymers made it possible to choose some polymers from a large group of polycrystalline materials, which possess such physical properties satisfying the laser ablation propulsion as: high dispersibility, relative chemical stability at standard atmosphere conditions, non-hydroscopicity, and non-toxicity. In the experiments of Ref. [28], the following types of polycrystalline materials were studied: metaldehyde C_2H_4O, carbamoyl hydrazine CH_5N_3O, oxybenzoic acid $C_7H_6O_3$, and dihydroxybenzene $C_6H_6O_2$. All these four substances are colorless crystals with a mid-fusion temperature, which are used in the pharmaceutical and food industries.

To determine K-factor in every case, N coefficients of delayed burning of CHO-polymer products in atmospheric oxygen were calculated. This procedure was performed by analyzing experimental results of polymers detonation.

Computed thermo-chemical parameters, relative afterburning coefficients N, K-factors from (3.30) equation, and momentum coupling coefficients C_m obtained from the experimental data of the ablation of polycrystalline polymers are listed in Table 3.4.

From analysis of the data presented in Table 3.4, we can conclude that the ablating coupling coefficient of polycrystalline polymers is determined more by their components released at detonation first of all. And maximum C_m is observed at minimum of the polymer specific ablation energy Q^* in accordance with the following relation:

Table 3.4 Computed thermo-chemical parameters of polycrystalline polymers

Polymer	Q_{det}, $\frac{kg}{kJ}$	Q_{db}, $\frac{kg}{kJ}$	N	Q^*, $\frac{kg}{kJ}$	Q_{det}/Q^*	$N \times Q_{db}/Q^*$	K	K_{av}	C_m, $\times 10^{-5}$ N/W	C_{mav}, $\times 10^{-5}$, N/W
Polyoxymethylene	2692	14,614	1.0	1200	2.24	12.18	6.09	5.96	20.0	20.25
- » -	- » -	- » -	- » -	1314	2.05	11.12	5.82		20.5	
Metaldehyde	698.9	24,391	2.04	3925	0.18	6.10	4.09	3.93	15.0	15.5
- » -	- » -	- » -	- » -	4733	0.15	5.15	3.77		16.0	
Carbamoyl hydrazine	397.6	10,685	3.86	2750	0.14	3.89	3.33	3.08	15.5	15.25
- » -	- » -	- » -	- » -	3971	0.10	2.69	2.82		15.0	
Oxybenzoic acid	32.7	20,937	0.71	947	0.03	15.26	6.26	5.42	10.0	12.0
- » -	- » -	- » -	- » -	1867	0.02	7.96	4.58		14.0	
Dihydroxy benzene	277.6	24,475	0.74	1425	0.19	12.71	5.77	5.42	7.2	8.9
- » -	- » -	- » -	- » -	1860	0.15	9.74	5.07		10.6	

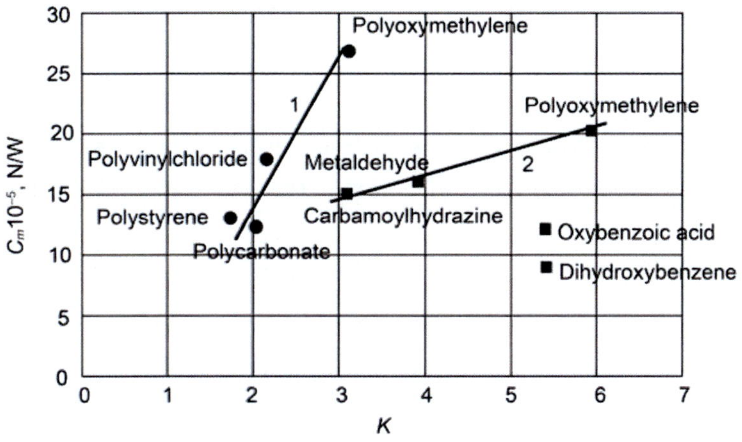

Fig. 3.9 Dependencies of C_m on K-factor for (1) the polymeric and (2) polycrystalline CHO-polymers

$$K \approx \left[1 + 2.5(Q_{det}/Q^* + NQ_{db}/Q^*)\right]^{1/2}$$

Figure 3.9 illustrates experimental results for C_m as functions of the K-factor for polymeric propellants (see line 1) within its range of change: $1.75 \leq K \leq 3.12$. It is obvious, the mentioned above relation is fulfilled at atmospheric conditions regardless of the oxygen presence in the polymer composition. The dependence of C_m on K-factor of polycrystalline propellants, the values of which for polyoxymethylene ($N = 1$), metaldehyde and carbamoyl hydrazine ($N > 1$)—see line 2—are obtained when K is changed within the range of the $3.08 \leq K \leq 5.96$. But, C_m dependence for dihydroxybenzene and oxybenzoic acid ($N < 1$) showed a large spread, and it did not coincide with any function of the K-factor.

The results of these experiments allow us to draw the conclusion that the high-temperature process of delayed burning of polymer detonation products proceeds as well as chemical reactions of the oxidation of atomic carbon and hydrogen with creation of the carbon dioxide and water molecules. Moreover, the delayed burning follows sequentially to the detonation reaction, and it is restricted by limited oxygen content in the atmospheric air.

So, the experiments on the laser ablation propulsion carried out with the use of CHO-polymers as propellants deposited on a plane target shows a maximum C_m of 27×10^{-5} N/W, and it is obtained only for a polyoxymethylene that is a linear polymer containing oxygen and possessing of a high detonation component. The polymer has a pronounced spectral region of the CO_2 radiation absorption feature due to the C-O-C stretching vibrations.

Polycarbonate, that is another oxygen-containing polymer with a linear aromatic chain, shows a smaller value of $C_m = 12.5 \times 10^{-5}$ N/W because of a weak absorption of IR-radiation via the deformation oscillations of C-H bond. However, such linear polymers as polystyrene and polyvinylchloride do not demonstrate high

Table 3.5 Comparative coefficients of delayed burning N, efficiency factor K, and experimental data on C_m for polycrystalline materials

Substance	Q_{det}, $\frac{kJ}{kg}$	Q_{db}, $\frac{kJ}{kg}$	N	Q^*, $\frac{kg}{kJ}$	Q_{det}/Q^*	K	K_{av}	C_m $\times 10^{-5}$ N/W
Polyoxy methylene	2692	14,614	1.0	1200	2.24	6.09	5.96	20.0
- » -	- » -	- » -	- » -	1314	2.05	5.82		20.5
Metaldehyde	698.9	24,391	2.04	3925	0.18	4.09	3.93	15.0
- » -	- » -	- » -	- » -	4733	0.15	3.77		16.0
Carbamoyl hydrazine	397.6	10,685	3.86	2750	0.14	3.33	3.08	15.5
- » -	- » -	- » -	- » -	3971	0.10	2.82		15.0
Oxybenzoic acid	32.7	20,937	0.71	974	0.03	6.26	5.42	10.0
- » -	- » -	- » -	- » -	1867	0.02	4.58		14.0

C_m value because of a lack of detonation components during the combustion process that results in the fact that main contribution into the momentum coupling coefficient is made by a laser breakdown of the polymer vapors and delayed burning of their components in the atmospheric air. In addition, it should be mentioned that the deformation oscillations of C-H and C-Cl bonds result in an insignificant absorption of the CO_2 laser radiation.

As it is seen also from Table 3.5, Q_{det} which is a detonation part of the internal energy reaches a maximum in polyoxymethylene. This energy exceeds the detonation components in metaldehyde and carbamoylhydrazine by an order of magnitude, and it exceeds the detonation components of oxybenzoic acid and dihydroxybenzene by many orders of magnitude.

It follows from the experiments that the delayed burning of polymers ablation products has not appreciably influence on the momentum coupling coefficient. In the case of polycrystallines, the losses of vapor particles are possible at the moment of the laser pulse action due to the particles diffusing out of the plasma volume. As a result, a noticeable part of the polycrystalline and free carbon (detonation product) do not burn completely under the laser radiation.

The results of the experiments also show the K-factor can be used to choose the CHO-polymer with respect to the laser ablation propulsion as a common criterion. For the example, the middle value of K is achieved for polymers with a large Q^* (5000–50,000 kJ/kg), and the factor K represents sufficiently magnitude of C_m. In the case of the polycrystalline materials, Q^* parameter are significantly lower (1200–4000 kJ/kg), which results in a considerable increase of K, by factor two for polyoxymethylene. But, for polycrystalline propellants, C_m is 2.5 times less than the forecasted increase. Nevertheless, C_m is proportionally depended on K (if $N \geq 1$) for both polymeric and polycrystalline polymers, respectively (see Fig. 3.9).

Despite the simplicity of the proposed physical model of the laser ablation propulsion based on CHO-polymers, evident choice of optimal CHO-polymer composition is determined by its specific ablation energy Q^*. Identification of Q^* for every polymer undertakes special experiments on ablation for these substances.

3.4.2 Similarity Laws of Laser Ablation Propulsion Based on Polymer Propellants

As for the laser ablation propulsion produced by a direct ablation of solid target (see case (a) in Fig. 3.1), Dr. Claude Phipps [5] proposed a similar theory of the momentum coupling coefficient with respect to laser radiation characteristics. In accordance with the theory, the direct laser ablation propulsion includes a characteristic invariant of basic laser characteristics, which is $(I\lambda\sqrt{\tau})$. Here, I is the laser radiation intensity, τ is the laser pulse length, and λ is the laser radiation wavelength (see Sect. 3.3). Basic statements of this theory are the following:

- all solid propellants for the laser ablation possess a strong adsorption of the laser radiation, and the propellant thickness exceeds $1/\mu\lambda$ where $\mu\lambda$ is the absorption coefficient of radiation;
- laser radiation intensity on the propellant surface corresponds to the intensity at which maximum specific recoil impulse is achieved at vacuum ambient conditions.

In accordance with numerous experimental data of the laser ablation of solid propellants [5], it follows that

$$I_{\max}\sqrt{\tau} = \text{const} = B_{\max} \tag{3.31}$$

where, $B_{\max} = 8 \times 10^4$ W$\sqrt{}$ s/cm^2, and it represents a law of the radiation intensity change as a function of the laser pulse duration at which maximal coupling coefficient is observed.

It is also assumed that a plasma ignited under the laser pulse absorbs the laser power due to the inverse Bremsstrahlung effect [11] so that the absorption coefficient is determined as follows:

$$\mu_\lambda = (4/3)(2\pi/3kT_e)^{1/2} n_e n_i Z^2 e^6 g_g [1 - \exp(-hv/kT_e)]/hcm_e^{3/2} v^3 \tag{3.32}$$

where Z is the degree of plasma ionization. Vapor density that corresponds to a neutral component of the plasma is determined as:

$$\rho_a = (Am_p/Z)n_e \tag{3.33}$$

Then, the pressure in vapor is equal to:

$$p_a = [(1 + \gamma M^2)]\rho_a c_a^2 \tag{3.34}$$

where, c_a is an adiabatic speed of sound in the vapor.

The basic parameters of this semi-empirical ablation theory are listed in Table 3.6.

Table 3.6 The basic parameters of the semi-empirical ablation theory from [5]

Parameter	Formula	Dimension
Ablation pressure, p_a	$5.83 \times 10^{-5}\mathrm{A}^{-1/8}\Psi^{9/16}\,\mathrm{I}^{3/4}(\lambda\sqrt{\tau})^{-1/4}$	$\mathrm{N/cm^2}$
Momentum coupling coefficient, $C_m = p_a/I$	$5.83 \times 10^{-5}\mathrm{A}^{-1/8}\Psi^{9/16}\mathrm{I}^{3/4}(I\lambda\sqrt{\tau})^{-1/4}$	N/W
Mass ablation rate, $\dot{m} = \rho_a \times c_a$	$2.66 \times 10^{-6}\mathrm{A}^{-1/4}\Psi^{9/8}\,(I/\lambda\tau)^{1/2}$	$\mathrm{g/(cm^2 \cdot s)}$

In accordance with the semi-empirical theory, the momentum coupling coefficient is determined as follows:

$$C_{\mathrm{mLPE}} = B \times \left(I_i \lambda_i \sqrt{\tau_i}\right)^{-1/4} \tag{3.35}$$

Formula (3.35) means similarity in all processes running in the laser ablation propulsion if the (I_i λ_i τ_i)parameter is constant. But, as it is follows from the experiments made in Refs. [21, 28], this similarity is valid only in the case when the laser power is absorbed completely.

We can extent this model for the CHO-polymers case as propellants for the laser ablation propulsion. Keeping similar conditions of solid targets ablation, the mass rate of the polymer evaporation can be expressed in the form:

$$\dot{m} = \rho_a \times c_a = B \times \left(I_i / \lambda_i \sqrt{\tau_i}\right)^{1/2} \tag{3.36}$$

Assuming also the lifetime of the laser detonation of CHO-polymers and afterburning of detonation products is much shorter than the time interval required to produce a thrust, the pressure of vapor t in the region located close to the target surface can be written as follows:

$$p_a = \left[\left(1 + \gamma M^2\right)\right]\rho_a c_a^2 + \rho_a \times c_a \times Q \tag{3.37}$$

where Q is a specific combustion energy of polymers. In accordance with the expressions listed above, we have:

$$p_a = A \times I_i^{3/4}\left(\lambda_i \sqrt{\tau_i}\right)^{-1/4} + B \times \left(I_i / \lambda_i \sqrt{\tau_i}\right)^{1/2} \times Q \tag{3.38}$$

where the A and B coefficients depend on the polymer composition. Then, the momentum coupling coefficient is expressed by the formula:

$$C_m = A \times \left(I_i \lambda_i \sqrt{\tau_i}\right)^{-1/4} + B \times \left(I_i \lambda_i \sqrt{\tau_i}\right)^{-1/2} \times Q \tag{3.39}$$

And finally, we have the expression for K-factor as follows:

$$C_{\mathrm{mLAP}} / C_{\mathrm{mLPE}} = 1 + B/A \times \left(I_i \lambda_i \sqrt{\tau_i}\right)^{-1/4} \times Q \text{ and } K = 1 + (B/A \times Q) \times \left(I_i \lambda_i \sqrt{\tau_i}\right)^{-1/4} \tag{3.40}$$

It follows from (3.40), K-factor depends on the polymer structure as well as laser radiation characteristics (compare this with the data presented in Fig. 3.9). Hence, this factor can be used to predict laser-propulsion characteristics applying ablation of CHO-polymers. But, to make it correct, it requires a complete conformity of the experimental data to assumed theoretical model.

3.5 Semi-empirical Models of Laser ablation Propulsion Based on CHO-Polymers

It is shown in Sect. 3.2.4, there are specific gas-dynamics effects following to generation of the various laser-sustained waves at the laser ablation of solid propellants, which depend on the laser power flux, q. The latter means that such parameters of the laser ablation propulsion as the recoil impulse imparted to a target, specific impulse of ablated jet, ablated mass rate, and so on are defined by the power flux. In that sense, the consideration of various models of the laser ablation propulsion is based on a detailed description of these gas-dynamic effects accompanying thrust generation.

3.5.1 Gas-Dynamics of the Laser Ablation Propulsion

There are a few gas-dynamic processes occurring during laser irradiation in the ablated vapor jet, which have been investigated recently with regard to the laser ablation propulsion [30–32]. For example, of these studies, the Schlieren's instant pictures of a vapor jet expanding from POM surface (PolyOxyMethylene or POM) into the surrounding gas are shown in Fig. 3.10 with a 6.7μs delay after the shot of CO_2 laser pulse with incident radiation flux was of 75 J/cm^2 [30]. In experiments, the ambient air pressure was changed from 50 Pa (for 3.9 (a)) to 3500 Pa (for 3.9 (b) and 3.9(c)). The elliptical black pattern in Figs. 3.10(a) and (b) is interpreted as nontransparent expanding dense plasma. The circular shock wave initiated in the

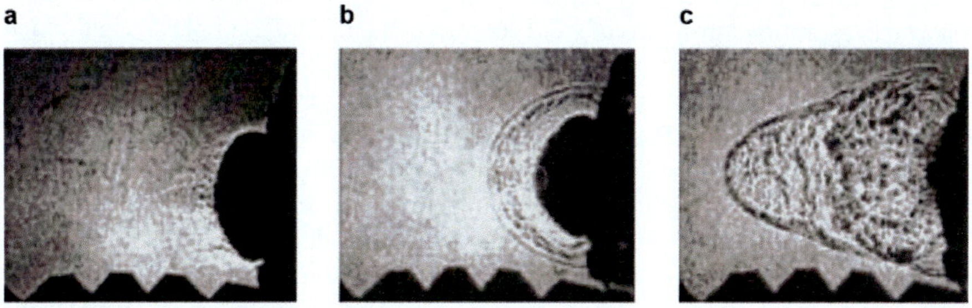

Fig. 3.10 Schlieren pictures of the POM ablation

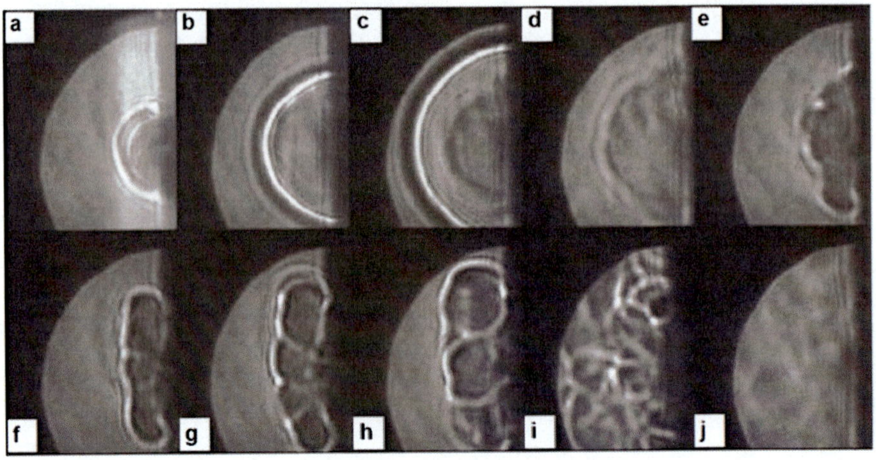

Fig. 3.11 Schlieren pictures of the ablation gas-dynamics effects on POM at various time moments after the laser pulse end

Table 3.7 Experimental evaporation thresholds of solid polymers and thresholds of plasma ignition in its vapor

Polymer	Radiation wavelength, μm	Energy threshold, J/cm^2		Intensity threshold, MW/cm^2	
		Advanced evaporation threshold	Plasma threshold	Advanced evaporation threshold	Plasma threshold
(C_2F_4)	1.064	0.55	7.48	61.1	871
$(CH_2O)_n$	1.064	2.32	9.15	257.0	1020
	0.532	0.84	5.89	93.6	655

air is seen in Figs. 3.10(b) and (c) s. There is one more detail of these experiments that the POM sample in Fig. 3.10(c) was seeded with a 20% aluminum powder.

Similar gas-dynamics phenomena were observed also in Ref. [31] (see photographs in Fig. 3.11). In the images of Fig. 3.11, we can also see the shock wave and a region occupied by dense plasma at different time moments since a start of the laser pulse.

More detailed treatment of experimental results [31] allowed the extraction of data on evaporation thresholds of solid polymers and thresholds of plasma ignition in the vapor (see Table 3.7).

The time-integrated photographs of laser-ablated jets for various laser power fluxes are also shown in Fig. 3.12 [32]. The data indicates the fact that the energy threshold of plasma ignition was of 10 J/cm^2 at a 101 kPa pressure of ambient air. All of these images were registered during a single pulse shot on a clean surface of POM target that was cleared before by it washing with water. It is incomprehensible at the present, whether the lower-than-expected threshold value was reduced due to the previously noted [13] conditioning effect, or if 10 J/cm^2 does in fact characterize the threshold of the plasma ignition at 101 kPa.

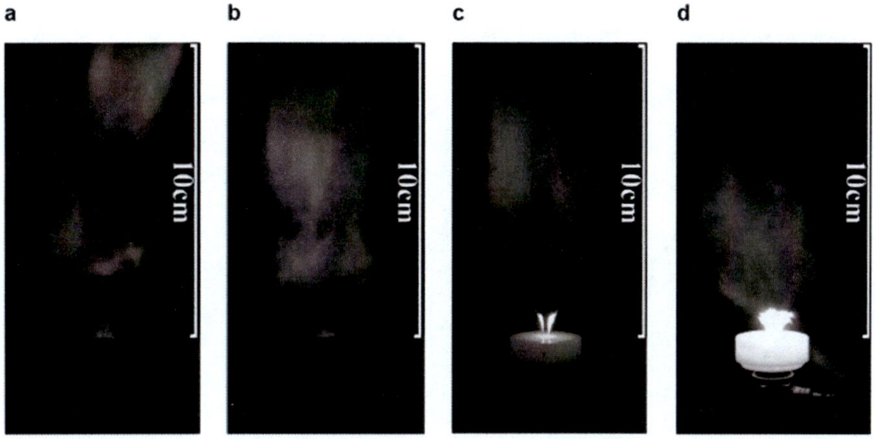

Fig. 3.12 Photos of vapor jet, illustrating the onset of the plasma generation threshold on new POM targets at 101 kPa: (**a**) 8.6 J/cm^2; (**b**) 9.1 J/cm^2; (**c**) 9.9 J/cm^2; (**d**) 10.7 J/cm^2

It is also seen from Fig. 3.12, POM vapor jet possesses two marked regimes of its propagation in the air, which include consequently: (a) a zone with a narrow jet that is transformed sharply into (b) an over-expanded jet at a certain distance from the POM surface. If the power flux is less than the plasma ignition threshold, the vapor jet is transformed into a neutral vapor flow being cooled because of its expansion (see Fig. 3.12(c)). And finally, the narrow high-speed jet is transformed into a flow of a weakly ionized over-expanded plasma when the power flux exceeds the energy threshold (Fig. 3.12(d)).

As previously discussed (see Sect. 3.3), the threshold intensity of the plasma ignition indicates the fact that deposition of the laser power in the vapor predominate over electron neutral atom collision process of the jet power dissipation into the temperature of ambient gas. At low pressure values, the threshold characterizes a point when the electric field induced by the laser pulse exceeds the ionization threshold of target molecules sufficiently to ionize gas molecules before plasma electrons diffuse out of the plasma volume (see also Ref. [33]).

One more interesting thing to observe from the photographs above is that the microparticles tracks are formed in the high-speed vapor jet up to the over-expanded jet flow. And the over-expanded part of the jet starts from a Mach disk (1) (Fig. 3.13) that is usually formed in supersonic flows. This assumption, in particular, is proved by the gas-dynamic model of the vapor jet, which was proposed in Ref. [34] (see also Fig. 3.13).

In the figure, a is the vortex generated due to an interaction of the jet with a surrounding gas, the dashed line b is the trajectory of a point on the boundary of the jet; c are the areas with an increased gas pressure in a front of the jet, and d is the shock wave moving in front of the jet.

Figure 3.13 illustrates the gas-dynamic picture of a vapor-plasma jet expanding into an ambient gas, constituting a complex pattern of jet flow structure with a head

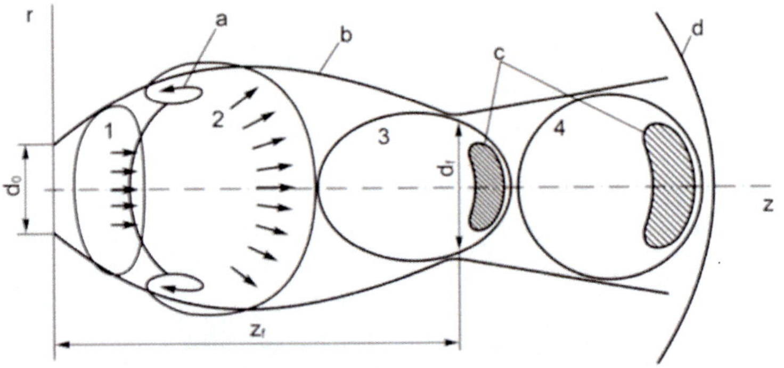

Fig. 3.13 Qualitative model of the laser-induced vapor-plasma jet in air

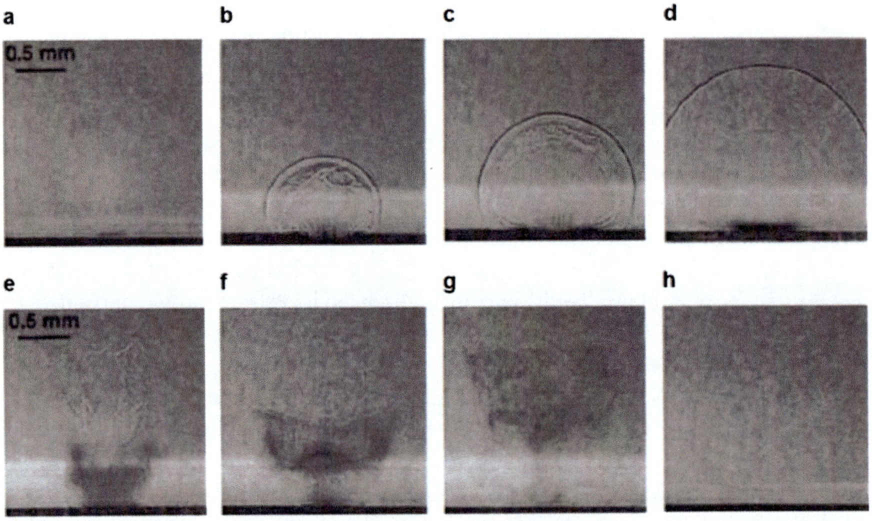

Fig. 3.14 Sequence of photographs showing the ablation of a PMMA surface by a single pulse from a CO_2 laser (9.17 J) (**a**) 60 ns; (**b**) 350 ns; (**c**) 700 ns; (**d**) 1100 ns; (**e**) 3μs; (**f**) 5μs; (**g**) 10μs; (**h**) ∞ s

shock wave. Here, 1–4 regions are consecutive stages of jet expansion, d_0 is the laser beam spot diameter, d_f is a minimal diameter of the vapor jet which is reached after focusing of the vapor jet, z_f is a distance up to the jet focus.

Similar pictures of an ablated jet are observed every time when the jet development is fixed after the short of a laser pulse. An experimental example demonstrating evolution of the jet structure is shown in Fig. 3.14 from Ref. [36].

It is also found experimentally that evolution of plasma jet induced at laser power around of the plasma threshold at ambient vacuum conditions differs significantly from the jet evolution ablated into ambient air. Nevertheless, both threshold power fluxes are nearly the same. The behavior of plasma jet is shown in Fig. 3.15 for

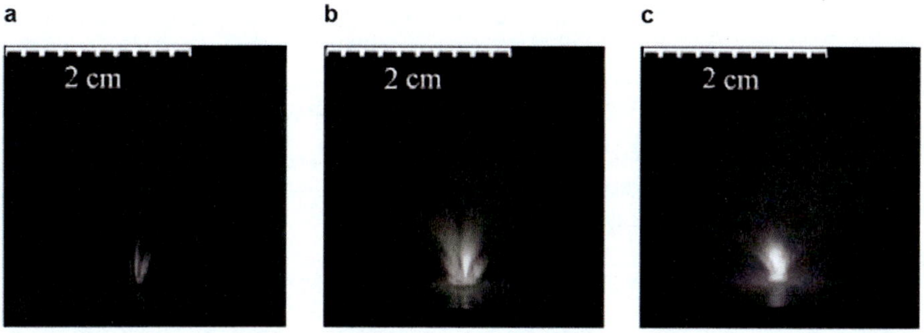

Fig. 3.15 Photos illustrating the onset of plasma on POM targets at 100 Pa at various laser power fluxes: (**a**) 10 J/cm^2; (**b**) 30 J/cm^2; (**c**) 60 J/cm^2

pulses of about 4.7 J on new POM targets prepared by the same way as described above.

Thus, we see that the structure of an ablation jet from a POM target depends on the laser power flux, and one of the criteria of the jet structure is a threshold power flux at increasing of which enhances the plasma generation. That is why there is an interest to consider the laser ablation propulsion models based on the critical laser power flux.

3.5.2 *Vapor and Plasma Models of the Laser ablation Propulsion Using Critical Laser Power Flux*

There are some theoretical attempts to explain the experimental data of the laser ablation propulsion obtained at various laser power-plasma conditions based on the model of the laser power flux as a crucial parameter that determines special conditions of thrust production via ablation of solid targets [21].

For example, the theory developed in Ref. [32] expresses the ablated mass areal density, μ, and impulse areal density, σ, as follows:

$$\mu = \rho z_t, \quad \text{and} \sigma = \mu \langle v_e \rangle \approx \sqrt{2\rho z_t (1 - \phi_R)(1 - \alpha_p)(\Phi - \Phi_t)} \tag{3.41}$$

where ρ is a solid density, z_t is a threshold depth (convenient models include photo thermal and photo chemical varieties), φ_R is the target surface reflectivity, α_p is plasma absorptivity, $<v_e>$ is a first moment of velocity distribution of exhaust jet, and Φ_t is a laser power flux of the ablation threshold. To confirm, polyoxymethylene (POM) as a perspective propellant for high-power laser propulsion is considered in this case.

Using these formulas and experimental data on the laser ablation propulsion, optimal laser power flux as applied to getting maximum thrust with the polyoxymethylene (POM) polymer can be defined. However, it is necessary to determine z_t that follows from two models of solid target evaporation near to the ablation threshold, namely: (a) photo chemical model based on a Bouguer-Lambert-Beer absorption law, and (b) photo thermal model based on one-dimensional theory of thermal diffusion derived by Bäuerle [32].

Let us consider both evaporation models in more detail, following Ref. [32].

Photo chemical model results in:

$$z_t = \frac{1}{\alpha} \ln \frac{(1 - \phi_R)\Phi}{\Phi_t} \tag{3.42}$$

where α is the radiation absorption coefficient of the target material. If the polymer consists of $(CH_2O)_n$ chains, then major of the laser power is absorbed via destroying C-O bonds which retain monomeric molecules. At the same time, a CO_2 laser photon energy is only 0.12 eV and the C-O bond energy of the molecular is 3.7 eV[33]. So, if we consider only a photo chemical model of POM ablation then it will require about 32 CO_2 molecular photons to destroy C-O bonds of the polymer.

Photo thermal ablation model results in:

$$z_t = \sqrt{4D_T\tau \ln \frac{(1 - \phi_R)\Phi}{\Phi_t}} \tag{3.43}$$

where D_T is the coefficient of thermal diffusion, τ is the laser pulse length, T_v is the evaporation temperature that corresponds to a boiling point of the polymer, and C_p is the thermal capacity coefficient (see (3.44)). It is physical fact that the absorption coefficient of the laser power in photo thermal case, α, depends on the relative polarization of the laser beam, and it can change within a 1000–9000 cm^{-1} range [33]. As for laser characteristics, the laser pulse length is usually known, and D_T can be found in the literature [29]. Only one parameter that is not determined in this formula is threshold of the power flux, Φ_t, at which the ablation process is started. It follows from the photo thermal model, Φ_t is determined from approximate equation:

$$\Phi_t \approx 2C_p\rho T_v\sqrt{\pi D_t\tau} \tag{3.44}$$

In the literature, the power fluence threshold is reported to range from 0.2 J/cm^2 [19] to 1.3 J/cm^2 [5].

As is known, the POM polymer is destroyed under high temperatures [28]. The destruction of the polymer molecules is observed as the polymer chains are broken, starting from end of the chain. This process results in production of water, formaldehyde monomer (CH_2O), and deblocking of some larger polymer chain fragments. The polymer can be destroyed into monomer molecules as well as in to simple molecules (e.g., CO, H_2) or up to atoms (C, H, and O) with a moderate laser power flux. Further increase of the laser power flux generates ions of such atoms as (C+, H+, O+) which will absorb the laser power in accordance with an inverse

Bremsstrahlung process and with number of ion density. Thus, we have another question concerning correct determination of the laser power flux threshold of plasma ignition in vapor jet consisting of C, H, and O atoms.

The Plasma Threshold Fluence

The photographs of the vapor-plasma jet shown in Figs. 3.12 and 3.15 indicate a crucial threshold of plasma formation of 10 J/cm^2 at 101 kPa of ambient air pressure. The vapor jet consisting of neutral particles is formed at low power flux on POM target surface, which is then transformed into a region of unstable flow at a certain distance from the target surface. It can be assumed that the region starts directly after formation of the Mach disk in a supersonic jet that is injected into the ambient gas of terminal pressure. In particular, the jet structure is seen clearly due to scattering of light by small solid particles evaporated from the target.

In his earlier paper [5], Dr. Phipps stated the following expressions for C_m and I_{sp} of the laser ablation propulsion, corresponding to a single-ionized plasma:

$$\begin{aligned} C_m &\approx \frac{(1.84 \times 10^2)}{A^{1/8}\left(I\lambda\tau^{1/2}\right)^{1/4}}\left(\frac{A}{2\left(Z^2(Z+1)\right)^{1/3}}\right)^{9/16} \\ I_{sp} &\approx 442A^{1/8}\left(I\lambda\tau^{1/2}\right)^{1/4}\left(\frac{2\left(Z^2(Z+1)\right)^{1/3}}{A}\right)^{9/16} \end{aligned} \tag{3.45}$$

where A is the averaged atomic mass, and Z is the averaged ionization potential of the plasma ($Z > 1$) (for comparison, see Sect. 3.2.3). It should be mentioned the results obtained by using these formulas will be strongly depending on the chemical composition of the target.

Now, we see that the laser ablation propulsion based on polyoxymethylene and other CHO-polymers looks very complicated because of numerous physical processes accompanying thrust production. But, these propellants are promising ones for the atmospheric and space applications of the laser propulsion. Moreover, it is expected to increase efficiency of the thrust production by propellants using for laser ablation propulsion.

3.6 Efficiency of the Laser Ablation Propulsion Based on CHO-Polymers

Proper definition of the laser ablation propulsion efficiency that uses conventional propellants to produce a thrust can be determined by general rule of the efficiency, η, as a ratio of the exhaust jet power P_{jet} to power P_Q of the total energy released in a combustion chamber of jet engine, namely:

$$\eta = \frac{P_{\text{pc}}}{P_{\text{Q}}} = \frac{g^2}{2} \cdot \frac{I_{\text{sp}}^2}{Q} = \frac{g}{2} \cdot \frac{I_{\text{sp}}}{C_{\text{T}}} \tag{3.46}$$

where Q is a specific heat of the rocket fuel combustion, $C_{\text{T}} = P_{\text{Q}}/T$ is a thrust coefficient.

By the analogy with this formula, the efficiency of the laser-propulsion η_{LPE} operating at space conditions is defined as a ratio of the momentum of the exhaust jet to the laser power released in the engine nozzle [3, 35]:

$$\eta_{\text{LPE}} = \frac{\dot{m}v^2}{2P} = \frac{T \cdot I_{\text{sp}} \cdot g}{2P} \tag{3.47}$$

Taking into account the fact that the momentum coupling coefficient C_{m} is the parameter that is most frequently used to estimate the conversion efficiency of the laser power into a thrust, this formula is transformed into the following:

$$\eta_{\text{LPE}} = \frac{C_{\text{m}} I_{\text{sp}} \cdot g}{2} \tag{3.48}$$

At the same time and in accordance with the general definition (3.46), the LPE efficiency is determined by such processes as:

- conversion of the laser power into the internal energy of a propellant, α, that can vary noticeably in dependence on laser radiation characteristics and chemical composition of the propellant
- conversion of the internal energy into a momentum of the exhaust jet, β, depended on the nozzle geometry and the ambient conditions (gas pressure, temperature, etc.).

In this case, the following general definition of the LPE efficiency can be introduced as [35]:

$$\eta_{\text{LPE}} = \frac{1}{2} C_{\text{m}} \langle v \rangle = \alpha\beta\Phi \tag{3.49}$$

Here, $<v>$ is the average flow velocity in the exhaust jet, Φ is the ratio of the average velocity square to the mean square of the jet velocity:

$$\Phi = < v >^2 / < v^2 > \tag{3.50}$$

As usually, the parameters $\alpha = 0.4$ (see Chap. 2) and $\Phi \sim 1$ are close to the actual values. As for the β parameter, its value varies over the range of 0.5–1. Then without significant error, we have:

Table 3.8 Thrust characteristics of the rocket and laser-propulsion engines

Propulsion engine	C_T, W/N	I_{sp}, s	η
IUS (International Upper Stage)—solid-propellant rocket engine	1.45×10^3	200	0.69
Centaur Upper Stage	2.04×10^3	300	0.74
Rocket engine with hydrogen-oxygen fuel	4.03×10^3	500	0.62
Rocket engine with fluid hydrazine	3.85×10^3	200	0.26
LPE with $C_m = 50 \times 10^{-5}$ N/W	2.0×10^3	160	0.4
LPE with $C_m = 10 \times 10^{-5}$ N/W	10^4	800	0.4

$$\eta_{LPE} = \alpha\beta \tag{3.51}$$

Then we can deduce the result that: $0.2 \leq \eta_{LPE} \leq 0.4$. For comparison, efficiencies of different rocket engines are listed in Table 3.8.

We can see from Table 3.8 the laser propulsion based on non-chemical propellants possesses an efficiency that is less than the efficiency of rocket engines, operating with chemically active propellants. In that sense, the laser ablation propulsion based on CHO-polymers could be comparable with chemical rockets. In accordance with (3.46), the total efficiency of the LAP is determined as a ratio of the momentum of an exhaust jet to the total power of energy sources released in the LPE nozzle, i.e.,

$$\eta_{LPE} = \frac{\beta(\alpha P + \dot{m}Q_{det})}{P + \dot{m}Q_{det}} \tag{3.52}$$

where Q_{det} is a specific heat of polymer combustion, which is released by two successive chemical reactions under the laser radiation, and $\dot{m}$ is ablation mass rate of the CHO-polymer (see Sect. 3.4).

After simple conversion, we have:

$$\eta_{LPE} = \frac{\beta(\alpha E + \dot{m}Q_{det})}{E + \dot{m}Q_{det}} \text{ or } \eta_{LPE} = \frac{\beta(\alpha + \dot{m}Q_{det}/E)}{1 + \dot{m}Q_{det}/E} \tag{3.53}$$

Here, E is the laser pulse energy.

The efficiency of the LAP with Delrin and polymethylmethacrylate, η_{LPE}, which have been obtained in various experiments are listed in Table 3.9.

As it is seen from Table 3.9, CHO-polymers used as propellants for the laser propulsion have to possess higher internal detonation energy, Q_{det}, in comparison with polyoxymethylene ($Q_{det} = 2690$ kJ/kg) to achieve more efficient propulsion. The latter will require specific technology to develop a new polymer composition [15, 28].

So, as for the laser ablation propulsion, we have a few ways of getting high efficiency of the thrust production, characterized by C_m parameter of several hundreds of dynes per Watt. Confined laser ablation in multilayer targets combines such advantages as availability of maximum values of C_m at medium radiation power and

Table 3.9 Efficiency η_{LPE} for investigated CHO-materials ($\lambda = 10.6\mu m$, $\alpha = 0.4$; $\beta = 0.9$)

Propellant	$\Delta m \times 10^{-6}$, kg	E, J	$\dot{m}\ Q_{det}$, J	$\dot{m}\ Q_{det}/E$	η_{LPE} space conditions
Delrin® [14]	15.0	250.0	40.5	0.16	0.44
Delrin® [27]	12.33	61.0	33.19	0.54	0.55
Delrin® [27]	8.0	57.1	21.60	0.38	0.51
PMMA [17]	37.80	90.0	18.51	0.21	0.45
PMMA [17]	40.30	130.0	19.73	0.15	0.43

operation in a vacuum. Long duration of a recoil momentum exceeded the laser pulse duration by a factor of 2 or 3 can be considered also as merit of the combined ablation. But, this type of the laser-propulsion technique has a disadvantage, namely, the specific impulse of the exhaust jet is up only several tens of seconds.

Laser ablation propulsion based on CHO-polymers used as propellants is a most promising technology of efficient production of thrust. This type of the laser propulsion provides a higher momentum coupling coefficient due to the additional chemical energy released in a vapor jet through laser irradiation. This energy is released via exothermic reactions proceeding in the polymer vapor via ignition of high-temperature plasma [37]. Moreover, the use of CHO-polymers allows achieving a total efficiency of the laser ablation propulsion higher 50%.

Nevertheless, the development of LAP with maximum thrust for prospective high-power laser-propulsion systems assumes achieving the propulsion efficiency of 70%. Choice and/or construction of the perspective propellant composition and engine device are the critical tasks of future investigations on the laser propulsion. We hope the scientific experience accumulated by now at the laser ablation propulsion study will allow solving these tasks in the near future.

References

1. Kantrowitz, A.: Propulsion to orbit by ground-based lasers. Astron. Aeron. **10**, 74–76 (1972)
2. Askaryan, G.A., Moroz, E.M.: Pressure impulse generated at material evaporation under the action of laser beam, light induced pressure. J. Exper. Theor. Phys. **43**, 27–25 (1962)
3. Ablation, L.: Principles and applications. In: Miller, J.C. (ed.) . Springer (1994)
4. Bunkin, F.V., Prokhorov, A.M.: Laser power source application to produce a thrust. Adv.Phys. Sci. (Rus.) **119,** 425-446 (1976)
5. Phipps, C.R., et al.: Impulse coupling to targets in vacuum by KrF, HF, and CO_2 single-pulse lasers. J. Appl. Phys. **64**, 1083–1096 (1988). https://doi.org/10.1063/1.341867
6. Phipps, C.R., Dreyfus, R.W.: Ch. 4: Laser ablation and plasma formation. In: Vertes, A. (ed.) Laser ionization mass analysis. Wiley (1993)
7. Phipps, C.R. (ed.): Laser ablation and its applications. Series: Springer Series in Optical Sciences, p. 129 (2007)
8. Danilychev, V.D., et al.: Investigations on plasma dynamics, generated close by a solid target under CO_2-laser pulse of microsecond length. J. Quant. Electron. (Rus.). **7**, 2599–2604 (1980)
9. Protasov, Y.Y., Stepanov, O.G.: Study oh radiative gas-dynamic processes and generation of laser detonation waves. Bull. MSTU. Ser.: Nat. Sci. **4**, 69–77 (2004)
10. Raizer, Y.P.: Laser-ignited spark and development of charges. Nauka, Moscow (1974)

11. Krokhin, O.N.: Modern physical principles of laser ablation. Proc. SPIE. **4065**, 6–16 (2000). https://doi.org/10.1117/12.407344
12. Zvorykin, V.D.: Comparative analysis of gas-dynamic regimes of high-power UV and IR gas lasers interaction with solids in atmosphere. Proc. SPIE. **4065**, 128–139 (2000). https://doi.org/10.1117/12.407369
13. Sinko, J.E., Sasoh, A.: Review of CO_2 laser ablation propulsion with Polyoxymethylene. Int. J. Aerosp. Innov. **3**, 93–129 (2011). https://doi.org/10.1260/1757-2258.3.2.93
14. Sinko, J.E., Phipps, C.R.: Modeling CO_2 laser ablation impulse of polymers in vapor and plasma regimes. Appl. Phys. Lett. **95**, 131–105 (2009). https://doi.org/10.1063/1.3234382
15. Lippert, N., et al.: Fundamentals and applications of polymers designed for laser ablation. Appl. Phys. A Mater. Sci. Process. **77**, 259–264 (2003). https://doi.org/10.1007/s00339-003-2111-y
16. Urech, L.: Design and characterization of energetic polymers applied in laser space propulsion. For the degree of Doctor of Sciences: DISS.ETH NO. 17068, ETH Zürich (2007)
17. Salvador, I.I.: Static and hypersonic experimental analysis of impulse generation in air-breathing laser-thermal propulsion. A Thesis Submitted Rensselaer Polytechnic Institute. Troy, New York (2010)
18. Pakhomov, A.V., Thompson, M.S., Gregory, D.A.: Ablative laser propulsion: a study of specific impulse, thrust and efficiency. AIP Conf. Proc. **664**, 194–205 (2002). https://doi.org/10.2514/2.1567
19. Phipps, C.R., et al.: Coupling coefficient at low laser flounce with a structured target. High-Power Laser Ablation III. Proc. SPIE. **4065**, 931–938 (2000). https://doi.org/10.1117/12.407330
20. Yabe, T., Nakagawa, R., Yamaguchi, M.: Simulation and experiments on laser propulsion by water cannon target. 1-st Int. Symp. Beam. Energy Prop. AIP Conf. Proc. **664**, 185–193 (2002). https://doi.org/10.1063/1.1582107
21. Phipps, C.R., et al.: Overview: laser-ablation propulsion. J. Prop. Pow. **26**, 609–649 (2010). https://doi.org/10.2514/1.43733
22. Remo, J.L., Hammerling, P.: Experimental and computational results for 1054 nm laser induced shock effects in confined meteorite and metallic targets. Proc. SPIE. **4065**, 635–643 (2000). https://doi.org/10.1117/12.407384
23. Phipps, C.R., et al.: Very high coupling coefficient at low laser fluence with a structured target. Proc. SPIE. **4065**, 931–938 (2000). https://doi.org/10.1117/12.407330
24. Fabbro, R., et al.: Physical study of laser-produced plasma in confined geometry. J. Appl. Phys. **68**, 775–784 (1990). https://doi.org/10.1063/1.346783
25. Yabe, T., et al.: Laser-driven vehicles—from inner-space to outer-space. Proc. SPIE. **4760**, 867–878 (2002). https://doi.org/10.1007/s00339-003-2125-5
26. Liukonen, R.A.: Efficiency of energy transfer into a recoil impulse for laser propulsion engine. Lett. J. Tech. Phys. (Rus.). **18**, 76–79 (1992)
27. Schall, W.O., et al.: Lightcraft experiments in Germany. Proc. SPIE. **4065**, 472–481 (2000)
28. Ageichik, A.A., et al.: Laser detonation of CHO-polymers as applied to laser propulsion. J. Tech. Phys. **79**, 76–83 (2009). https://doi.org/10.1117/12.407369
29. Kikoin, I.K. (ed.): Tables of physical parameters. Atomizdat, Moscow (1976)
30. Schall, W.O., Eckel, H.-A., Bohn, W.L.: Laser propulsion thrusters for space transportation. Phipps, C.R. (ed.) Laser ablation and its applications. Springer (2007)
31. Loktionov, E.Y., Protasov, Y.Y.: Investigation of the dynamics and microstructure of laser-induced optical discharges with ablating target. High Temp. **48**, 766–778 (2010)
32. Sinko, J.E., et al.: Critical fluence and modeling of CO_2 laser ablation of polyoxymethylene from vaporization to the plasma regime. AIP Conf. Proceed. **1230**, 395–404 (2010). https://doi.org/10.1063/1.3435456
33. Bäuerle, D.: Laser processing and chemistry. Springer-Verlag, Berlin (2000). Ch. 2.2.2: The dimensionality of heat flow, p. 20; and Ch. 11.6: Plasma formation, pp. 206–212

34. Bulgakova, N.M.: Investigation of the dynamics and processes of laser ablation under the action of milli-, nano-, and femtoseconds pulses. Doct. Thes. Novosibirsk (2002)
35. Larson, C.W., et al.: Laser propulsion and the constant momentum mission. AIP Conf. Proc. **702**, 216–227 (2004). https://doi.org/10.1063/1.1721002
36. Srinivasan, R.: Ablation of polymethyl methacrylate films by pulsed (ns) ultraviolet and infrared (9.17 μm) lasers: a comparative study by ultrafast imaging. J. Appl. Phys. **73**, 2743 (1993). https://doi.org/10.1063/1.353048
37. Zeldovich, Y.B., Raizer, Y.P.: Physics of shock waves and high-temperature gas-dynamic processes. Nauka, Moscow (1966)

Chapter 4
Aerospace Laser-Propulsion Engine

Abstract Development of the airspace laser-propulsion engine operating both in RP and CW modes of thrust production is validated. The engine is considered as a basic unit for creation of HPLP. The aerospace laser device differs sufficiently as compared with other laser-propulsion engines by its design, and it consists of two principal units, namely, laser beam concentrator and jet nozzle. Such type of the device allows optimizing both units of the engine independently on other one, taking into account only laser beam parameters and gas-dynamic effects.

The ASLPE engine was tested experimentally by the use of the RP and CW types of CO_2 lasers. The engine model demonstrated thrusts of 1.5–2.0 N at the laser power of 5–20 kW, which corresponded to momentum coupling coefficient of 10–20 dynes/W.

In the chapter, thermal and mechanical loads on the device construction are analyzed to optimize design of the engine units, too. The proposals for increasing the efficiency of thrust production by the engine are considered.

Keywords Pulsejet and air-breathing propulsion · Onboard engine · Two-mirror beam concentrator · Jet nozzle · Beam wave front aberrations · Continuous wave and pulsed modes of operation · ASLPE device · Thermal aberrations

4.1 Introduction

Basic design principles of high-power laser-propulsion (HPLP) systems are assumed to be based on some engineering requirements to vehicle with a laser-propulsion engine (LPE) and especially:

- The LPE device has to allow the engine to operate both in pulsejet and air-breathing propulsion modes during the vehicle flight in atmosphere as well as in rocket mode when it reaches the low Earth orbits.
- The vehicle device has to arrange optical equipment on a board to collect the laser radiation, to transform it into LPE.

Y. A. Rezunkov, *High Power Laser Propulsion*, Springer Series on Atomic, Optical, and Plasma Physics 116, https://doi.org/10.1007/978-3-030-79693-8_4

- The vehicle system has to allow co-adjusting of both the optical equipment and the remote laser so that the vehicle can move independently on a mutual orientation of the vehicle and laser.

A specific problem for the laser propulsion is the vehicle's transition from a subsonic regime of flight in the atmosphere up to supersonic flight, when the laser-propulsion engine has to operate efficiently as well.

In principle, the vehicle named as Lightcraft Technology Demonstrator LTD satisfies two of the three listed requirements, excluding namely dependence of its flight direction on a spatial location of the laser beaming system [1]. The laser-propulsion engine of LTD can produce thrust both at subsonic flight in close-to-ground atmosphere and at supersonic flight of the vehicle in the upper atmosphere, as well as finally in a rocket mode using the hydrogen or helium as onboard propellants. In our opinion, one of the important steps in designing the LTD was made when the vehicle was developed as a single-stage vehicle for sub-orbital and orbital flights in the close-to-Earth space. The LTD looks like a very compact and perfect engineering device being adapted for high-power laser-propulsion launch system.

Nevertheless, as progressed from the experimental investigations, the development of LTD was slowed down by the problems of thrust generation during supersonic modes of the air-breathing laser propulsion [2]. Hence, special attention should be given to the LTD air inlet and absorption chamber shapes to provide lossless braking of incoming supersonic flow and efficient generation of air-breathing propulsion [3]. Both problems are considered in Chap. 5 more in detail.

Here, we propose Aerospace Laser-Propulsion Engine (ASLPE) [4], developed especially as the onboard laser-propulsion engine providing vehicle flight independently the mutual orientation of the vehicle and laser. Moreover, the ASLPE design corresponds to all conditions to be satisfied for the multipurpose engine of high-power laser-propulsion systems. An ASLPE design is developed by matching the receiver optics of a vehicle with a beam collection system onboard the laser-propulsion engine. Moreover, the ASLPE is assumed to be used for orbital maneuvers of space vehicles.

The basic idea of the ASLPE device is based on the special design of the laser beam focusing system that allows the laser radiation be directed into the engine nozzle by around an exhaust jet. This type of the beam concentrator got the name of a tractor beam conception [5] a little bit later. In science fiction, a tractor beam is a hypothetical beam of energy that can be used to move objects such as spaceships or hold them stationary. But, the ASLPE conception possesses a wider field of applications, including the supersonic airborne and high-orbit space-based vehicles.

To investigate the features of the ASLPE, some test experiments were required. These experiments included:

- proof of the ASLPE conception that is based on a separation of the optical and gas-dynamic thrust units of the engine so that the laser radiation enters into the engine nozzle through a small slit in the nozzle wall

- verification of the momentum coupling coefficient dependence on laser pulse energy and estimation of the slit size influence on the C_m magnitude
- testing of CHO-polymers as additional propellants to produce additional thrust

The choice of grade of material for the ASLPE device is defined by mechanical resistance of the materials to the shock and thermal loads as well as by high resistance to intensive laser radiation, and so on. These and other questions are considered in this chapter.

4.2 The Aerospace Laser-Propulsion Engine Conception

As usual laser-propulsion engines operating under a CW laser radiation are based on the conventional gas-dynamic design to accelerate gas flow in a jet nozzle and to produce a thrust [6, 7]. Here, laser radiation is injected into a combustion chamber through an optical window. One of the characteristic features of thrust production under CW laser radiation is an instability of the laser radiation interaction with a plasma ignited in the chamber, which results in an instability of the thrust produced.

A pulsejet LPE operating under RP laser radiation is a simplest design. A jet nozzle of the RP LPE is combined as a laser beam concentrator and jet nozzle (see Refs. [8, 9]). In this case, laser radiation is directed into the nozzle from the exhaust section side of the nozzle. This simplification of the LPE design results in some restrictions on efficiency of laser propulsion, namely:

- Laser power is partially absorbed by the gas and plasma of the exhaust jet.
- Optical quality of nozzle inner surface is also destroyed by the hot gases and plasma as well as products of various chemical reactions, happening in the nozzle.
- Vehicle flight direction depends on a mutual spatial orientation of the vehicle and laser.

Therefore, the airspace laser-propulsion engine (ASLPE) is specially designed without these defects. The principle of the structural and functional separation of the beam concentrator and the jet nozzle allows the received laser radiation to produce a thrust independently on a mutual orientation of the vehicle and laser beam. Moreover, the laser beam comes to the ASLPE from a position that is opposite to exhaust jet [10]. That means, the ASLPE device may be considered as a multipurpose laser-propulsion engine operating both in continuous and repetitively pulsed modes depending on the emission characteristics of the laser radiation [11].

The ASLPE device consists of the following units, namely (see Fig. 4.1), receiver optics adjusting receiver laser beam aperture with the beam concentrator geometry, laser beam concentrator optics, and gas-dynamic unit (nozzle) that are arranged to be co-axial with the beam concentrator. The beam concentrator is designed as two optically matched mirrors: the first one R1 is a short focus parabola, the second reflector RO2 is designed as an annular mirror, axis of which coincides with the axis of symmetry of first reflector. Hence, the focus regions of these two reflectors are

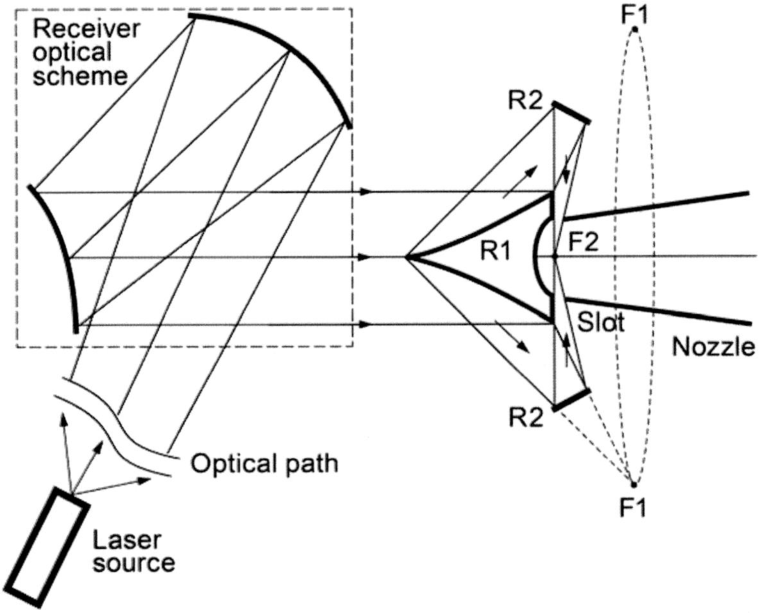

Fig. 4.1 Schematic diagram of airspace laser-propulsion engine units R1—first (parabolic) reflector, R2—second (elliptic) reflector; slit is arranged between R1 and nozzle wall, F2 is laser beam focus located inside the nozzle

combined with each other. The nozzle consists also of two parts, namely, the pressure impulse receiver PR that is arranged closely to the rear of first reflector and the received jet nozzle C. There is a slit between the pressure receiver and nozzle wall, being designed to transmit laser power into the nozzle.

A critical feature of the ASLPE design principles is decreasing of the slit influence on the pressure losses because of lateral flow of the working gas through the slit. To minimize these losses, the slit width is minimized by arrangement of the laser beam intermediate focus in the slit. In this case, the slit width is determined by the beam divergence and common focus of the R1 and R2 reflectors. The slit width of 100μm was achieved in the experiments. More experiments with the ASLPE device made in Ref. [12] proved the fidelity of the ASLPE conception.

4.2.1 Designing of Two-Mirror Beam Concentrator

The beam concentrator of ASLPE consists of two off-axis reflectors, optical surfaces of which are formed as a curve of a second order. Below, the principles of designing this beam concentrator are presented.

The condition of matching the mirrors surfaces means coincidence of the mirrors conjugate points by such a way that an image of any object located in a conjugate

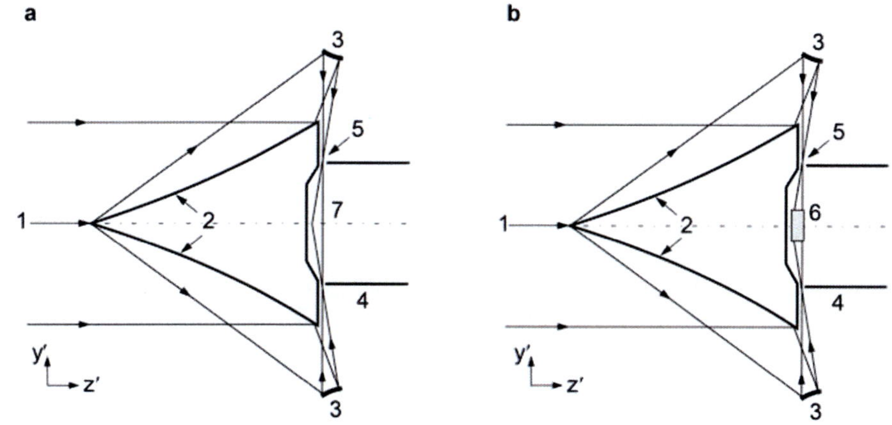

Fig. 4.2 Optical schematics of the ASLPE two-mirror reflector: (1) laser beam; (2, 3) optical surfaces O_1 and O_2; (4) nozzle; (5) a slit for the laser beam; (6) solid propellant or (7) gaseous propellant

point of first mirror is transmitted into a conjugate point of second mirror by both the reflectors without aberrations [13]. The choice of the mirror shape of a second order for every reflector is determined by the requirement of designing the concentrator packaged device with a real focus point existence but only as an imaginary focus. This type of the spatial coordination of the conjugate points is required to control the laser beam phase structure and its intensity distribution at a real focus of the combined concentrator.

In Fig. 4.2, the ASLPE beam concentrator scheme is shown in a meridian section. In the scheme, toroidal optical surface mirror components 2 and 3 are presented. The optical surface forms can be chosen from the following mirrors with a shape of second order, namely, elliptical, hyperbolic, spherical, parabolic, and even rectilinear.

A particular feature of the radial symmetric focusing of a laser beam on the nozzle axis is that a high intensity region is formed in a local area in the vicinity of the nozzle axis, so that the local beam intensity exceeds the threshold value for a gaseous (Fig. 4.2(a)) or solid (Fig. 4(b)) propellants only in this region. The focused beam has an intermediate circular focus in a slit between the first beam concentrator and the nozzle. But, the radiation intensity in the small slit is such as to avoid a breakdown of gas in the slit because of lengthy circular focus of the beam.

The optical scheme of the ASLPE beam concentrator looks like a conical beam transformer that has been developed and applied to resonators for high-power lasers [14, 15]. For the examples, the conical transducers serve to transform a beam aperture inside laser cavities into a compact aperture or to focus the laser beam at long distances out of the laser outlet aperture as applied to laser drilling technology or similar metal working [16]. But, the ASLPE beam concentrator has to satisfy some additional conditions, namely:

1. formation of a minimal waist of the laser beam that comes through the nozzle slit
2. guaranteed laser breakdown of the propellant on the nozzle axis

To satisfy these conditions, the minimal width of a slit has to be of a size defined by the diffraction divergence of the transformed laser beam, namely:

$$d_{\mathrm{dif}} = 2\lambda(R_1 - r)/D \tag{4.1}$$

where λ is laser radiation wavelength, r is the nozzle radius, R_1 is average radius of mirror 2, and D is the laser beam size on the second mirror surface. For example, the ASLPE concentrator from Ref. [10] was designed with the following geometry parameters: $R_1 = 80$ mm, $r = 30$ mm, $D = 7.5$ mm, and minimal size of the slit equal to 13.3 λ.

But in a reality, a high-power laser beam is usually distorted in phase and amplitude because of misalignment of the laser resonator mirrors, inhomogeneity of the laser active medium, mechanical and thermal loads on the laser optics, and so on. All of these defects result in broadening and wandering the laser beam focus. Therefore, these effects will limit the laser power delivered into the nozzle and will decrease the thrust produced if the slit is of a diffractively limited width. Besides, the slit raises the requirements for accuracy of adjustment of the beam concentrator mirrors as a whole system.

It is expected that the radiation intensity in the beam focus point will depend on a deflection of incoming laser beam axis the total optical axis of the beam concentrator, defocusing (spherical distortion) of incoming beam and similar beam distortions. For example, the laser beam deflection with the concentrator axis of symmetry in the order of 2 mrad causes an astigmatic aberration of the beam in the focus region, which is presented like couple lines in the beam focus. At that, the radiation intensity falls twice in the focus. If the incoming beam possesses a spherical aberration in its wave front, then the radiation intensity decreases by a factor of two because of formation of the coma aberration.

Therefore, the detailed analysis of the two-mirror concentrator optical schemes is performed by considering various combinations of the mirrors of four types of reflecting surface, which are listed in Table 4.1. As it follows from this table, the optimal configurations of the mirrors are with an imaginary intermediate focus to

Table 4.1 Feasible configurations of the two-mirror beam concentrator

		1	2	3	4	5	6	7	8	9	10	11	12	13	14	15	16
Plane mirror	M1	+	+	+							+						
	M2				+	+	+				+						
Elliptical mirror	M1				+			+						+		+	
	M2	+						+							+		+
Hyperbolic mirror	M1					+			+				+		+		
	M2		+						+			+		+			
Parabolic mirror	M1						+			+		+					+
	M2			+						+			+			+	

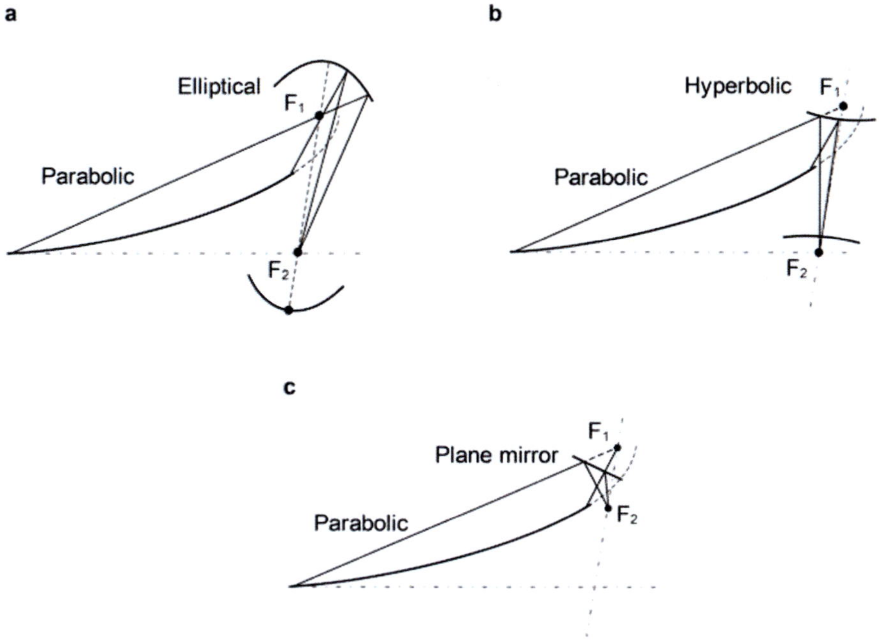

Fig. 4.3 Optical schematics of the ASLPE beam concentrators

exclude a laser breakdown of the medium filling up the ASLPE concentrator volume (see Fig. 4.3).

Configuration under number 6 is illustrated in Fig. 4.3(a), version 11 is in Fig. 4.3 (b), and version 16 is in Fig. 4.3(c).

It should be noted that the use of a plane mirror in the beam concentrator (see Fig. 4.3(c) simplifies the ASLPE design. But, as is shown numerically in Ref. [10] this version of the concentrator has a few practical difficulties, namely, the outer mirror has a large diameter and lengthened generatrix of the mirror, which result in increased mirror sizes. Geometric limitations on the relative arrangement of the two mirrors of the concentrator resulting from the prescribed distance from second mirror to beam focus point do not allow any decrease in the diameter of the second mirror. That is why the second mirror should be a hyperbolic mirror because this type of optical surface can refocus spherical beams transforming the beam wave front curvature from large to small magnitude and vice versa.

Thus, the basic statements of the ASLPE system are the following:

1. Laser power input into the laser-propulsion engine is realized from the side of the engine that is opposite to exhaust jet.
2. Laser radiation is directed into the engine nozzle through a small slit between the first mirror of the beam concentrator and engine nozzle.
3. Laser breakdown of a propellant is initiated on the axis of the engine nozzle.
4. Laser breakdown of the propellant in a slit is avoided by the choice of optical concentrator.

4.2.2 Optical Model of the Two-Mirror Beam Concentrator

To simplify the procedure of collecting the incoming laser beam, first mirror of the beam concentrator, O1, is chosen as a figure of rotation with a parabolic profile. Its basic geometric parameters are shown in Fig. 4.4.

Canonical equation describing the paraboloid profile can be written in the following form of rectangular coordinate system XOY [15]:

$$(y - y_A)^2 = -2p(x - x_A) \tag{4.2}$$

where $p = 2f$ is focus parameter of the parabola, $p > 0, f$ is parabola focus. Negative sign "–" in a right part of the equation denotes the fact that first mirror generatrix O1 is a lower branch of the parabola.

Substituting coordinates:

$$x_A = x_F + f \quad \text{and} \quad y_A = y_F \tag{4.3}$$

of the parabola tip, coordinates $(-x_c, 0)$ of the first mirror top C, as well as coordinates $(0, r)$ of the point B, corresponding to the location of the parabola bore edge $OB = r$, into Eq. (4.2), the following system of equations is derived:

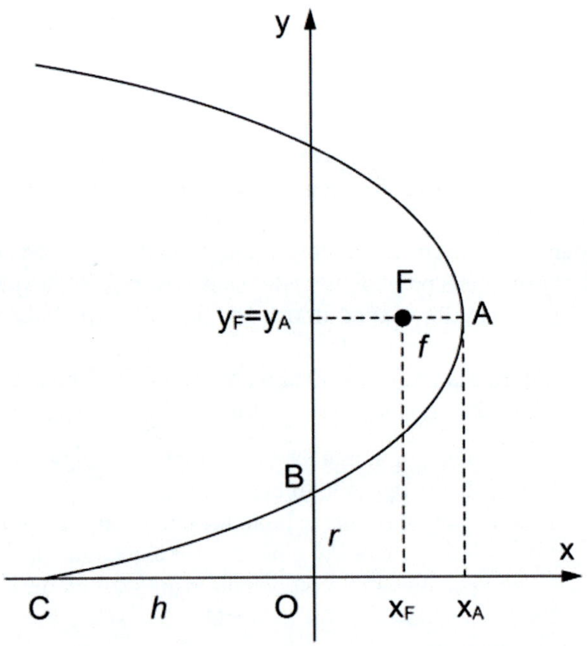

Fig. 4.4 Profile of the first mirror, O1, namely, $A(x_A, y_A)$ and $F(x_F, y_F)$ are the tip and focus of the parabola, f focus, r and h radius of the parabola basis OB and height OC of the paraboloid

$$(r - y_F)^2 = 4f(x_F + f) \tag{4.4}$$

$$y_F^2 = -4f(x_c - x_F - f) \tag{4.5}$$

Using Eq. (4.5) to determine x_F and substituting it into Eq. (4.4), the focus coordinates of the reflector O1 is found to be:

$$y_F = \frac{r}{2} - \frac{2fx_c}{r}, \quad x_F = x_c - f + y_F^2/4f \tag{4.6}$$

Taking into account the expression (4.6), one can obtain the equation for the parabolic profile by using such parameters as height x_c, r as a radius of parabola base, and its focus f:

$$y_1(x_1) = -\sqrt{4f(x_F + f - x_1)} + y_F \tag{4.7}$$

or in more common form [15]:

$$Ax_1^2 + 2Bx_1y_1 + Cy_1^2 + 2Dx_1 + 2Ey_1 + F = 0 \tag{4.8}$$

where $A = 0$, $B = 0$, $C = 1$, $D = 2f$, $E = -y_F$, $F = -4fx_c$ when x_1 varies in the range of $-x_c - 0$.

As one can see, the focus of the first reflector is a circular line in the form of a ring with a y_F radius, arranged at x_F plane.

As for the second reflector O_2, its mirror is a segment of toroidal optical surface that is formed by a hyperbolic profile. Equation describing the second reflector optical surface O2 will depend on a location of the second focus (total focus) F_2 (Fig. 4.5). The reflector generatrix O2 has a symmetric character relative to the OX′-axis in the X′O′Y′ coordinate system, located at α angle to OX-axis of the main coordinate system, XOY. The second reflector surface is formed by a rotation of the GH segment about the OX-axis.

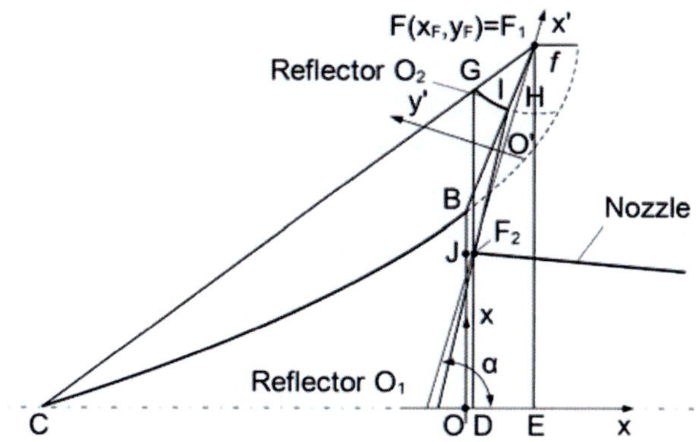

Fig. 4.5 Schematic diagram of the two-mirror beam concentrator with coordinate systems to design

Canonical equation of the hyperbolic profile of the O2 reflector written in rectangular coordinate system X′O′Y′, is as follows [17]:

$$\frac{x'^2}{a^2} - \frac{y'^2}{b^2} = 1, \; b = \sqrt{c^2 - a^2} \tag{4.9}$$

To get the hyperbolic equations describing the second mirror's optical surface in the XOY system, it is necessary to make a dual transformation of the X′O′Y′ system, namely, turning the X′O′Y′ system clockwise at an α angle—about O′ point, corresponding to using a X″O′Y″ as the new reference system, then parallel transforming the X″O′Y″ system to achieve a coincidence of its coordinate origin O′ with the coordinate origin of O of the basic system XOY [18]. The consequent mathematical expressions are as follows:

$$\begin{aligned} &x' = x'' \cos\alpha + y'' \sin\alpha, \quad y' = -x'' \sin\alpha + y'' \cos\alpha \\ &x'' = x - OD - O'F_2 \cdot \cos\alpha, \; y'' = y - OJ - O'F_2 \cdot \sin\alpha \end{aligned} \tag{4.10}$$

All notations correspond to those that are shown in Fig. 4.5. As a result, general expressions of the second reflector coordinates will be:

$$\begin{aligned} x' &= (x - OD)\cos\alpha + (y - OJ)\sin\alpha - O'F_2, \\ y' &= -(x - OD)\sin\alpha + (y - OJ)\cos\alpha \end{aligned} \tag{4.11}$$

Substituting (4.11) into (4.9), one can get the equation for the second reflector optical surface O2 written in basic coordinate system, XOY:

$$y_2(x_2) = K_1\sqrt{x_2^2 + K_2 \cdot x_2 + K_3} + K_4 \cdot x_2 + K_5 \tag{4.12}$$

where K_i coefficients are written in the following forms:

$$\begin{aligned} K_1 &= \frac{-O'H \cdot \sqrt{O'F_1^2 - O'H^2}}{O'H^2 - O'F_1^2 \sin(\alpha)^2}, \\ K_2 &= -2 \cdot (OD + O'F_1 \cdot \cos(\alpha)), \\ K_3 &= OD^2 - O'H^2 + O'F_1^2 + 2 \cdot OD \cdot O'F_1 \cdot \cos(\alpha), \\ K_4 &= \frac{O'F_1^2 \cdot \sin(\alpha) \cdot \cos(\alpha)}{O'H^2 - O'F_1^2 \sin(\alpha)^2}, \\ K_5 &= \frac{(O'H^2 - O'F_1^2) \cdot (OJ + O'F_1 \sin(\alpha)) + O'F_1^2 \cdot \cos(\alpha) \cdot (OJ \cdot \cos(\alpha) - OD \cdot \sin(\alpha))}{O'H^2 - O'F_1^2 \sin(\alpha)^2} \end{aligned} \tag{4.13}$$

From Fig. 4.5 and taking into account that $OE = x_F$, $EF_1 = y_F$, $OC = x_c$, $OJ = DF_2 = r_c$, $OD = \Delta$, we have:

- angle between OX and O′X′: $\alpha = \text{arctg}\left(\frac{y_F - r_c}{x_F - \Delta}\right)$
- eccentricity of hyperbola: $e = {}^{O'F_1}/_{O'H}$
- hyperbola tip coordinates:

$$\begin{aligned}
&x_H = r_H \cdot \cos(\alpha) + \Delta, \;\; y_H = r_H \cdot \sin(\alpha) + r_c, \\
&\text{where } r_H = HF_2 = O'F_2 + O'H, \\
&O'H = 0.5 \cdot (GF_2 - GF_1), \\
&O'F_1 = O'F_2 = 0.5 \cdot \sqrt{(x_F - \Delta)^2 + (y_F - r_c)^2}, \\
&GF_1 = \sqrt{(x_F - \Delta)^2 + (y_F - DG)^2}, \\
&GF_2 = DG - r_c, \\
&DG = y_F \cdot \frac{|x_C| + \Delta}{|x_C| + x_F}
\end{aligned} \tag{4.14}$$

It is seen that if the expressions (4.13) and (4.14) are taken into account, then Eq. (4.12) for the reflector profile O2 depends on the overall sizes of the first mirror O1 (x_c height), on displacement of the mirror O2 relative to base line of first reflector O1, Δ, as well as on radius r_c of the intermediate ring focus.

But in reality, there are a few optical distortions of the beam concentrator which originated during the manufacture of the concentrator mirrors, its misalignment during the assembly into the concentrator unit, and so on. Moreover, the laser beam may be with phase and amplitude distortions which affect the beam size in a focus region. All of these defects should be taken into account during the development of two-mirror beam concentrator.

4.2.3 Numerical Techniques to Develop the Two-Mirror Beam Concentrator

Most technological defects, which are originated during the manufacture and assembly of the beam concentrator, consist of (a) misalignment of the concentrator mirrors one relative to the other, including displacement of its axis, ΔL, (b) longitudinal displacement of the mirrors, Δz, and (c) reciprocal inclination of its axis, $\Delta\alpha$. We must consider also the phase distortions of the incoming laser beam, including (a) inclination of the beam with respect to the beam concentrator axis, $\Delta\theta$, and (b) spherical aberrations of the beam wave front illustrated as a beam radius of curvature, R. Moreover, thermal deformation, $\Delta\tau$, of the concentrator mirrors can occur through heating by the high-power laser radiation.

To test the influence of most of these defects on the laser beam optical quality, the numerical techniques are applied to specify the technology requirements of both the optical quality of the concentrator mirrors and its misalignment with respect to the incoming laser beam [10]. The factor that the concentrator consists of decentralized mirrors with a non-paraxial bundle of rays should be taken into account. Moreover, the numerical techniques used have to operate as applied to an arbitrary optical scheme of the beam concentrator [13].

The principal subject of the proposed technique consists in the following Ref. [18]. Let us assume that the linear plane-polarized laser beam with a given aperture (see Fig. 4.6) is incident on the entrance mirror of the beam concentrator and parallel to the concentrator axis of symmetry. Moreover, the laser beam diameter coincides with the mirror diameter. It is also assumed that the entrance mirror profile, O1, is described by Eq. (4.7), and the second mirror profile O2 is defined by Eq. (4.12).

To apply the ray tracing technique to analyze the beam concentrator optical properties, the laser beam is divided into a number of elementary rays. The spacing between neighboring elementary rays is chosen to be equal to $\lambda/4$. Initial coordinates of every ray are specified at the entrance aperture of the beam concentrator. Elementary ray vectors are defined by direction cosines of the angles with respect to the x, y, z spatial axes. In the plane of the entrance aperture, all laser beam rays are parallel to the z-axis, and the direction cosines are put into a matrix $M_{\rm i}$, $M_{\rm i} = (0, 0, 1)$, where the i index denotes a ray number. If it is required, we can specify a laser beam quality in the entrance plane by varying components of the $M_{\rm i}$ cosines matrix.

The matrix of direction cosines of the i-th ray polarization vector has the $P_{\rm i} = (0, 1, 0)$ form. Intersection of the light rays with a mirror surface is determined by the Newtonian iteration process at using the following set of equations:

$$\frac{x_{\rm j} - x_{\rm i}}{(M_{\rm i})_0} = \frac{y_{\rm j} - y_{\rm i}}{(M_{\rm i})_1} = \frac{z_{\rm j} - z_{\rm i}}{(M_{\rm i})_2} = L_{\rm i} \tag{4.15}$$

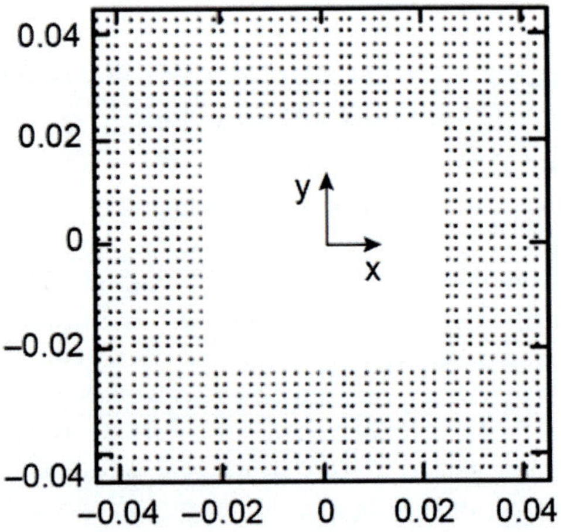

Fig. 4.6 Assumed aperture of laser beam

where (x_i, y_i, z_i) and (x_j, y_j, z_j) are the coordinates of initial space point of a ray and intersection point of the ray with consequent mirror surface. A sum of L_i^k determines a whole optical ray length in the concentrator, and k is a number of the concentrator mirrors.

Reflection of an arbitrary laser ray from a mirror surface with an arbitrary normal to the surface is defined by the following vector ratio if $n_2 = n_1$:

$$M1_i = M_i - (sign(M_i \cdot N_i) + 1) \cdot (M_i \cdot N_i) \cdot N_i \quad (4.16)$$

where $M1_i$ is reflected ray vector, N_i is normal vector of the mirror surface at the point of the ray incidence.

In order to determine laser radiation intensity distribution in a slit and inside the ASLPE nozzle, the proposed numerical ray tracing method is applied by using a ZEMAX-specialized software (Radiant Zemax LLC) [19], taking into account the linear polarization of incoming laser beam. To analyze the proposed optical scheme of the concentrator, one can assume that toroidal laser beam is focused into a ring with a radius of R_0. To estimate the radiation intensity in the focus area, let us consider an axial section of the ring beam in the y, z plane (see Fig. 4.7).

In Fig. 4.7, coordinate system $y'Oz'$ is a local spatial coordinate system, where the point of origin coincides with a slit center. Here, d_0 is the slit width counted along the nozzle wall line, R is actual distance along y' axis, D is the circular beam aperture on the second reflector O2, F is its focus. In accordance with the diffraction law, we have $R_0 \sim \lambda/(D/F)^2$ where R_0 is a distance from the nozzle axis up to the slit (nozzle

Fig. 4.7 Laser beam scheme right after reflection from a second mirror of the ASLPE beam concentrator

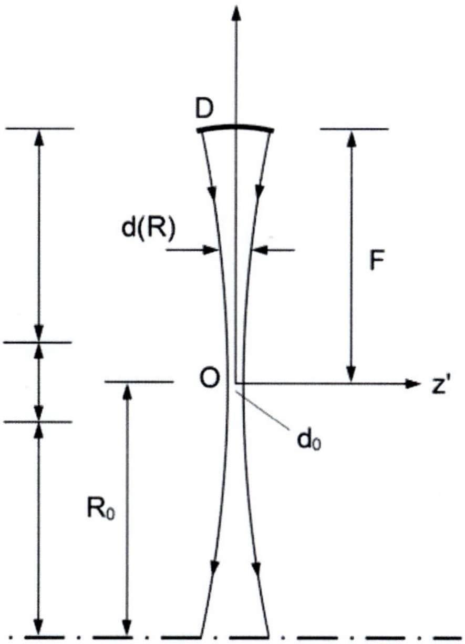

diameter). The intensity distribution in the slit will be determined by the following relation as function also of the linear power density, P/d_0:

$$I(R) = \frac{2\pi}{\lambda}\frac{P}{d_0}J_0^2(k \cdot R) \tag{4.17}$$

where J_0 is Bessel function of a zero order, k is the wavelength number.

Following (4.17), the asymptotic expression for the radiation intensity may be obtained as:

$$I(0) = \frac{2\pi}{\lambda}\frac{P}{d_0}, \quad \text{at } R = 0 \tag{4.18}$$

$$I(R) = \frac{1}{2\pi R}\frac{P}{d_0}, \quad \text{at } \lambda < R < R_0 \tag{4.19}$$

Combining (4.18) with (4.19), one can obtain the simple expression for the radiation intensity:

$$I(R) = P/\left[d_0 2\pi\left(R + \frac{\lambda}{4\pi^2}\right)\right] \tag{4.20}$$

Taking into account these formulas, one can deduce simple expressions for the radiation intensity in a slit (I_{sl}):

$$I_{\text{sl}} = \frac{D}{2\pi\lambda}\frac{1}{(F - R_0)\cdot F}P \tag{4.21}$$

and at the nozzle axis (I_{00}):

$$I_{00} = \frac{2\pi}{\lambda}\frac{F}{D(F - R_0)}P \tag{4.22}$$

$I_{\text{R}} = \frac{I_{00}}{I_{\text{sl}}} = 4\pi^2(F/D)^2$ is the parameter that depends on the laser beam divergence.

So one of the principal criteria defining optimal design of the ASLPE beam concentrator is the laser radiation intensity I_0 on the nozzle axis must be twice as large as the breakdown threshold of a propellant (I_{th}) and the radiation intensity I in the slit must be less than the threshold intensity by a factor of ten, namely, $I_0 \leq 0.1 \times I_{\text{th}}$.

Figure 4.8 illustrates the two-dimensional distribution of radiation intensity in a slit, which was calculated numerically for the case of focusing a diffractive-limited beam (4.8 (a)) and for the case of focusing the beam reflected by the concentrator mirrors being under a thermal load of 10 kW (4.8 (b)). These figures scales are 2 cm

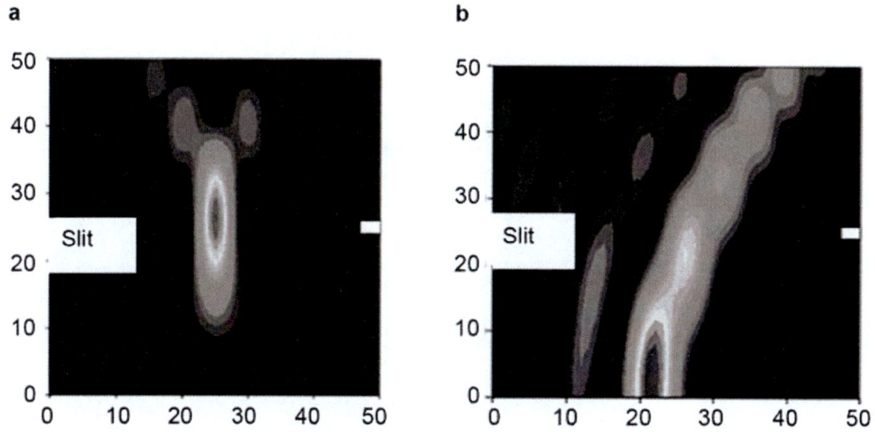

Fig. 4.8 Two-dimensional distributions of radiation intensity in a slit for the experimental model of ASLPE

for vertical axis and 0.1 cm for horizontal axis. The peak of the radiation intensity for the case 4.8 (a) is located in the slit center.

In the case 4.8 (b), the average thermal phase distortions of the laser beam reached 20μm. Consequently, the laser beam focus is shifted by 8.5 mm in the directions of the nozzle center and by 0.08 mm into the slit center. Nevertheless, the radiation intensity on the nozzle axis exceeds the threshold for air breakdown in this case, that the incoming laser power is assumed to be more than 9 kW [12].

The same analysis of a two-mirror beam concentrator is considered theoretically in Ref. [20]. Basic combinations of the beam concentrator mirrors studied are presented in Table 4.2 (see also Fig. 4.9). A specific feature of this model of the beam concentrator is the focusing of incoming laser beam at a certain angle in respect to a nozzle center line.

It is assumed in Ref. [20] that the diameter of the incoming laser beam is within the first mirror aperture. One should note that considered aberrations of the beam are important to consider when a vehicle maneuvers during the laser radiation transmission from a remote laser. And the efficiency of thrust production will depend on the radiation intensity distribution on the nozzle center line.

In Fig. 4.10, we see a distinctly dissimilar evolution of the beam focus spot defining a region of plasma ignition as a function of a misalignment angle, $\Delta\Theta_\Gamma$, for three different combinations of the concentrator mirrors. Generally, modifications of the ignition region are shown for A, B, and C concentrators. If $\Delta\Theta_\Gamma = 0°$, there are no phase aberrations in the focused beam, and a round ignition region centered on the focus and radius of nearly 0.4 mm is displayed in the three cases. If $\Delta\Theta = 1°$, the ignition region is shaped with two joined annuli, and disk of high intensity in the center has a deviation of almost 6 mm from the focus in case A. In case B, an annulus ignition region with a radius about 1.1 mm has a deviation of nearly 5 mm. In case C, the ignition region is also an annulus with radius of 2 mm, but its center is located on

Table 4.2 Geometric parameters of three groups of combinations of concentrator mirrors

Focusing mode	Reflecting surface	Equation for reflecting surface profile	Start point (z, y)	End point (z, y)
A	R1: paraboloid	$(y-550)^2=-500\ (z-350)$	(−225, 0)	(−201.3, 25)
	R2: hyperboloid	$\sqrt{(z-225)^2+(y-550)^2}-\sqrt{z^2-y^2}=404.26$	(−138.4, 102.5)	(−169.6, 97.9)
B	R1: cone	$y=z+180$	(−180, 0)	(−162.3, 25)
	R2:paraboloid	$z^2=-580(y-145)$	(−180, 89.1)	(−162.3, 99.6)
C	R1: paraboloid	$(y-270)^2=-500(z+18)$	(−163.8, 0)	(−138.1, 25)
	R2: cone	$y-135=0.53(z+143)$	(−156.9, 89.8)	(−139.6, 98.9)

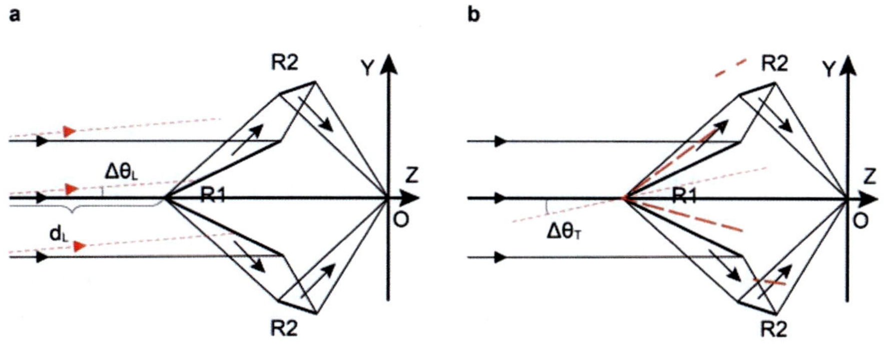

Fig. 4.9 Two-mirror beam concentrator with focused laser beam on the first mirror geometric center in the cases of (**a**) beam outcoming aberrations and (**b**) flying aberrations of ASLPE

the concentrator center line. If $\Delta\Theta = 3^\circ$, in case A we have two joined annuli that are stretched to two separated tilting elliptical loops at the bottom of which, the intensity is much higher, while lower at the edges and the deviation is almost 16 mm. In case B, deviation of nearly 11 mm, two inter-tilting annuli are also shown and radiation intensity is relatively high on their overlapped bottom. In case C, the ignition region cannot be formed and the focusing spot is still a center-in-focus annulus but the area is broader with radius of nearly 7 mm.

It follows from Fig. 4.10, focusing the laser beam directly on to nozzle axis is not an effective technique because of the strong influence of various technological and operational constraints on a thrust production. That is why, the proposed two-mirror beam concentrator of ASLPE, operating with an intermediate focusing of laser beam on to a ring focus in a slit of the nozzle, is a more preferable way allowing to keep the efficiency of laser propulsion high even in the case of technological limitations on the mirrors surfaces optical quality, thermal loads of the mirrors, and so on [11].

4.3 ASLPE Thrust Characteristics in a Pulsed Mode of Operation

To test the ASLPE thrust characteristics, experiments with a light model of the engine were carried out by using both RP and CW CO_2 lasers [10–12]. The ASLPE model (see Fig. 4.11) was made of a D16 aluminum alloy, except that the second mirror O2 of the beam concentrator which was made of a copper. There was a special secondary frame in the center of the pressure receiver, which allowed placing a solid propellant on the center line of the engine nozzle.

The purpose of the experiments was to obtain the data on the dependencies of the momentum coupling coefficient, C_m, of the thrust produced on the laser pulse energy (or laser-radiated power) and on geometric characteristics of the engine nozzle.

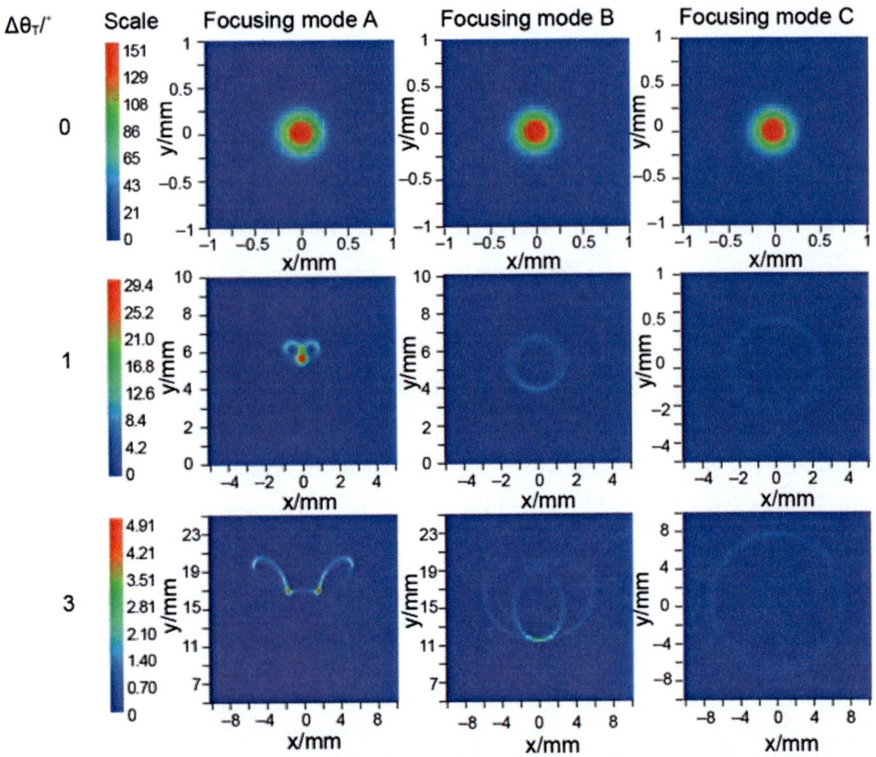

Fig. 4.10 Influence of mismatching of concentrator mirrors on distribution of radiation intensity in the concentrator focus region [20]

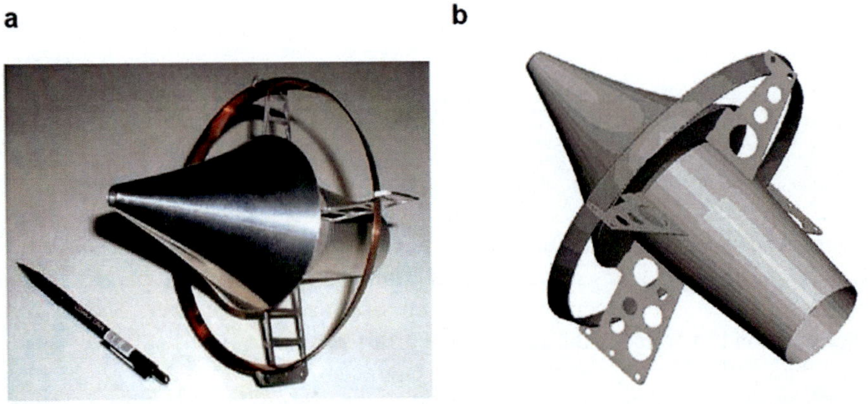

Fig. 4.11 Photo of (**a**) the ASLPE experimental model and (**b**) its computerized view are presented

Recoil impulse produced by the ASLPE was measured by a pendulum technique [10].

It is shown experimentally, laser breakdown of air in the region located closely to the nozzle axis has the appearance of a thin disk of 20–40 mm in diameter, which is oriented perpendicular to the center line. Figure 4.12 shows the image of the plasma that was seen from the side of the nozzle exit. In these experiments, the efficiency of the laser energy conversion into the thermal energy of the atmospheric air was measured to be 20–25%.

An E-beam sustained CO_2-laser was used in the pulsejet laser-propulsion experiments [21]. The near-field energy distribution across the laser beam aperture is shown in Fig. 4.13. The beam aperture size was 80×90 mm^2.

The experiments allowed the following thrust characteristics of the engine:

- dependence of C_m on laser pulse energy
- influence of the slit width on C_m
- thrust produced at using additional solid polymers arranged in the nozzle focus region

In the experiments, the following ASLPE geometric parameters were varied, namely, width, *D*, and length, *L*, of the engine nozzle, nozzle shape, slit's width, *d*, and position of the engine pressure receiver, *l*. Certain experimental results concerning C_m are presented in Table 4.3, which show the nozzle expanding into a direction of its exit section and demonstrates a smaller momentum coupling coefficient than C_m obtained when using a cylindrical nozzle. Moreover, the convergent nozzle with a 9 cm length demonstrates the same C_m, as the cylindrical nozzle.

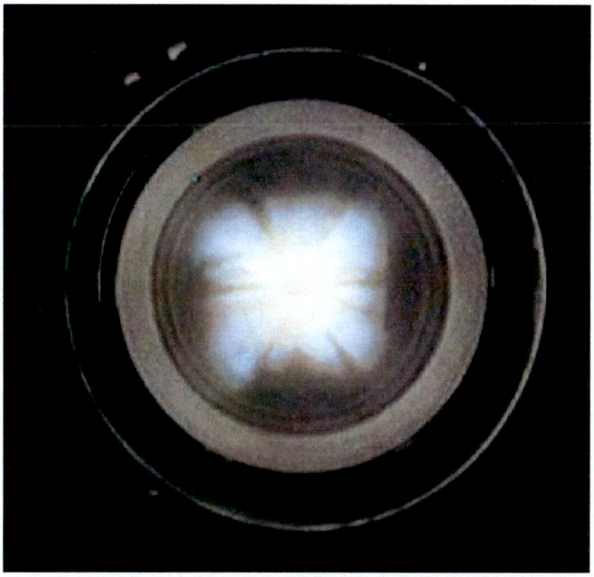

Fig. 4.12 Photo of plasma luminescence, ignited in atmosphere air

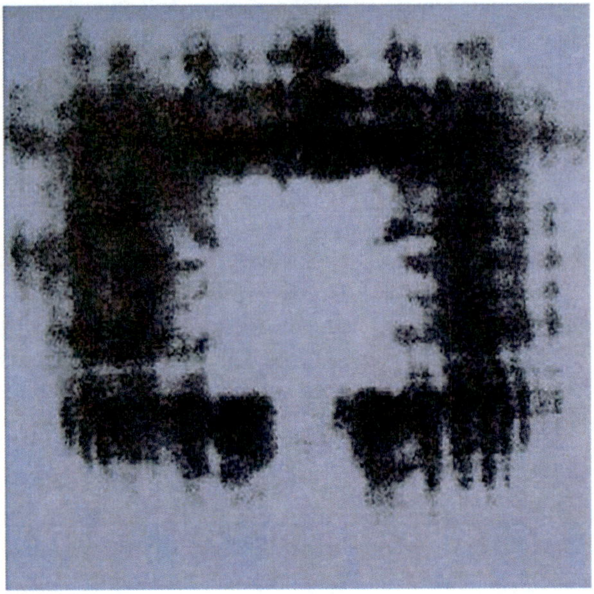

Fig. 4.13 Typical pattern of the CO_2 laser beam (near-field) on heat-sensitive paper

The profile of the pressure receiver also affects C_m parameter, namely, the parabolic receiver showed less coupling coefficient than a plane receiver (see Table 4.3).

As one can see from Table 4.3, the momentum coupling coefficient is changed if the ASLPE nozzle design is varied. Maximum and minimum C_m differ from each other by a factor of two. Besides, maximum C_m of ~1.4×10^{-4} N/W is achieved for cylindrical nozzle with a length of 9–15 cm.

Taking into account the fact of a strong influence of the slit width on a thrust produced in the ASLPE, some experiments were carried out by using a simplified model specially designed to represent the ASLPE engine to satisfy the following conditions:

- Slit width of the model is to be varied from zero to the ASLPE engine slit width.
- Characteristic sizes of the model nozzle are to be close to its geometric parameters to the ASLPE nozzle.
- Beam concentrator of the model has to be made in the form of a conical reflector with a diameter of a cone with a diameter being equal to the nozzle exit diameter.

In Fig. 4.14, photographs of the plasma which is ignited at a laser breakdown of pure air (left photographs) and after ablation of Delrin polymer (right photographs) arranged on the nozzle center line in the vicinity of the beam concentrator axis are presented.

Experimental dependencies of C_m on CO_2 laser pulse energy are presented in Fig. 4.15 for two experimental cases, namely, (a) engine model with a slit between the beam concentrator and nozzle, and (b) without the slit. Moreover, the data were

Table 4.3 Experimental results concerning the thrust characteristics, C_m, for the ASLPE nozzle with different structural units

No	Structure modification of the engine nozzle	Variable parameter	Laser pulse energy, J	$C_m \times 10^{-5}$, N/W
1	Divergent cone with $\theta = 20°$ pressure receiver in the form of a bowl, slit width of 10 mm	$L = 14$ cm	134	7.9
2		$L = 10$ cm	134	9.5
3		$L = 5$ cm	134	7.8
4	Divergent cone with $\theta = 45°$, pressure receiver in the form of a bowl, slit width of 10 mm	$L = 7$ cm	136	5.4
5		$L = 5$ cm	132	4.6
6	Cylindrical nozzle pressure receiver in the form of a bowl, slit width of 10 mm	$L = 15$ cm	129	12.3
7		$L = 12$ cm	134	11.5
8		$L = 9$ cm	137	11.5
9		$L = 5$ cm	134	9.0
10	Cylindrical nozzle, $L = 9$ cm, pressure receiver in the form of a bowl, slit width of 10 mm	$L = 20$ mm	140	10.9
11		$L = 12$ mm	134	9.7
12		$L = 3$ mm	139	10.4
13	Cylindrical nozzle, pressure receiver in the form of a cup, slit width of 7 mm	$L = 9$ cm	141	13.3
14	Cylindrical nozzle pressure receiver in the form of a cup, slit of 7 mm	$L = 5$ cm	87	12.0
15	Taper cone pressure receiver in the form of a cup, slit of 7 mm	$L = 12$ cm	105	12.0
16		$L = 9$ cm	112	10.4

also obtained at changing propellant in a focus region of the engine nozzle. In these experiments, a Delrin polymer rod of 7 mm diameter was arranged on axis of the conical beam concentrator. The experiments were made with the use of E-beam sustained CO_2 laser. As it was recorded experimentally, the mass rate of Delrin was of 0.2 mg/J at laser pulse energy of 60–90 J.

The experimental results confirm certain laser-propulsion characteristics of ASLPE predicted theoretically, namely:

- The presence of a slit in the ASLPE device decreases momentum coupling coefficient about 40%.
- C_m increases with laser pulse energy.
- C_m increases monotonically until the laser pulse energy reaches 60–90 J whereupon it keeps constant value.
- The use of Delrin inside the nozzle increases the momentum coupling coefficient by a factor two.

Nevertheless, these experiments proved the ASLPE device to be a prototype of laser-propulsion engine for future flights of a vehicle being driven by laser radiation. Flight experiments were carried out by using a laboratory inclined range [12]. A CO_2 laser generated pulses of a 120–130 J energy with a 50 Hz repetition rate so that the laser average power was of 6 kW. A laser pulse burst lasted about 3 s in every

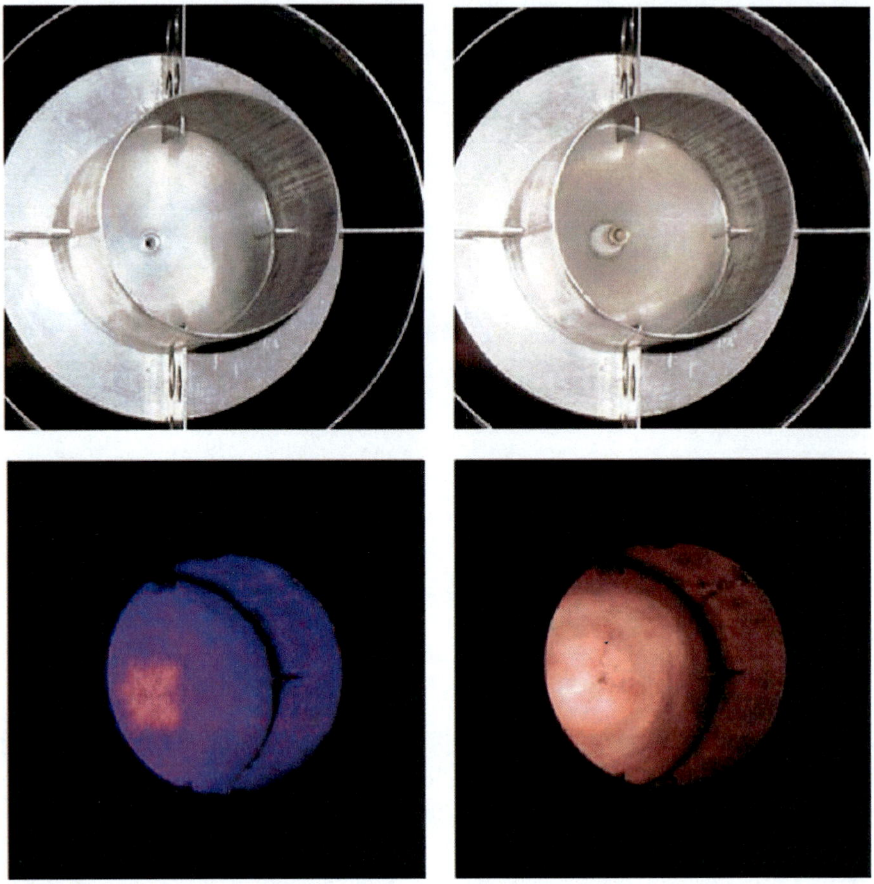

Fig. 4.14 Photographs of plasma luminescence ignited experimentally in the air (left) and in the air + Delrin (right)

experimental run. In accordance with the ASLPE design, the vehicle moved towards the laser beam at a distance of 7 m. Total mass of the vehicle was 150 g.

The experimental momentum coupling coefficient of the thrust as a function of the laser pulse energy is presented in Fig. 4.16. As one can see, C_m reaches (30–42) $\times$ 10^{-5} N/W, and it proves possibility of the ASLPE device to be used as a prototype for high-power laser propulsion.

The flight experiments demonstrated the following maximum characteristics of the ASLPE, namely:

- flight velocity: 3–4 m/s
- acceleration: 0.4–0.5 g
- momentum coupling coefficient: $4.0 \cdot 10^{-5}$ N/W
- thrust achieved in the flight: 1.3–1.5 N

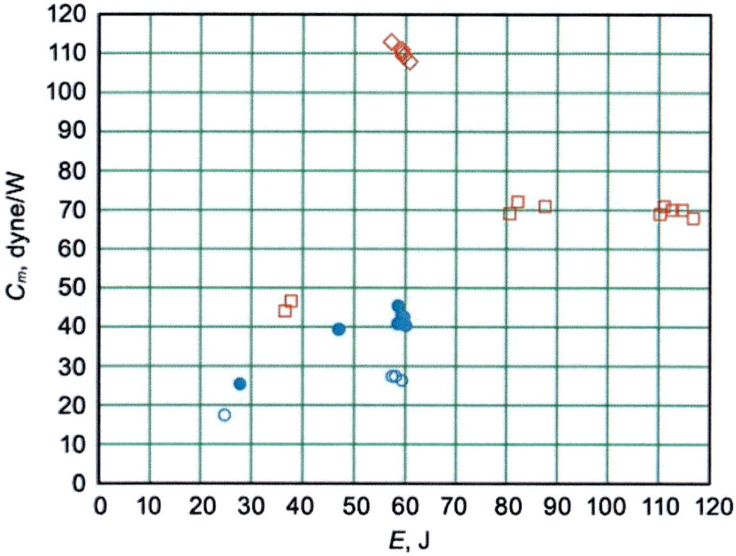

Fig. 4.15 Momentum coupling coefficient C_m dependences on laser pulse energy for various experimental conditions. Air: white circle—*nozzle with a slit*, block circle—*nozzle without a slit*, Delrin: white square—*nozzle with a slit*, white diamond—*nozzle without a slit*

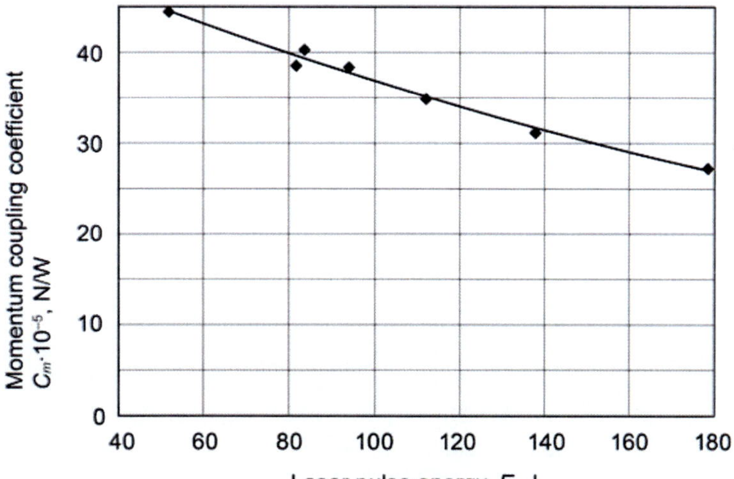

Fig. 4.16 Experimental data on C_m of thrust produced by ASLPE

These experimental results showed the necessity to modify the ASLPE design by refining the optical quality of the beam concentrator mirrors, strengthening mechanical units of the vehicle device as well as increasing the thermal resistance of its elements under a load of high-power laser radiation.

4.4 Adaptation of ASLPE for Continuous Wave (CW) Laser Propulsion

4.4.1 Principles of CW Laser Propulsion

CW laser propulsion is based on ignition of plasma in a gases medium by a CW laser radiation and followed conversion of the laser power into the plasma temperature via interaction of the radiation with the plasma to produce a thrust. The plasma is also called like an optical continues discharge—OCD. The phenomena were discovered and demonstrated in the 1970s in Ref. [22] for the first time. Prof. Yu.P. Raizer was the first researcher who proposed using an optical plasmatron as for laser propulsion. This idea was extended for launching vehicles into low Earth orbits [23] in the form of a first technological project of a new rocket system. The concept of a laser thermal engine (LTE) based on OCD phenomenon was developed this time, too. LTE is assumed to produce a thrust via to laser power conversion into a kinetic energy of gas molecules with a subsequent conversion of this energy into an exhaust jet power. This concept looks very attractive because it allows using gases which possess a small molecular mass as propellants to produce a high specific impulse of an exhaust jet.

Theoretical and experimental investigations on CW laser propulsion were studied in a Tokyo University [6, 24] using CW CO_2 lasers. A schematic diagram of LTE that was developed in the University is shown in Fig. 4.17. Here, the laser beam comes into a combustion chamber through a co-axial optical window in the chamber. To input the laser power into a gas propellant, the OCD is ignited in a region that is located close throat of the combustion chamber. After heating of the gas-plasma mixture, this mixture flows through the chamber throat into a nozzle, reaching supersonic speed.

In the experiments [6], efficiency of laser power conversion into a thrust reached 15–17% when laser power was of 1.2 kW, with a mass flow rate of gas of 1.5 g/s, and an initial gas pressure in the chamber of 7 atm. The LTE produced a 0.55 N thrust. What is more important is the fact that maximum momentum coupling coefficient of CW laser propulsion reached $C_m \sim 10^{-3}$ N/W in these experiments. This was available due to a high initial pressure of working gas in combustion chamber.

But, the experiments displayed strong spatial oscillations of the OCD along the chamber axis, observed under the high-power laser radiation. The oscillation amplitudes reached 0.1–0.9 mm depending on the initial location of the discharge with respect to the nozzle throat. The OCD oscillations modulated the exhaust jet power with a frequency of 400–500 Hz and with an intensity of a few decibels.

The OCD oscillations might be caused by acoustical, entropy, vortex, and thermo-chemical instabilities of ignited plasma. Moreover, everyone from these processes could be initiated independently from other ones and could interact with other in a nonlinear way. To suppress these discharge instabilities, consideration of the laser radiation and working gas parameters, as well as thorough designing of the LTE, are required. A few technological techniques need to be used to stabilize the

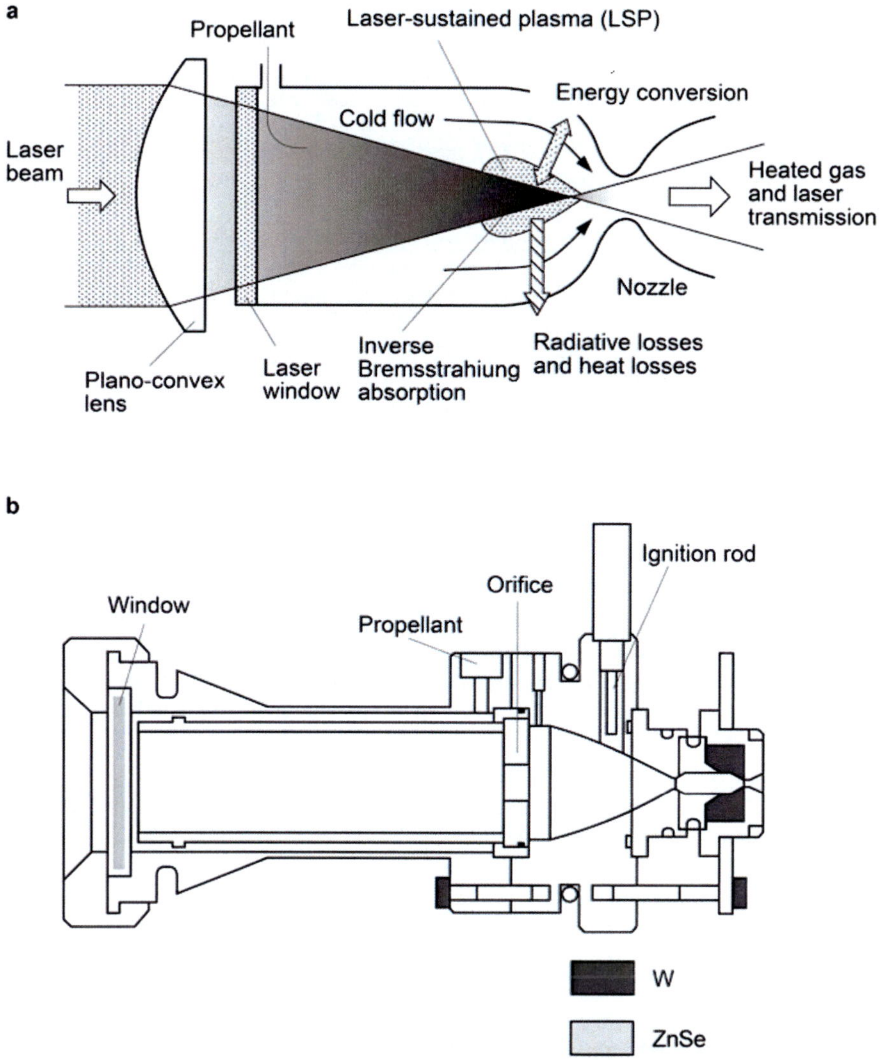

Fig. 4.17 Concept of (**a**) LTE to produce laser propulsion and (**b**) experimental model of the engine

laser power release in the gas medium, as well. One of the techniques is spinning of the injected gas.

In this regard, such topics as transferring of the laser radiation into combustion chamber, ignition of plasma, stabilization of the OCD, protection of the chamber walls against a high-temperature plasma, and so on were considered in Ref. [25, 26] in some detail. For the example, spatial stabilization of the OCD is achieved by using a counter propagating axial swirling flow of gas into the combustion chamber (see Fig. 4.18).

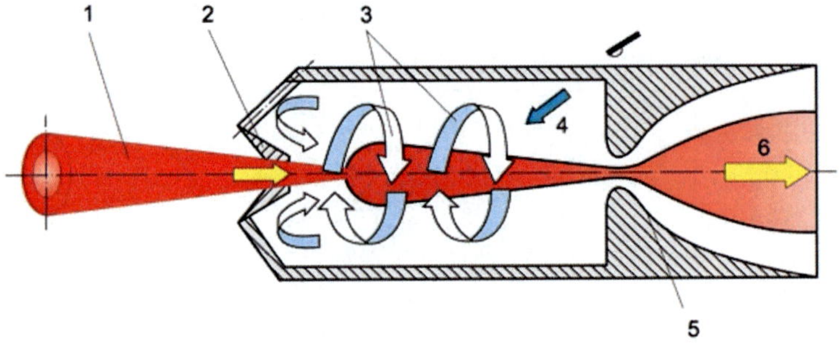

Fig. 4.18 CW LTE schematic diagrams with stabilization of the OCD by counter propagating axial swirling flow of working gas [25]

As is usually proposed, that nitrogen might be used as a gaseous propellant for CW LPE. Additionally, ammonia is possible propellant because it allows achieving a high specific impulse up to 600 s or higher. Both of the gases do not require special conditions to be stored.

The experiments made in Ref. [26] confirmed the fact that OCD can be stabilized along the axis of a combustion chamber due to the formation of a vortex flow from the injected gas. It showed the maximum specific impulse of CW laser propulsion depended on a kind of gas used in the experiments, namely, I_{sp} of 1200–1500 s was obtained with hydrogen, 300–500 s with ammonia, and 250–300 s with argon. And the laser power per unit thrust produced by the LTE was 2×10^2 W/N. For example, this parameter for an electric rocket engine is equal approximately to 4.5×10^3 W/N, it is 4.5×10^2 W/N for ion engines, and 4.5×10^4 W/N for plasma engines.

As we can see from the brief review of CW laser-propulsion engines using an OCD, there is a similar technical ability of adapting the ASLPE to operate in CW laser-propulsion mode after some modification of the ASLPE design. This adaptation is considered in the following section in detail.

4.4.2 CW ASLPE Thrust Characteristics

It has been proved theoretically and experimentally that a transverse flow increases stability of an axial OCD [27, 28]. It is also theoretically shown in Ref. [28] that two crossed laser beams (transverse and longitudinal in respect to gas flow direction) when applied to initiate gas-plasma allows positioning the OCD in a fixed volume. Both of these techniques may be realized in an ASLPE.

To originate axial gas flow inside the ASLPE engine, a separate orifice is made in the top of the first beam concentrator mirror of the engine. The experimental CW model of ASLPE is shown in Fig. 4.19, which consists of first and second mirrors of the beam concentrator, engine nozzle as well as an air inlet assembly allowing to inject gas flow along the engine axis.

Fig. 4.19 Photograph of the CW ASLPE model: (1) first mirror of the beam concentrator; (2) second mirror; (3) nozzle; (4) air inlet assembly

Figure 4.20 illustrates the engineering design of the CW ASLPE device that was tested experimentally in Ref. [29]. The ASLPE CW nozzle unit differed from ASLPE nozzle developed for pulsed laser propulsion, when its shape corresponded to a Laval nozzle.

To make experiments with the ASLPE CW device, a gas-dynamic CO_2 laser with an output power of 50 kW was used. The experimental setup of the experiments is shown in Fig. 4.21.

In Fig. 4.21, the gas-dynamic nozzle assembly (1) is shown as a part of the CW CO_2 laser device. Unstable resonator (including mirrors 2 and 3) with magnification of $M = 1.2$ was used to generate CW radiation. The laser radiation was directed into a laboratory hall through a supersonic gas-dynamic window arranged close to the output mirror of the resonator. In the experiments, the laser power was of 20 kW. An auxiliary telescope formed with the mirrors 7 and 8 was used to adjust the beam aperture to the beam concentrator of the CW ASLPE.

The thrust produced by the ASLPE CW model was recorded by using a strain sensor (made by the Honeywell company), which was attached to the model by means of a special spring. The engine itself was fixed on a special mechanism that allowed the model to be shifted longitudinally with a minimal friction force. The initial force produced by the spring was considered when the detector signals were recording.

In the experiments, Delrin polymer was used in the form of a cylinder of 4 mm in diameter to increase propulsion efficiency. The cylinder was arranged on the nozzle axis close the pressure receiver right behind of first mirror of the ASLPE beam concentrator. To stabilize OCD inside the nozzle, atmosphere air was injected into the nozzle through a ring orifice made in the beam concentrator top. Characteristic pictures of the ASLPE CW operation are shown in Fig. 4.22 at different regimes of thrust production.

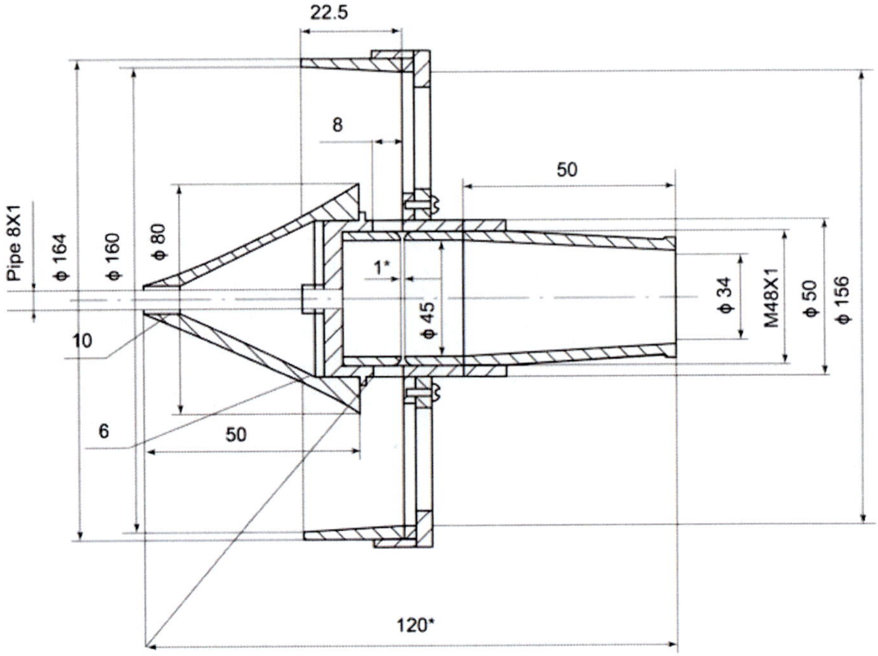

Fig. 4.20 Experimental CW ASLPE model (all dimensions are given in centimeters)

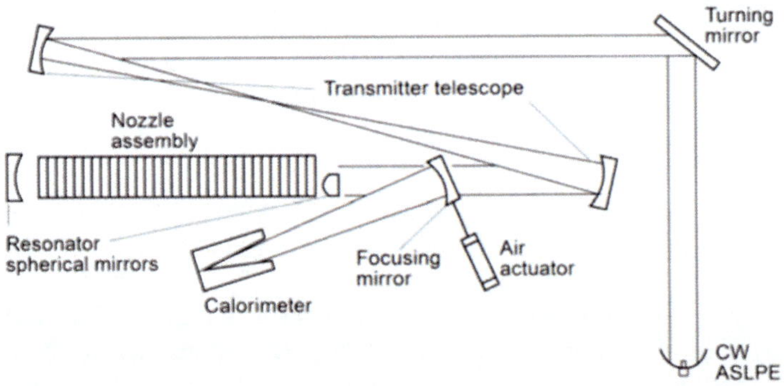

Fig. 4.21 Optical setup of experimental facility

It is shown experimentally the plasma that blows out through a slit at start moment of engine operation is observed in the case when the flow mass rate of air through the nozzle is rather small and doesn't exceed 0.4 g/s (see Fig. 4.22(a)). The plasma ejection is distributed along the circular slit non-uniformly that may be caused by a non-uniform distribution of the laser beam focused into the slit. Conversely, if the air mass rate exceeds 2 g/s, then the ejection of plasma isn't observed (see Fig. 4.22(b)).

a b

Fig. 4.22 Various structures of the exhaust jet pattern, obtained at various air mass rate

a b

Fig. 4.23 Photos of the exhaust jet at air consumption of $G_a = 0.4$ g/s

It is observed in the experiments that the flow mass rate of air flow defines also produced the thrust when laser radiation power fixed. For example, maximum thrust of $F = 2.0$ N was achieved when the mass rate was $G_a = 0.2$–0.4 g/s. Therefore, the momentum coupling coefficient C_m was of 10.2 dyne/W. But, further increase in the air mass rate resulted in decreasing thrust and it was only of 0.8 N at $G_a > 0.2$–0.4 g/s.

Analysis of series of photographs being similar to those shown in Fig. 4.23 displays a visible length of the exhaust jet to be rather long, and the length depends on the air mass rate through the nozzle. Maximum jet length of 300 mm was observed when the air mass consumption was of $G_a = 0.4$ g/s (see Fig. 4.23(a)). Moreover, its length changed a little with time, so that the jet possessed a ripple character (Fig. 4.23(b)).

The produced thrust also varied in a similar oscillatory character that was detected by strain sensor, and the thrust pulsation frequency was of 80–100 Hz. As shown above, instability of optical continuous discharge in gases may be caused by plasma screening of laser radiation being focused into the nozzle. The effect of a small mass rate of air flow may result in this type of thrust instability, too.

4.5 Analysis of Available Technologies as Applied to ASLPE Development and its Engineering Constraints

4.5.1 Effects of Slit on Thrust Production

To determine the influence of slit width on the thrust produced by ASLPE under laser radiation, a numerical simulation of the gas-dynamic processes in the ASLPE nozzle, assuming equilibrium in the ignited plasma by the laser radiation, was carried out. The numerical techniques based on the Godunov numerical scheme of high numerical resolution on the basis of non-structured dynamic adaptive numerical grid were applied [30]. Figure 4.24 illustrates some results of these calculations in the form of gas flow patterns in a cross section of the ASLPE nozzle at various points of time after the end of the laser pulse. It is seen that a part of a bow shock wave generated in axial region of the nozzle comes through slit and interacts with second mirror of the ASLPE beam concentrator. Moreover, a part of the current flow of the wave comes through the slit. Both of these effects result in decreasing of the thrust.

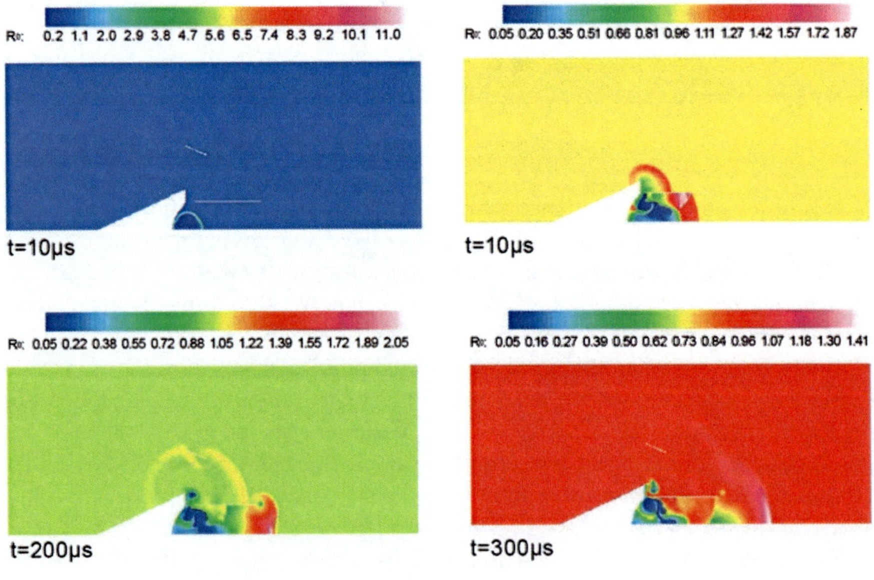

Fig. 4.24 Density contours of a gas inside the ASLPE model at some moments of time after the laser pulse end, *t*

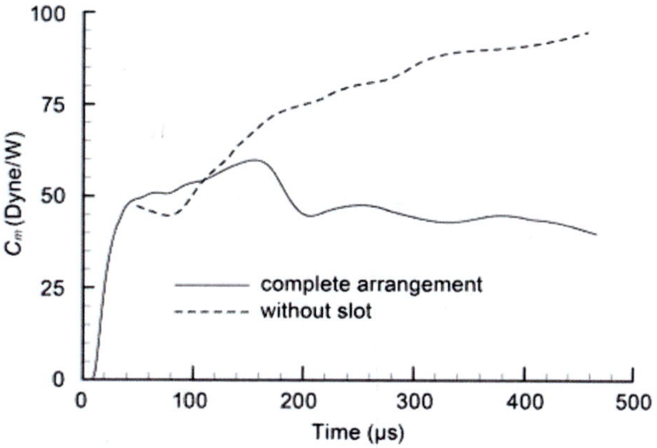

Fig. 4.25 Coupling coefficient for the ASLPE (complete arrangement with co-axial concentrator and arrangement without slit)

Figure 4.25 shows the time-behavior of the momentum coupling coefficient C_m of a generated thrust both in the presence of slit (solid line) and without slit (dashed line) with time. It is seen that instant C_m for the nozzle with a slit becomes less than C_m for the nozzle without a slit starting from such moment of time when a part of bow shock wave reaches second mirror of beam concentrator. This effect is also associated with a decrease of gas pressure inside the ASLPE nozzle, being caused by the slit [10].

The last statement is also confirmed by similar studies of gas-dynamic effects in pulsejet laser propulsion with the use of a nozzle with slit which were made in Ref. [31]. Figure 4.26 illustrates the velocity vector and pressure contours for tested nozzles at $t = 7.57$μs. It is seen, that a spiral vortex in the flow is generated close to the slit, when the bow shock wave is reflected from the nozzle wall. Moreover, we can see a high pressure area that is formed outside the nozzle wall which also results in thrust decreasing.

This study also demonstrates generation of a low pressure regions located close to both the nozzle slit and exit edge of the nozzle. Figure 4.26 shows the velocity vector and pressure contours for these two cases. In Fig. 4.27(a), two vortexes appeared near the slit and pressure gradient of gas changes here sharply so that is was captured in the nozzle. In Fig. 4.27(b), the pressure outside the nozzle is higher than the internal pressure. The different pressure areas are labeled as "P_{max}" and "P_{min}". It is also evident that the negative pressure causes drop in the thrust.

So, one of the promising ways of increasing of thrust produced by an ASLPE is to decrease the slit width between pressure receiver and nozzle walls, which has to be of the order of the diffraction size of a focused laser beam (see Sect. 4.1). But, there are also other effects which may limit thrust production by ASLPE if high-power radiation loads on the ASLPE mirrors will take place.

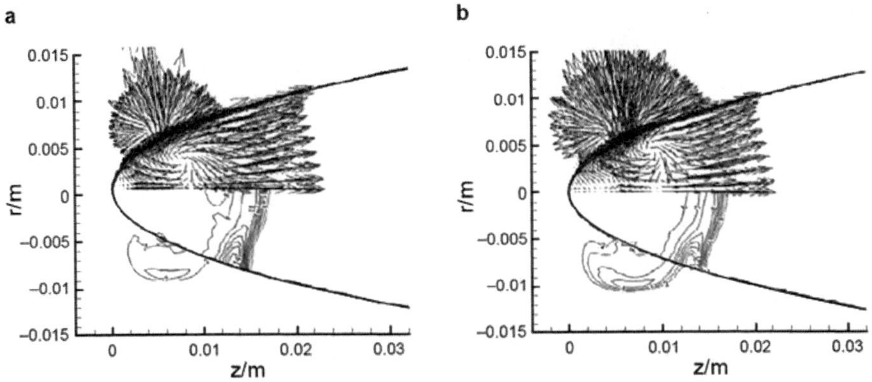

Fig. 4.26 Velocity vectors (top half of nozzle) and pressure contours (bottom half of nozzle) (**a**) nozzle with a 1 mm slit, (**b**) nozzle with a 2 mm slit

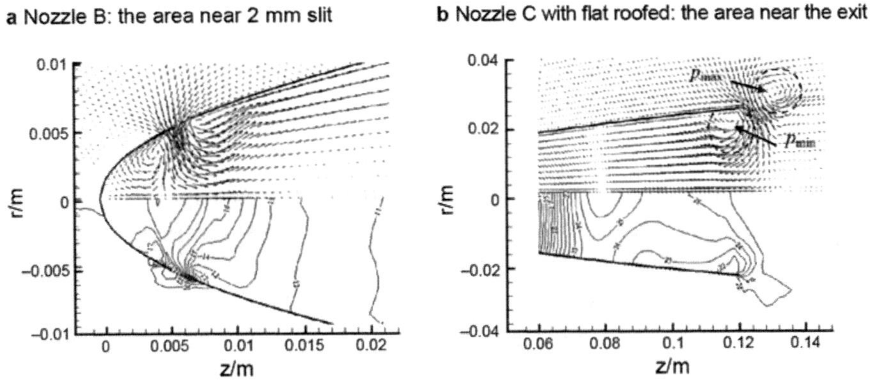

Fig. 4.27 Velocity vectors and pressure contours for two kind of nozzles, $t = 600\mu s$

4.5.2 Thermo-physical Model of the ASLPE Device

To consider exploitation of ASLPE under real conditions, we examine such effects as thermal loads on the ASLPE mirrors and the mechanical loads on the engine nozzle walls. These loads may cause both thermal aberrations and vibrations of the engine mirrors which will result in misalignments of the laser beam focus with respect to the slit position. A common result is that these loads can reduce efficiency of thrust production in ASLPE.

As for the thermal aberrations of its mirrors, we can use a thermo-physical model to estimate these loads, taking into account axial symmetry of the mirrors structure. The thermal equation describing temperature of the mirrors is written at cylindrical coordinates as follows:

$$c(T) \cdot \rho \cdot \frac{\partial T}{\partial \tau} = \frac{1}{r} \cdot \frac{\partial}{\partial r}\left(r \cdot \lambda(T) \cdot \frac{\partial T}{\partial r}\right) + \frac{\partial}{\partial z}\left(k(T) \cdot \frac{\partial T}{\partial z}\right) + q(r, z, \tau) \quad (4.23)$$

where ρ, $c(T)$, $\lambda(T)$, $k(T)$ are thermal parameters of the materials which from the mirrors are made, namely' ρ is the material density, $c(T)$ is the specific thermal capacity, T is temperature, and $\lambda(T)$, $k(T)$ are the material thermal conductivities along r- and z-axis, respectively. We propose $k(T) = \lambda(T)$ because of the homogeneity and isotropy of the material are assumed (Fig. 4.28).

To solve Eq. (4.23), some initial and boundary conditions have to be formulated for every mirror, which have to satisfy to the experimental conditions. The basic purpose of the numerical simulation of the mirror thermal loads is to insure the correct choice of mirror material. Moreover, the following requirements for the ASLPE device have to be satisfied, namely:

1. ASLPE has to be developed as a lightweight device and its mirrors will be very thin.
2. Selected mirror materials have to possess high thermal conductivity to exclude great temperature gradients induced under high-power laser radiation.
3. Every ASLPE's mirror and other elements can be created of various kinds of materials but linear expansion coefficient of every material must be matched, approximately.
4. One should be also considering the laser breakdown thresholds of every mirror surface under high-power laser radiation.
5. Moreover, ASLPE materials have to possess a high mechanical resistance.

To illustrate the ASLPE mirrors behavior under the loads of high-power laser radiation, let us consider the ASLPE device that was tested in flight experiments in Ref. [11]. The experiments were carried out with CO_2 laser radiation, operating in an RP mode with a 50 Hz pulse repetition rate (PRR) and pulse energy of 200 J. So that, the average laser power was of 10 kW.

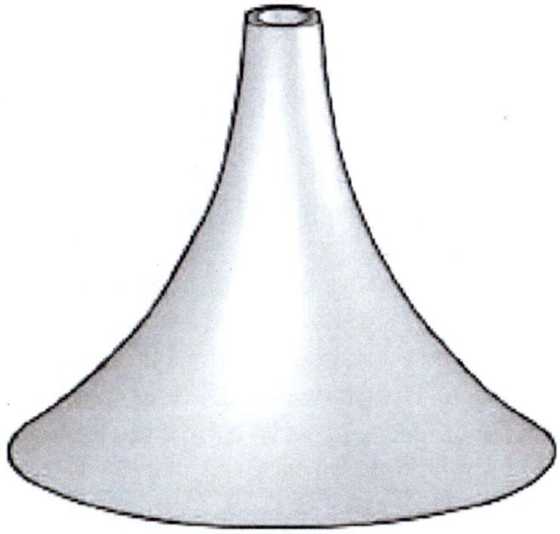

Fig. 4.28 ASLPE first mirror selected for numerical calculations of the mirror thermal loads

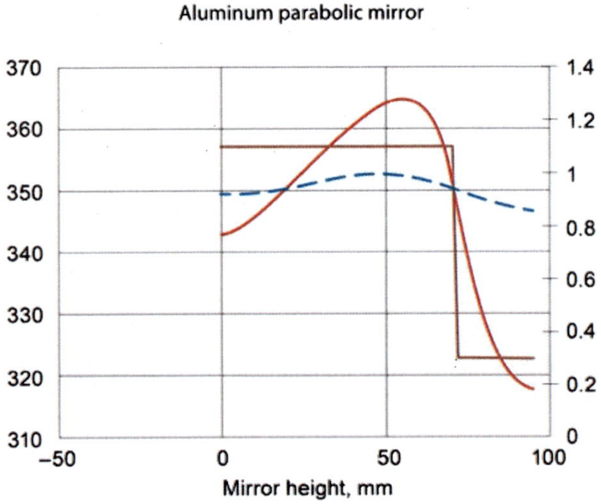

Fig. 4.29 Variations of the mirror surface temperature for given distribution of laser power along the surface (brown line)

Figure 4.29 illustrates dependences of the mirror surface temperature along X coordinate counting the mirror height starting from the bottom. Maximum height when $X = 95$ mm corresponds to the top of the mirror. Curve 1 (red line in the figure) corresponds to the temperature rise in the surface to 2 s after start of the mirror illumination, and curve 2 (blue line)—to a 10 s. The temperature variations shown in the figure correspond to a face surface of the mirror because the rear surface temperature of the mirror will be practically the same because of a small thickness of the mirror.

As is evident, second reflector of ASLPE beam concentrator will be under higher thermal load than the first mirror because of the increase of the radiation intensity (power density) that is incident on its surface. That is why the mirror must have a very high thermal conductivity to exclude large temperature gradients of its surface and the resulting aberrations.

In the calculations, two versions of the mirror design were studied (Fig. 4.30).

The first structure consists of two parts, namely, a bottom part that is used to fix the mirror in the beam concentrator, and top part is used as a reflector. Temperature variations of the mirror surface along its surface are shown in Fig. 4.31 at the moment of time $t_3 = 40$ ms. It is also assumed that the mirror is made of aluminum that absorbs about 8% of the incident CO_2 laser power. The mirror thickness is 0.5 mm, and the average laser power density is $4.0 \cdot 10^8$ W/m^2.

The temperature variations along the mirror surface of the second structure (see Fig. 4.30(b)) are shown in Fig. 4.32, corresponding to a 2 s after the high-power laser irradiance was started. This mirror is made of copper; the mirror thickness is equal to 1 mm. Curve 1 (brown line) corresponds to the laser power absorption of 3% by the mirror surface, and curve 2 (blue line)—to 5%. The laser power density is $4.0 \cdot 10^8$ W/m^2 on the mirror surface.

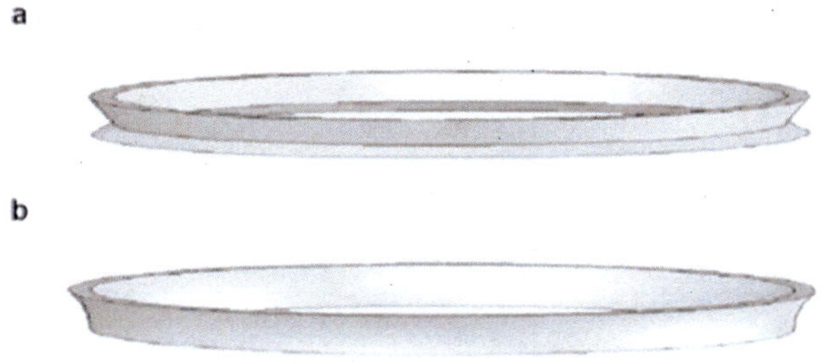

Fig. 4.30 ASLPE second mirror structure

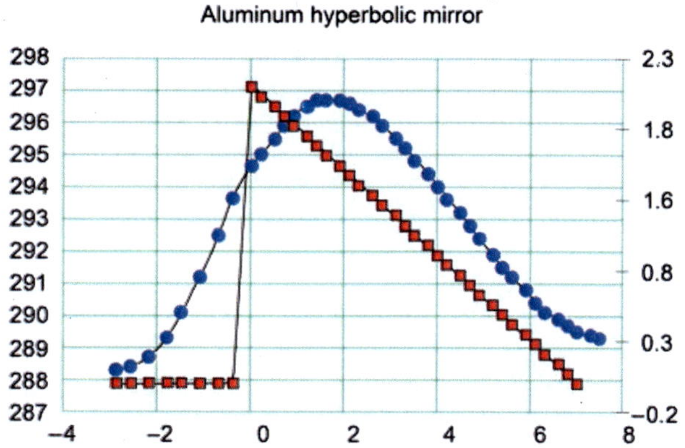

Fig. 4.31 Temperature distribution (*blue line*) along the ASLPE second mirror surface (see Fig. 4.30(a)), corresponding to the given radiation intensity distribution (*red line*)

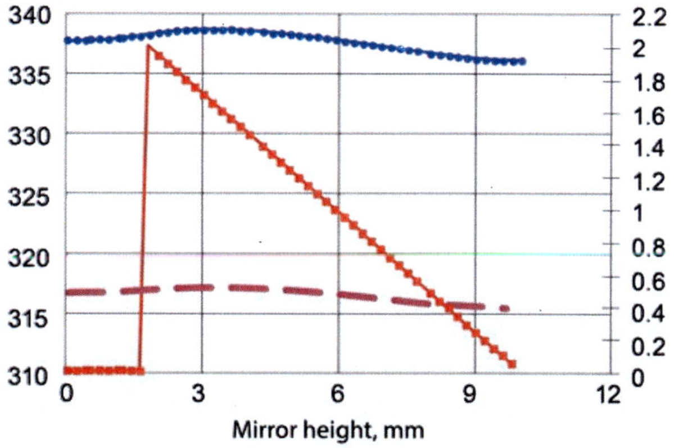

Fig. 4.32 Temperature of the second mirror surface, version of Fig. 4.29(b)

Analysis of these effects of thermal loads of the ASLPE beam concentrator allows some conclusions to be drawn, which have to be taken into account with any future design, namely:

- Mirror surfaces have to be covered by specific high reflecting coatings to decrease thermal loads on the mirrors.
- Mirror device has to include additional systems to cool the mirrors in the case of high-power laser irradiations.

Table 4.4 Fused quartz and sapphire thermal characteristics

Material	Density ρ, kg/m^3	Refractive index at the average spectral range	Melting temperature, °C	Temperature coefficient of linear expansion, 10^{-6}, K^{-1}	Coefficient of thermal strength, $M_f \times 10^{-4}$
Fused quartz	2650	1.43	1470	0.5	5.3
Sapphire synthetic	3980	1.67	2040	5.7	2.1

Table 4.5 Optical characteristics of materials that could be used as reflecting coatings for ASLPE mirrors

Coatings	Index of refraction	Spectral range of coatings, nm
Al_2O_3	1.69	200–7000
GeO_2	2.30	400–5000
MgF_2	1.38	200–5000
SiO_2	1.45	200–8000
Si_2O_3	1.55	300–8000
TiO_2	2.30	400–12,000
ZnS	2.30	140–15,000
ZrO_2	2.10	250–7000

As an example, fused quartz and sapphire are considered as substrates for beam concentrator mirrors which can be used under laser radiation with $\lambda = 1.06\mu$m wavelength. Its thermo-mechanical properties are presented in Table 4.4.

Materials such as aluminum dioxide (Al_2O_3), germanium dioxide (GeO_2), magnesium fluoride (MgF_2), silicon oxide (SiO_2), silica (Si_2O_3), and zinc selenide (ZnS) could be used as reflecting coatings, too. Optical properties of these components are shown in Table 4.5.

Some combination of materials can be used to make multilayer coatings. But, one of the basic parameter characterizing the coatings is its radiation resistance. Hence, the following factors have to be considered when choosing the coatings, namely:

- ambient conditions (temperature, components, and so on) and technique of substrate coating as well as structure of multilayer coating
- laser radiation wavelength, pulse length, radiation intensity, affecting coating strength
- storage conditions of coatings

4.6 Preliminary Conclusion

It may be concluded from this chapter that it is possible to make an ASLPE that operates both RP and CW modes of the pulsejet and CW laser-propulsion regimes with one and the same efficiency of thrust production. Moreover, the fact that the ASLPE device using a transverse focusing of laser beam into the engine nozzle with respect to axial gas flow allows stabilizing the thrust in a CW mode. For that, it will require matching of air mass consumption in the engine and applied laser power.

Exploitation of the ASLPE for high-power laser propulsion will require more detailed designing of its mirrors and nozzle to compensate for such effects as the thermal aberrations and vibrations induced into the mirrors, which result in a misalignment of laser beam axis in a respect to the ASLPE engine slit.

References

1. Myrabo, L.N.: Brief history of the Lightcraft Technology Demonstrator (LTD) project. Beamed Energy Propulsion. AIP Conf. Proc. **664**, 49–60 (2003). https://doi.org/10.1063/1.1582095
2. Salvador, I.I., et al.: Hypersonic experimental analysis of impulse generation in airbreathing laser thermal propulsion. J. Prop. Pow. **29**, 718–731 (2013). https://doi.org/10.2514/1.B34598
3. Harrland, A., Doolan, C., Wheatley, V.: Hypersonic inlet for a laser powered propulsion system and its interaction with an idealized laser induced detonation wave. In: Proc. of the 10th Australian Space Science Conference, 27–30 September 2010. Sydney NSW 2001, pp. 15–26 (2011). https://doi.org/10.1063/1.3657023
4. Ageichik, A.A., et al.: Aerospace laser propulsion engine. RUS Patent № 2266420 8 Oct 2003
5. Sinko, J.E.: Laser ablation propulsion tractor beam system. J. Prop. Pow. **26**, 189–191 (2010). https://doi.org/10.2514/1.46037
6. Toyoda, K., Komurasaki, K., Arakawa, Y.: An experimental research on a CW CO_2 laser thruster. In: 26-th IEPC, Oct 17–21, Kitakyushu, Japan (1999)
7. Bikmuchev, A.R., Sattarov, A.G.: Investigations of laser propulsion engine characteristics on the base of CW optical discharge. Man. Space Flight. **10**, 51–71 (2014)
8. Bunkin, F.V., Prokhorov, A.M.: Laser power source application to produce a thrust. Adv. Phys. Sci. (Rus.). **119**, 425–446 (1976)
9. Pirri, A., Simons, G.A.: The fluid mechanics of pulsed laser propulsion. AIAA J. **15**, 835–842 (1977)
10. Rezunkov, Y.A., et al.: Model tests of Aerospace Laser Propulsion Engine. AIP Conf. Proc. **766**, 183–194 (2005). https://doi.org/10.1063/1.1925142
11. Rezunkov, Y.A., et al.: Performance characteristics of laser propulsion engine operating both in CW and in repetitively-pulsed modes beamed energy. AIP Conf. Proc. **830**, 3–13 (2005). https://doi.org/10.1063/1.2203241
12. Rezunkov, Y.A., et al.: Experimental study on thrust characteristics of Aerospace Laser Propulsion Engine. AIP Conf. Proc. **702**, 49–60 (2003). https://doi.org/10.1063/1.1720985
13. Rusinov, M.M.: Non-spherical surfaces in optics. Nedra, Moscow (1973)
14. Nadorff, G., DeWitt, F.: Reflaxicon objectives for imaging. Proc. SPIE. **8486**, 84860C (2012). https://doi.org/10.1117/12.930010
15. Klimentiev, S.I., Kuprenyuk, V.I., Khloponina, I.V.: Numerical estimations of laser beam aberrations of optical conic transducer. Opt.-mech. Ind. (Rus). **7**, 38–41 (1991)
16. Drago, N., Villoresi, P., Bertogli G.: Laser technologies: a step forward for small and medium sized enterprises. In: International center for science and high technology. Trieste, High

Technology and New Materials, International Centre for Science and High Technology, ICS-UNIDO, AREA Science Park. Padriciano 99, 34012 Trieste, Italy (2008)
17. Braunecker, B., Hentshel, R., Tiziani, H.J. (eds.): Advanced optics using non-spherical elements. SPIE Press, Washington (2008)
18. Egorov, M.S.: Investigation and development of the onboard optical system for small satellites with laser propulsion. A Thesis Submitted to the Graduate Faculty of Saint-Petersburg ITMO University in Partial Fulfillment of the Requirements for the degree of Doctor of philosophy. S.-Petersburg (2016)
19. ZEMAX User Manual. Radiant ZEMAX LLC (2013)
20. Cheng, F.Q., Dou, Z.G.: Study on focusing performance of the twice reflecting laser focusing system. J. Phys.: Conf. Ser. **276**, 1–9 (2011). https://doi.org/10.1088/1742-6596/276/1/012014
21. Rezunkov, Y.A., et al.: Modeling of physical processes on the basis of high-energy laser facility. J. Opt. Tech. (Rus.). **1**, 84–94 (1994)
22. Raizer, Y.P.: Bases of modern physics of gas-discharges processes. Nauka, Moscow (1980)
23. Caveny, L.J.: Orbit-raising and maneuvering propulsion: research status and needs. Progr. Astron. Aeron. **89**, 160–171 (1984)
24. Komurasaki, K., et al.: Fundamental researches on laser powered propulsion. In: AIAA, 2002–2200, (2002). https://doi.org/10.2514/6.2002-2200
25. Sattarov, A.G.: Laser propulsion engine on the basis of continuous optical discharge. Proc. Univ. Aeron. Eng. N. **3**, 46–49 (2008)
26. Dregalin, A.F., et al.: Experimental and theoretical studies of laser propulsion engine characteristics of continuous optical discharge. Proc. Univ. Aeron. Eng. N. **4**, 39–43 (2010). https://doi.org/10.3103/S1068799810040094
27. Generalov, N.A., et al.: Temporal stability of continuous optical discharge in the atmosphere air flow. Comb. Deton. Phys. **22**, 91–93 (1986)
28. Kulumbaev, E.B., Lelevkin, V.M.: Numerical analysis of continuous optical discharge characteristics in atmosphere air. Plas. Phys. **27**, 12–15 (2001)
29. Rachuk, V.S., et al.: Experimental investigations of laser propulsion by using gas-dynamic laser. AIP Conf. Proc. **830**, 48–57 (2005). https://doi.org/10.1063/1.2203246
30. Golovachov, Yu.P., Kurakin, Yu.A., Schmidt, A.A.: Numerical investigation of non-equilibrium MGD flows in supersonic intakes. In: AIAA 2001-2883 (2001). https://doi.org/10.2514/6.2001-2883
31. Wen, M., Hong, Y., Song, J.: Investigation on multiple-pulse propulsion. AIP Conf. Proc. **1402**, 74–81 (2011). https://doi.org/10.1063/1.3657017

Chapter 5
Supersonic Laser Propulsion

Abstract Supersonic laser propulsion is aimed to accelerate supersonic gas flows in a laser-propulsion engine. The problem is that some gas-plasma-dynamic processes complicate the overall nature of a thrust production in a supersonic regime. To produce supersonic laser propulsion, original technique based on the interaction of laser-ablated jet with supersonic gas flow in a nozzle is proposed. It is shown that such parameters of the vapor jet as gas-plasma pressure and temperature in the ablation vapor as well as the mass consumption rate of the ablated propellant are important ones in this case. The results of numerical simulations of the supersonic laser propulsion are presented for two types of a nozzle configuration. The feasibility of momentum coupling coefficient C_m to be of 10^{-3} N/W is shown.

Keywords Laser breakdown · Continuous optical discharge · Supersonic gas flow · Focal region · Laser-ablated jet · Repetition pulse rate · Exhaust jet · Pulsejet propulsion · Ramjet propulsion · Conversion efficiency · Lightcraft vehicle · Parabolic nozzle · Gas-dynamics effects · Merging of shock waves · Ablation jet injection · Mach number · Propulsion efficiency

5.1 Introduction

Supersonic laser propulsion is considered in cases when (a) vehicle reaches a supersonic velocity in the atmosphere or (b) the gas flow in the engine nozzle reaches a supersonic speed [1, 2]. In both cases, the laser radiation interacts with supersonic flows, going with particular processes both in the case of RP and CW laser-propulsion systems.

These processes are as follows:

- pulsed laser breakdown of a gas flow in the focal region of a parabolic nozzle to produce ramjet laser propulsion, this is when power shock waves propagate up and down from the focal region [2, 3]
- laser breakdown of a gas flow in a focal region where using laser radiation with a high repetition pulse rate, when a series of low-power shock waves follow each

Y. A. Rezunkov, *High Power Laser Propulsion*, Springer Series on Atomic, Optical, and Plasma Physics 116, https://doi.org/10.1007/978-3-030-79693-8_5

other to form a strong common shock wave [4, 5]; this is the so-called merging of shock waves, which was proposed by Prof. Yu. Raizer in Ref. [6] as a technique to decrease air drag of vehicles

- as well as the creation of a laser ablation jet in supersonic flows to transform and to accelerate the total gas flow [7].

As a matter of fact, the supersonic laser propulsion has been considered since the development of the first projects on laser-propulsion applications. The basic problem of the supersonic propulsion concerns the inefficient conversion of applied laser power into a power of exhaust jet in supersonic engines.

One of the first studies on pulsejet supersonic laser propulsion was theoretically carried out by Prof. Anthony Pirri in Ref. [2]. In his paper, Pirri used a theoretical model of the laser propulsion produced due to the laser breakdown of supersonic gas flow in a parabolic nozzle that was also used as a beam concentrator. The theoretical model of RP pulsejet laser propulsion was assumed to be based on equations describing shock waves (SW) downstream propagation from the gas breakdown region. In addition, the semi-empirical approach to the laser propulsion was used to determine the thrust characteristics and to define principal requirements for the laser radiation parameters to achieve efficient thrust.

But, the semi-empirical model of supersonic propulsion does not consider some principal phenomena of the laser radiation interaction with the supersonic flows, which can affect the thrust. So, analysis of the whole process of the supersonic laser propulsion should consider:

- The RP laser radiation interaction with a supersonic flow is assumed to be realized through two interactions, namely: (a) restoration of a perturbed gas flow structure after the end of the laser pulse and (b) adaptation of the flow structure to the ambient atmosphere conditions.
- Such physical phenomena as "the continuous optical discharge" and "laser breakdown" of gas flows should be distinguished in the case of supersonic gas flows. As we can see in Chap. 2, there are two different scientific studies on laser propulsion existed because of these effects. Consequently, we can separate these two problems from each other, namely: (a) efficiency of the laser power conversion into a plasma temperature of the optical discharge and (b) transforming of the gas flow structure under the action of CW or RP laser radiation.
- Unfortunately, currently there is no technological ability of making experiments on the ramjet supersonic laser propulsion right now by using high-power lasers. Therefore, the main part of investigations on this subject is carried out only theoretically.
- Two separate types of the ramjet supersonic propulsion should be considered in way of vehicle flight speed difference with a Mach number equal to 5 related to supersonic propulsion and with Mach number exceeding 5 related to hypersonic propulsion. In both cases, a special attention is made to the vehicle forebody of laser-propulsion engine (LPE) because its configuration affects the supersonic flow inside the engine nozzle and efficiency of the laser power conversion into

exhaust jet power. The most interesting paradigm of such engine design type is a new configuration of the Lightcraft vehicle developed in Ref. [8].

5.2 Lightcraft Engineering Version Adapted to the Pulsejet Supersonic Laser Propulsion

Lightcraft (LTD) developed by Prof. Leik Myrabo & Co. [9, 10] is considered as a demonstrable vehicle to test the ramjet and pulsejet supersonic laser propulsion experimentally. As we note in the Chap. 1, one of the features of the vehicle design is a forebody with an attached cowl that is supposed to be one of jet nozzle walls (see Fig. 5.1). Other feature of the Lightcraft device is an LPE with a central body formed as an off-axis paraboloid of revolution to focus the laser beam on to the inner surface of the forebody cowl. Then the laser beam produces a laser breakdown of gas in the form of a ring located close to the cowl surface. The engine geometric parameters and laser pulse characteristics are chosen by such manner that the laser radiation interacts with the ignited plasma in the gas breakdown region at a regime of laser-sustained shock wave generation.

Let us reconsider the Lightcraft model, including the evolution of its design. One of the first designs of Lightcraft achieved ramjet laser propulsion by using a slit formed between the cowl and the forebody in the inlet section of the vehicle engine. The forebody shape and slit sizes were matched, so that a shock wave formed in the slit transforms incoming supersonic flow into a subsonic one to increase the static gas pressure in beam focus (see Fig. 5.2).

The gas pressure growth must decrease the laser pulse energy needed to initiate a plasma in this region. Simultaneously, to increase the laser-propulsion efficiency, the laser-sustained detonation wave needs to be generated. To test these conditions of the supersonic laser propulsion, specific studies of thrust produced by the Lightcraft model were carried out in Refs. [9–11]. These studies were based on a modification

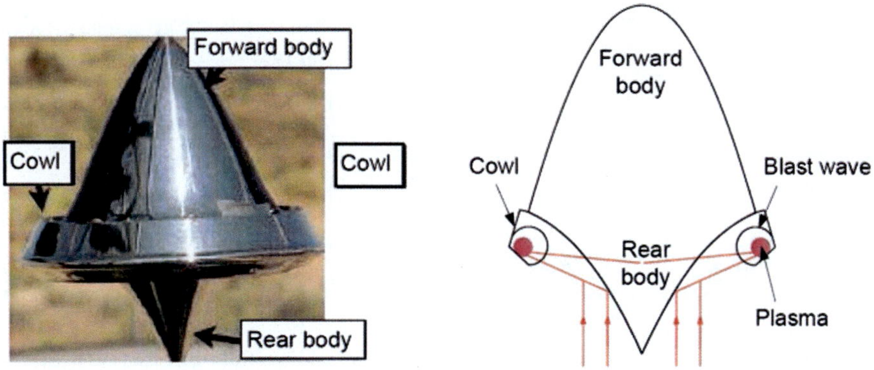

Fig. 5.1 Conceptual design of the Lightcraft vehicle

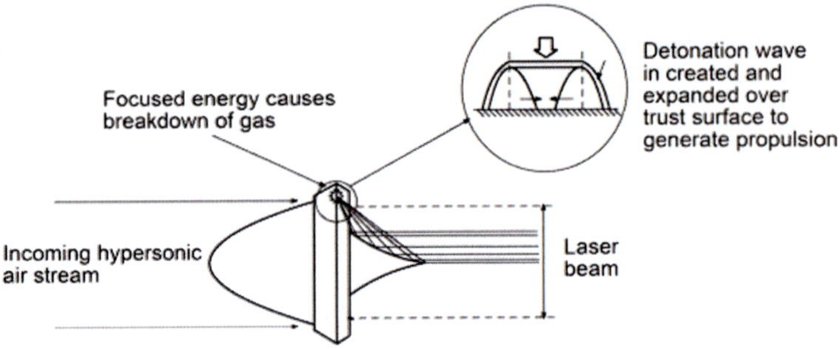

Fig. 5.2 Current Lightcraft configuration with the external laser-induced detonation

Table 5.1 Lightcraft vehicle shapes

Vehicle label	Description	Vehicle shape
A	Final flight configuration. Closed inlet and reverse-curved nozzle. Parabolic optic and rounded nose	
E	Original baseline, open inlet. Same optic and nose as configuration A	
E-C	Same as configuration E but closed inlet	
P	Closed inlet with cowl tangent to forebody contour	

of the Lightcraft model by varying the cowl and inlet geometric forms (see Table 5.1).

"A" version of the Lightcraft model was chosen for subsonic flight experiments. The forebody inlet of this model is closed to the incoming flow. As for a supersonic flight of the Lifhtcraft vehicle, one of the problems is an efficient compression of the incoming gas flow in the forebody inlet with minimal losses of dynamic velocity pressure in the flow. To overcome this problem, the construction of the model consisted of a forebody, an engine cowl, and a central body are considered jointly as a common gas-dynamic design of the Lightcraft (Fig. 5.3).

Following engineering principles of the Lightcraft device which are adapted for supersonic flights are established by summarizing of vehicle designs developed to the present time, namely:

- Lightcraft forebody must reduce minimal losses of gas pressure through compression of the gas in the inlet section.
- Forebody shape must not cause large hydraulic losses at a supersonic flight of the Lightcraft vehicle.
- Efficient compression of incoming gas flow by the forebody must exist within a wide range of the gas flow Mach numbers.

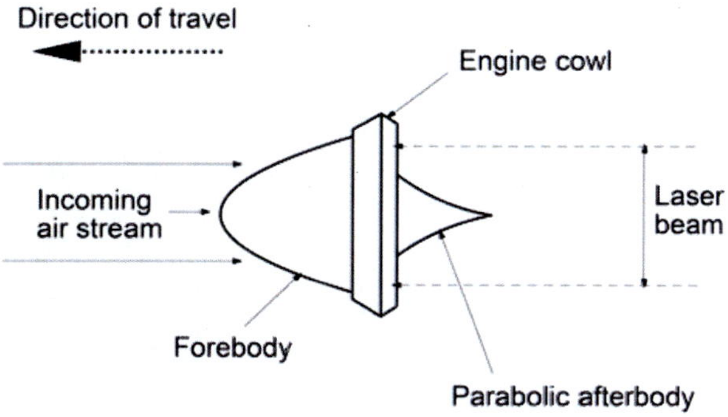

Fig. 5.3 Basic design of the Lightcraft vehicle

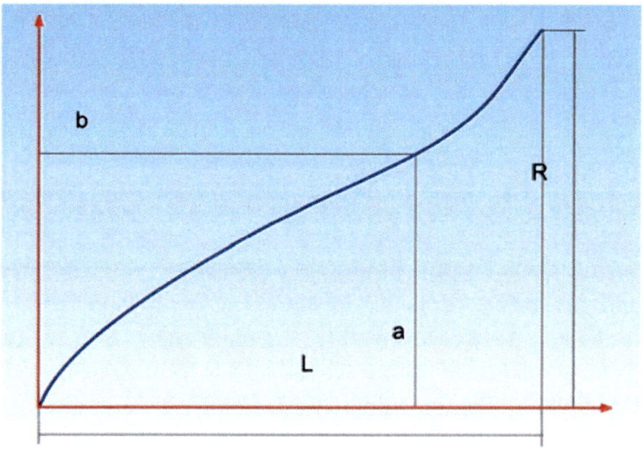

Fig. 5.4 Contour of the Lightcraft forebody

The vehicle developed in Ref. [12] to carry out the experiments on the hypersonic laser propulsion can be reviewed as an example of the Lightcraft model following the engineering principles. Forebody cone of the vehicle is designed to withstand the mechanical and thermal loads with the minimal increase in aerodynamic drag. In this case, its shape is described by the following power function:

$$y(x) = a\left(\frac{x}{b}\right)^n$$

where a is the nose cone length, b is the nose cone radius, and **n** is a power factor.

This type of an inlet elbow is shown in Fig. 5.4 with the nose cone followed by the compression ramp.

Next version of the Lightcraft basic model is shown in Fig. 5.5.

The cowl structure has to withstand an impact load caused by the strong shock waves generated at pulsed breakdown of the gas flow inside the engine nozzle. At the same time, the Lightcraft has to provide maximum coupling coefficient, C_m. All of Lightcraft models presented were developed to make technological tests on the

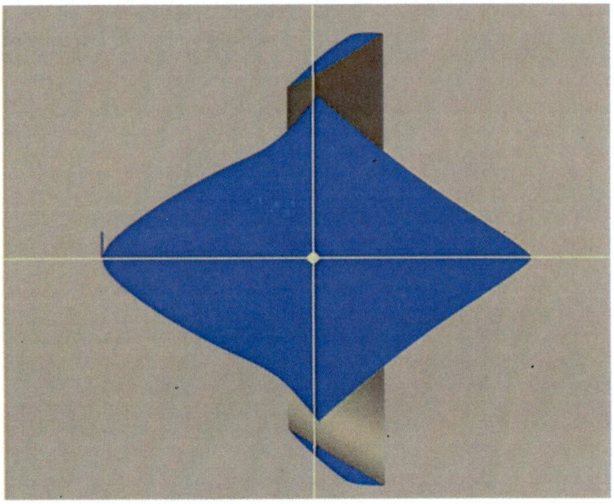

Fig. 5.5 Basic design of the Lightcraft vehicle

pulsejet or air-breathing laser propulsions at supersonic flights of the vehicle in the atmosphere. The main subject of investigations in these tests was experimental verification of thrust production efficiency which depend on both the vehicle engine and its forebody designs as well as on shock waves generated in supersonic gas flows in the engine. Let us consider the main aspects of this problem related to generation of laser-sustained shock waves and to propagation of the waves upstream the gas flow. The problem is possible choking of the engine flow inlet after the end of the laser pulse [13].

To be more precise study of the pulsejet supersonic laser propulsion produced in the Lightcraft model; it is assumed that the laser beam is focused at a local point with spatial coordinates, which correspond to the middle of the flow channel, formed by the afterbody and the engine cowl (see Fig. 5.6). In the figure, generation of a blast wave is observed immediately 1 μs after the end of 200 J energy pulse.

In Fig. 5.7, we see one of the principal effects of the pulsejet propulsion when shock wave is ejected from the engine inlet when the laser pulse energy exceeds 400 J, due to a propagating possibility of a strong shock wave upstream of a supersonic flow. Moreover, the wave acts as additional drag on the Lightcraft forebody. The calculations show that when the laser beam is focused on the engine cowl, then the shock ejection does not occur even if the laser energy exceeds 600 J. Therefore, to exclude generation of shock wave being ejected from the flow inlet, the laser beam focus has to be located correctly.

On the other hand, Fig. 5.8 shows the coupling coefficients estimated in the computations of Ref. [12], in the experiments of Ref. [11], and for various location of beam focus of Ref. [13]. We see, the coupling coefficients are fairly constant and equal to 150 N $\times$ s/MJ in the cases of the laser beam focusing on the cowl, being independent of the laser energy or slightly decreasing with the laser energy. This is a common tendency of all reviewed investigations.

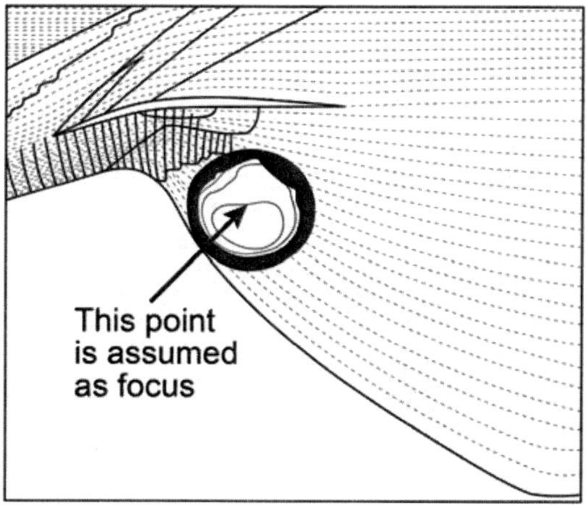

Fig. 5.6 Gas pressure counters in $t = 1$ μs after the laser pulse end

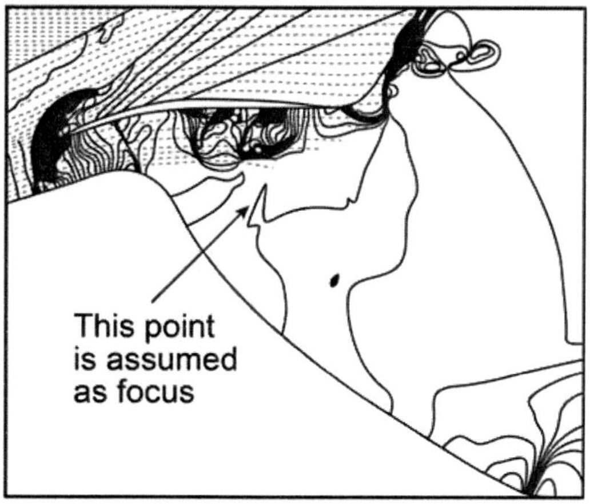

Fig. 5.7 Pressure counters in $t = 30$ μs after the end of the laser pulse with 400 J energy input

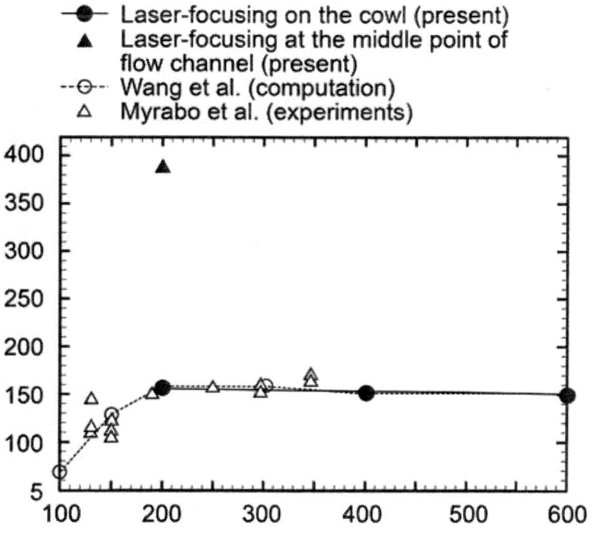

Fig. 5.8 Dependence of the coupling coefficient on laser pulse energy for various location of the laser beam focus

In case of laser beam focusing in the middle point of the flow channel, the coupling coefficient is about 2.5 times large in comparison with the one in case of its focusing on the cowl. So, the coupling coefficient is found to be sensitive to the location of the beam focus in the Lightcraft engine geometry.

The first experiments on a supersonic laser propulsion were made in Refs. [14–19] by using a simplified model of the Lightcraft vehicle as a quasi-two-dimensional model. The experiments took place in Henry T. Nagamatsu laboratory of supersonic aerodynamics and thermodynamics in Brazil, where a 1 kJ TEA CO_2 laser equipment is available. The supersonic shock tube is arranged in the laboratory close to the laser equipment. This favorable arrangement of the experimental equipment allows combining investigations on the supersonic laser propulsion.

Figure 5.9 shows the tested Lightcraft model, consisting of the cylindrical beam concentrator that has off-axis parabolic mirror and the cowl (shroud). Both units are arranged in a supersonic shock tube. The width of the model was equal to 24 cm. The initial static pressure of the air in the tube was change as 0.06, 0.15, 0.30, and 1 bar, and the laser pulse energy varied from 150 to 230 J.

To determine the coupling coefficient experimentally, the static gas flow pressure distribution along the inner surface of the model engine was measured and integrated. The experiments showed the coupling coefficient of the thrust was ~600–3000 Ns/MJ that is 2.5–5.0 times higher than the one measured in subsonic modes of the Lightcraft model operation [9, 10].

To present the experimental data adequately to vehicle flight in the atmosphere, the experimental conditions were reproduced using two basic gas-dynamic parameters, namely: Reynolds (Re) and Mach (M) numbers. Examples of the similarity from the calculations are presented in Fig. 5.10, showing the dependence of Reynolds number on Mach number for various atmosphere conditions.

These experiments are the first detailed investigations of the supersonic gas flow in the Lightcraft engine before and right after a laser-induced breakdown of the supersonic flow inside the engine. The examples of the supersonic flow patterns are shown in Fig. 5.11.

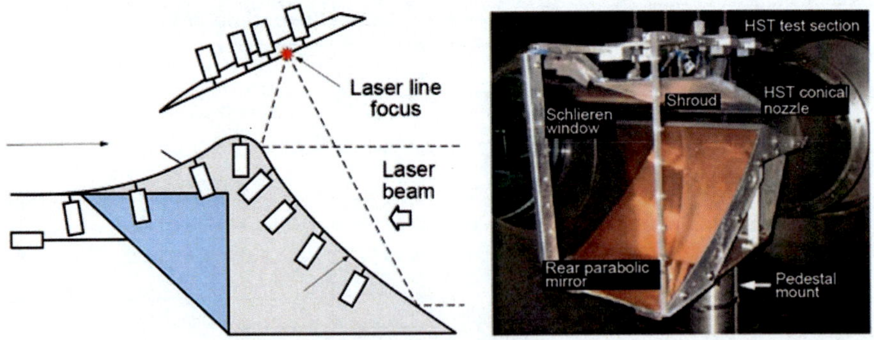

Fig. 5.9 One-dimension model of the Lightcraft engine developed for testing in supersonic flows

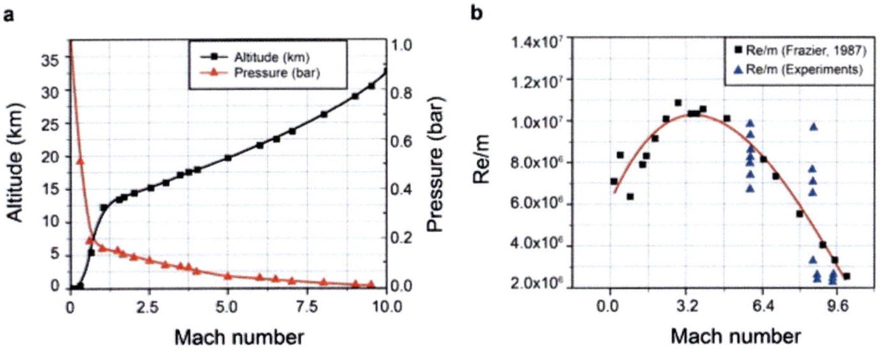

Fig. 5.10 The conditions of the Lightcraft experiments in comparison with its flights in the atmosphere

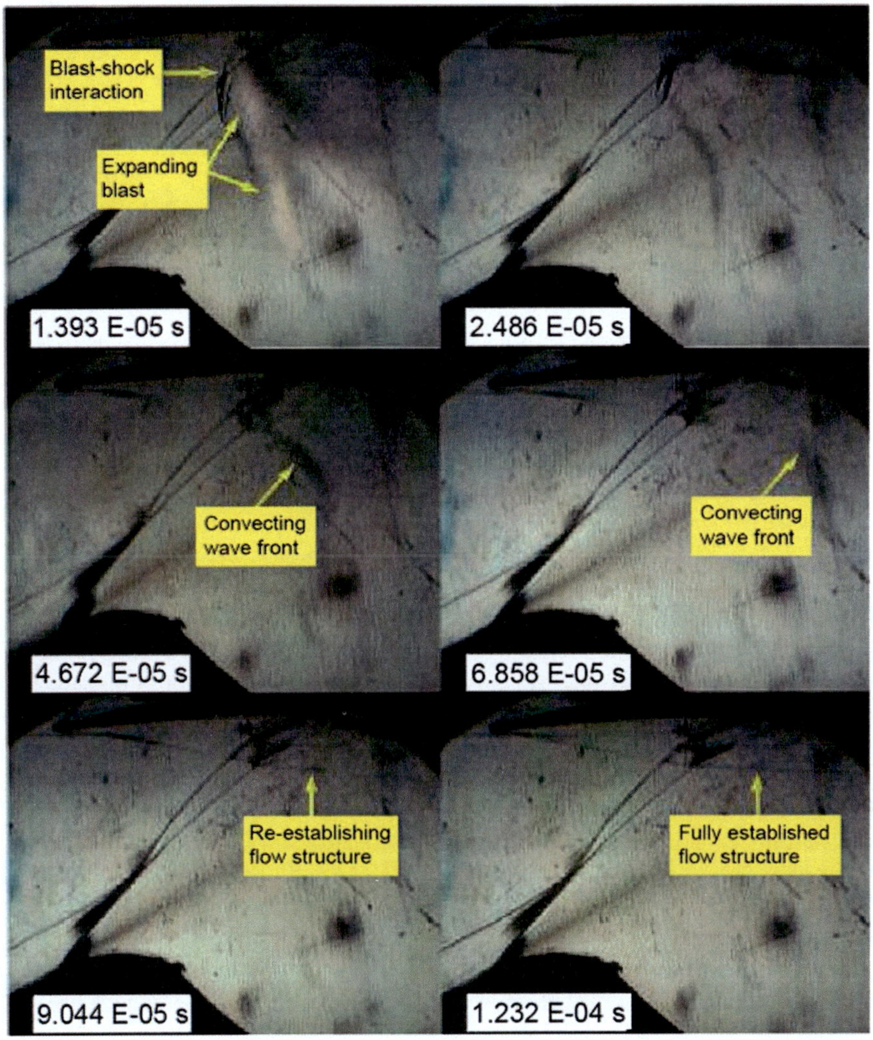

Fig. 5.11 Pictures of laser-induced blast wave interaction with oblique shocks and shroud undersurface

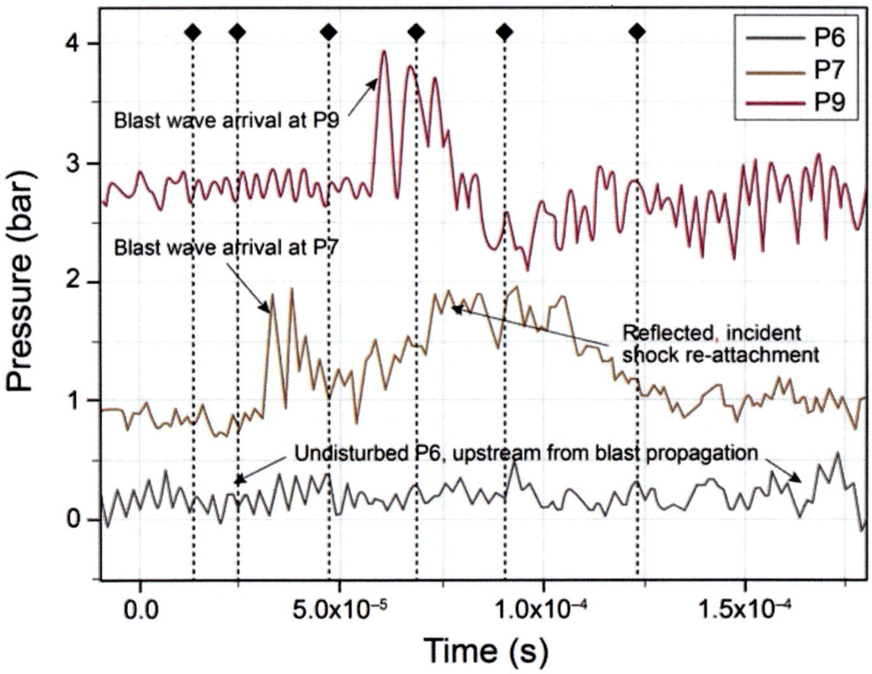

Fig. 5.12 Measured pressure temporal variation across the shroud under-surface. Diamond symbols indicate timing of Schlieren frames of Fig. 5.11

These pictures were obtained under the following experimental conditions: $M_\infty = 5.94$, $T_\infty = 256.8$ K, $P_\infty = 6.25$ kPa, and a laser pulse energy $E_P = 186 \pm 19$ J in the test region.

In these experiments, the processes of blast wave generation and interaction of the wave with the gas flow inside the channel were studied in detail. It is seen from Fig. 5.11 that the blast wave propagates upstream the laser breakdown region and then interacts with a bow shock wave that is formed due to the interaction of the incoming supersonic flow with the vehicle forebody. The blast wave also reaches a pressure sensor that is arranged nearby the leading edge of the shroud, increasing the static gas pressure on the inner side of the shroud. Temporal variations of static gas pressure registered by two pressure sensors are shown in Fig. 5.12.

As we can see from Fig. 5.12, there is a confined period of 100 μs of pressure growth. It is obvious that the blast wave doesn't reach the leading edge of the shroud because of the high velocity of incoming flow, estimated to be as high as $M_\infty = 9$.

Thus, the experiments discovered certain intricate processes of the ramjet supersonic laser propulsion, which will strongly affect the thrust production efficiency. One of the processes is supersonic flow instability inside the engine nozzle caused by the generated blast wave. Hence, the laser pulse energy must not exceed a certain magnitude in order to exclude the chocking effect in the nozzle inlet by the blast wave, and the wave must not be so weak that it doesn't reach the nozzle walls.

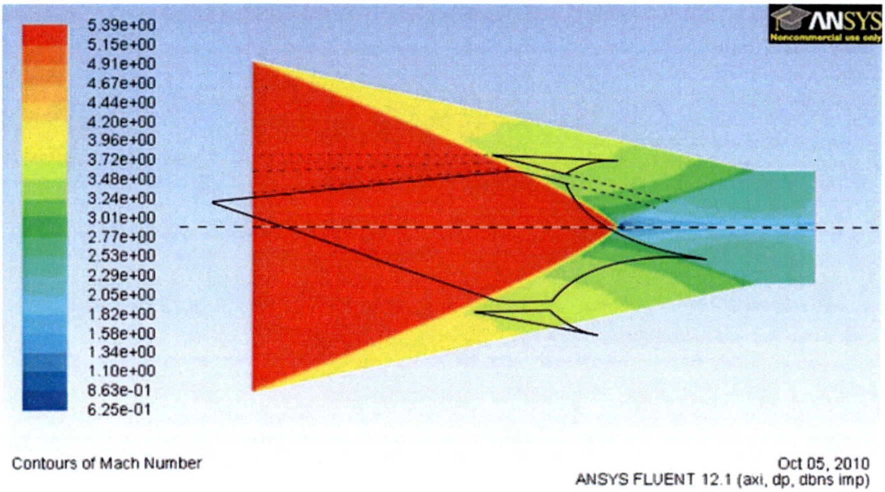

Fig. 5.13 Streamlines through a cross section of the axisymmetric flow, corresponding to Much number

5.2.1 Perspective Designs of the Lightcraft

Assuming further improvement of the Lightcraft design, a numerical technique was proposed in Ref. [3] to modify the Lightcraft forebody design complying with the ramjet supersonic laser propulsion. This technique is based on stream tracing method that provides a specified location of the incident flow stagnation point. Figure 5.13 is an illustration of the method as applied to designing of a new inlet cone of the Lightcraft vehicle.

The method is intended for optimization of gas flow static pressure at a specific Mach number of incoming flow to achieve the condition of laser-sustained blast wave generation. As an example, the novel design of a Lightcraft model corresponding to this condition is shown in Fig. 5.14.

The forebody is designed in such a way that a bow shock wave is inclined so the incoming supersonic flow at the channel entrance connecting the model inlet and model nozzle without unnecessary gas-dynamic losses in gas pressure. That type of the forebody is usually used to develop air intakes of supersonic airborne vehicles. There is another feature of the forebody which involves a lack of parasitic shock waves inside the inlet channel [20].

The atmosphere conditions corresponding to the proposed Lightcraft model with a speed of $M = 5$ and $M = 8$ are presented in Table 5.2.

Typical flow pattern of the model flight at $M = 5$ is shown in Fig. 5.15.

We see the supersonic flow structure has a homogeneous character in the inlet channel that connects the model inlet with its nozzle in spite of a complex flow structure in front of the vehicle forebody. However, if the ramjet laser propulsion is produced in the pulsejet regime by using the proposed engine design, shock (blast)

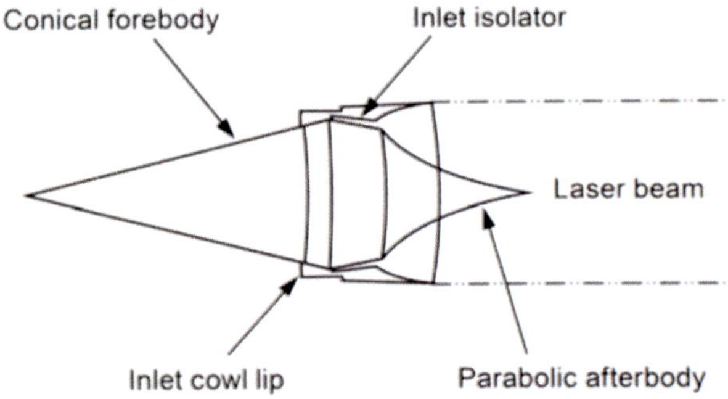

Fig. 5.14 New configuration of the Lightcraft with a stream traced inlet

Table 5.2 Flight conditions simulated for supersonic flights

Flight Mach number	5	8
Altitude, km	21.5	35
Static pressure, Pa	4500	575
Incoming flow temperature, K	220	237
Air density, kg/m^3	0.071	0.0085

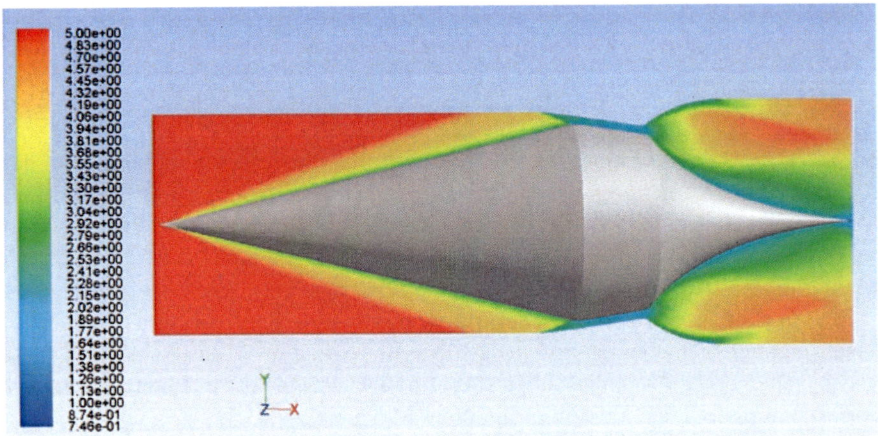

Fig. 5.15 Flow patterns of Lightcraft flight in the atmosphere at $M = 5$

waves generated inside the engine nozzle can destroy the flow homogeneity at some laser pulse energy, and the waves can also choke the engine inlet during fixed-time period [3]. Moreover, time of the choking period will depend on the Mach numbers of the shock wave and incoming flow. The choking effect may look like a surging phenomenon if the repetitively pulsed laser propulsion will be considered. To exclude such type of negative effects on the thrust production, the laser pulse energy should correlate with the Lightcraft flight velocity.

5.2.2 Intermediate Conclusion

We can conclude that the supersonic pulsed laser propulsion can be more effective if the shock waves generated by the laser pulses don't reach an exit of the inlet channel and don't choke the flow channel. This condition can be satisfied only in case if the pulsejet supersonic laser propulsion with a high repetition rate of laser pulses is developed by formation of a single strong shock wave located inside the engine nozzle [21]. To consider other techniques of supersonic laser propulsion, it will be useful to analyze the physical phenomena occurring with a laser-induced breakdown of gases and running at a supersonic speed.

5.3 Physical Phenomena Going with Ramjet Supersonic Laser Propulsion

To consider possible ways to develop efficient thrust with supersonic laser propulsion, we propose to consider some physical phenomena which are developed through interaction of laser pulses with supersonic gas flows [22, 23].

5.3.1 Gas-Dynamics Effects Induced by Lasers in a Supersonic Gas Flow

Remembering that a high-power laser pulse interacts with a plasma ignited in a gas flow of the form of (a) fast ionization wave (FIW) that propagates in the gas with velocity of $D_F = 20$–100 km/s, (b) laser detonation wave (LDW) that propagates with velocity of $D_F = 3$–10 km/s, and (c) radiative wave (RW) propagating with velocity of $D_F = 10 - 40$ km/s [22]. Here, D_F is a velocity of the wave front that propagates towards the laser beam. Every wave can be developed only at a specified laser radiation intensity, I (W/cm^2). For example, LDW is developed in an argon if $I = 10^7$ W/cm^2 [22]. Moreover, the flow volume occupied by the ignited plasma looks like a cylinder with a diameter equal to the laser beam waist in the focus region (see Fig. 5.16).

There are a number of physical features of the laser power deposition in the supersonic flow, which are observed theoretically and experimentally in Refs. [25, 26]. Detailed numerical analysis of the gas-dynamic phenomena accompanying the laser breakdown of gas flow was performed in Ref. [26]. The calculations show a complex destruction of a plasma core generated by a laser in a breakdown region (see Fig. 5.17). Moreover, reverse flow along the laser beam axis is observed to be formed behind the shock wave induced by the laser pulse. This flow generates complex vortex flow field that looks like toroidal ring whirlwinds. This vorticity is generated by various effects. One of these effects is a result of reverse streams

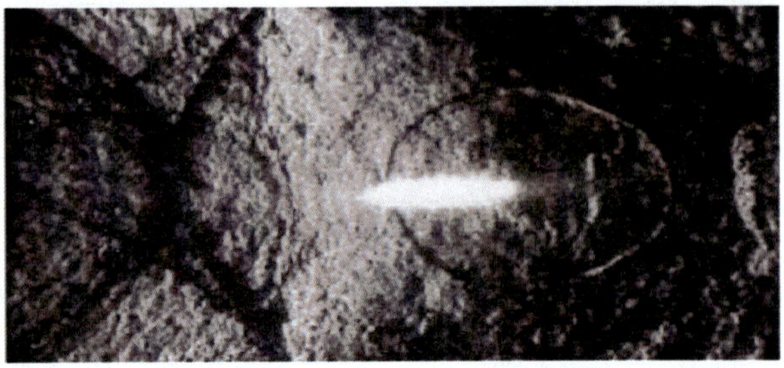

Fig. 5.16 Photograph of the optical discharge luminescence in argon [23]

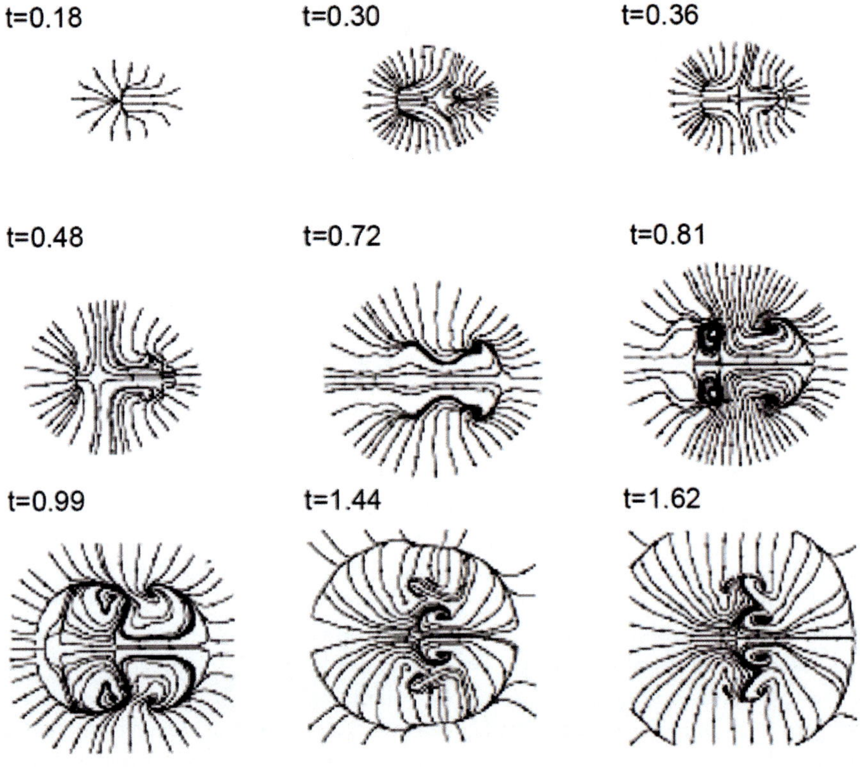

Fig. 5.17 Plots of velocity stream lines showing evolution of the flow field in time

generated in the plasma due to the plasma viscosity and laser power dissipation in the plasma.

Similar flow vorticity was experimentally observed in Ref. [4] (see Fig. 5.18) in laser-induced discharge in gases. Arrows at left side of Fig. 5.18 show the direction of the laser beam propagation.

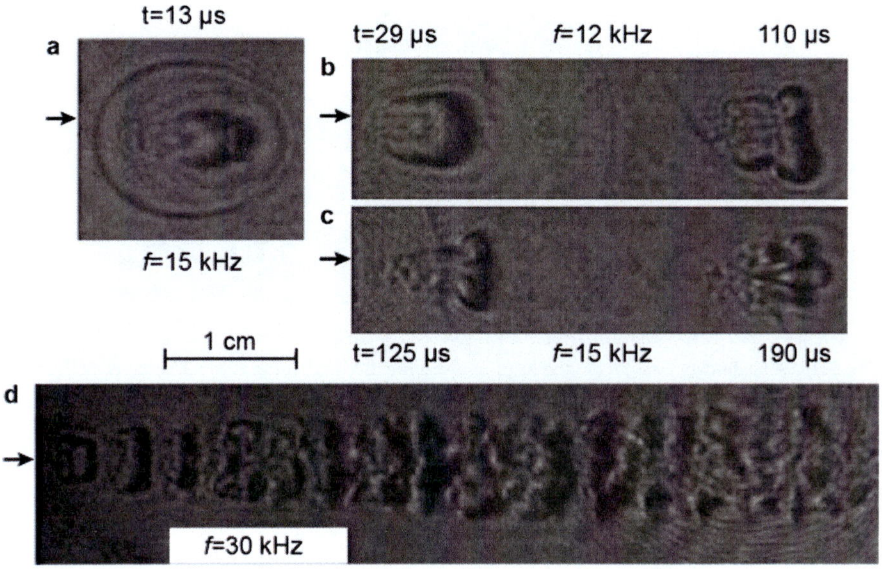

Fig. 5.18 Shadowgraphs of laser-induced flow vorticity in subsonic gas flow

It is interesting to mark the plasma "cavern" developed in time, including the region of the plasma ignition. As it is seen from the shadow pictures of the laser-induced gas discharge, gas vortexes are generated in the direction of laser beam propagation which are transformed into a turbulent jet if the gas flow velocity is less than the speed of sound in the gas.

Thus, both theoretical and experimental data on the supersonic laser propulsion show that this process is accompanied by a complex gas-dynamic phenomenon along with the laser-induced plasma after laser power conversion into the plasma temperature.

These gas-dynamic phenomena in supersonic gas flow were studied in Ref. [23] in some detail. Following this research, let us consider the results of numerical calculations of a supersonic flow of monatomic gas with an adiabatic constant of $k = 5/3$. It is also assumed that the "plasma" cylinder length is equal to 8.5 mm as in Ref. [34], and its radius is 1 mm (see Fig. 5.16). Besides, the pulsed energy is released in the plasma, with a pulse repetition rate of 45 kHz is $E = 0.04$ J. Then, the specific energy absorbed by the plasma is determined from $e = E/(m \times t)$ where m is the mass of the gas cylinder and t is the laser pulse length. The laser pulse length is equal to 1 μs, a pulse separation is $T = 22.2$ μs for PPR = 45 kHz and 10 μs if PPR = 100 kHz for the experimental conditions.

Both the supersonic and subsonic local shock waves are originated behind the developed bow shock wave, and the flow patterns change in dependence on the laser power. This phenomenon is illustrated in Fig. 5.19. Moreover, regular local spots of heated gas in the flow interact with each other or independently move downstream with time. In the case of PPR = 45 kHz, a quasi-stationary flow pattern is formed

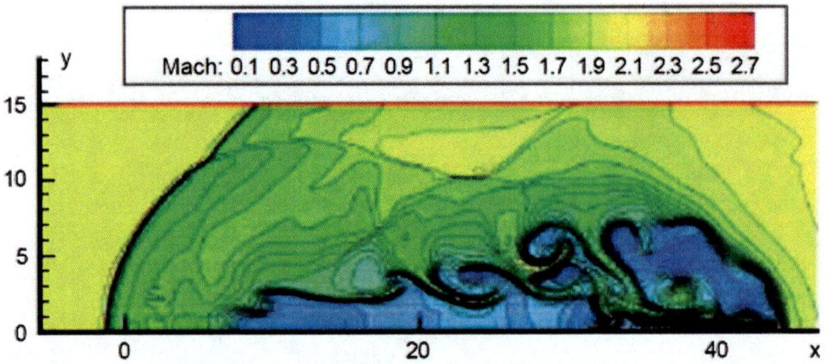

Fig. 5.19 Gas density contours in the laser-induced breakdown region at Mach number of 2 and at PRR of 45 kHz

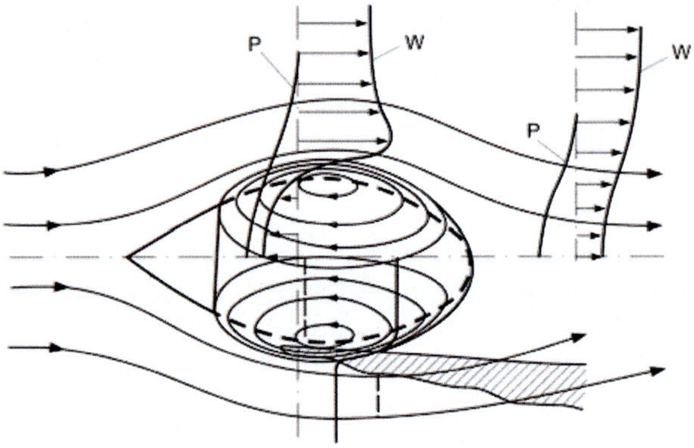

Fig. 5.20 The flow structure around the V-shaped stabilizer of ramjet engine

after 4–5 laser pulses. Thickness of the heated flow expands with time. In addition, a sharp spatial boundary between the heated area and other gas flow volume is formed, as well.

In principle, the flow structure in a breakdown region doesn't change strongly if pulse repetition rate will be changed with a kilohertz pulse repetition rates. A similar structure of heated gas trace is observed at PRR = 65 kHz, but at PRR = 20 kHz the heat trace consists of separate thermal spots and a quasi-stationary bow shock wave is not formed in that case.

To define the ways of stable development of supersonic laser propulsion, let us compare the supersonic flow patterns formed due to the laser radiation interaction with supersonic flow (Fig. 5.19) and similar flow in a ramjet combustion chamber (Fig. 5.20) [26].

The comparison allows finding out some common details. The combustion chamber of a ramjet engine is an engineered structure in which the heat of chemical reactions in a fuel is released into a gas due to fuel combustion by oxidation with the atmospheric air (Fig. 5.21).

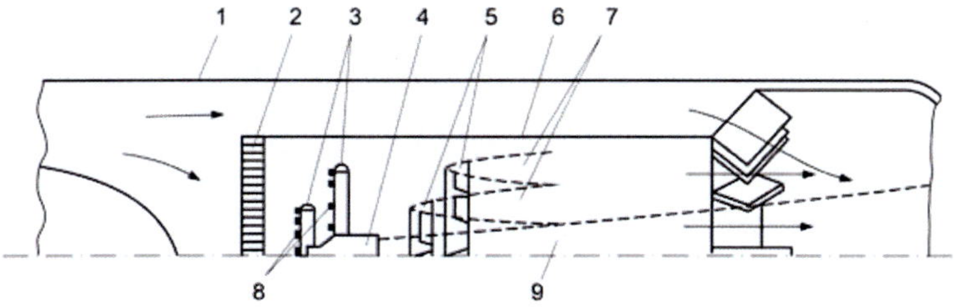

Fig. 5.21 Ramjet engine combustion chamber: (1) camera enclosure; (2) honeycomb; (3) fuel headers; (4) pre-combustion chamber; (5) flame ring holders; (6) screen; (7) flames down from the stabilizers; (8) fuel jets; (9) flame induced after the pre-combustion chamber

Four basic conditions have to be satisfied for operation of the combustion chamber, namely:

1. Liquid fuel should be evaporated, mixed with air, and heated.
2. The mixture of components must correspond to a stoichiometric composition needed for the oxidation of chemical reaction, which provides ignition of the mixture.
3. The gas mixture flow velocity is equal to the flame velocity at the point of the mixture ignition.
4. Initial temperature of the mixture is matched to the flame temperature.

To satisfy the first condition, liquid fuel is injected into the combustion chamber through sprayers. The fuel drops evaporate and mix with the hot air. As a result of these processes, a heated fuel-air mixture with a specific composition is produced. The second and third conditions are fulfilled in the flow regions located behind the flame holders. The stabilizers are made of bluff bodies in the form of cones, rings, and so on.

The regions of low gas pressure as well as reverse-direction flow are formed close by the stabilizer and behind it (see Fig. 5.20). Simultaneously, there is the recirculation local zone of the gas-fuel mixture flow, in which the flows in the forward and reverse directions are equal to each other. The area around the recirculation zone is called as the active region. The strong turbulent mass transfer takes place in the region located between the recirculation zone and the active region. In Fig. 5.20, the distributions of gas pressure p, flow velocity W, temperature T, and vapor concentration c_{T} are shown both in the recirculation zone and in the active region. One can see that the gas temperature reaches a maximum value in the recirculation zone, and the gas maintains a constant temperature along the gas flow.

So, comparing both pictures of the gas flow in a combustion chamber of a ramjet engine with the flow patterns formed by interaction of high repetition rate pulse laser radiation with a supersonic flow, we can conclude there is similarity in these flows. This physical similarity of the flows is connected with the generation of vortex zones caused by heat release in a confined gas volume as well as with the existence of boundaries between the heated gas and the incoming flow. But, the principal difference between laser-propulsion engine and a conventional ramjet device is an absence of such components as sprayers and pre-combustion chambers which complicate the engine device. But, supersonic laser-propulsion engine has to be specially designed to focus on a laser beam inside the engine nozzle and to provide a stable laser power release in a supersonic gas flow.

Consequently, let us consider one of the original techniques proposed in Ref. [24] in respect to supersonic laser propulsion to produce a thrust by using laser pulses which follow with a PRR of a few kilohertz.

5.4 Merging of Individual Shock Waves into a Quasi-Stationary Integrated Shock Wave

The process of generating a quasi-stationary bow shock wave from merging of a number of individual shock waves of low power with a laser with a PRR of a few kHz was studied in Refs. [4, 5]. The flow patterns corresponding to this phenomenon are shown in Fig. 5.22, which illustrates the influence of PRR on the bow shock wave formation in a trans-sonic gas flow.

It can be seen from the figure that merging effect is only achieved under certain condition when incoming gas flow reaches a supersonic velocity (Fig. 5.22(c)) [24]. It should be noted that every version of these pictures depends on the gas-dynamic model for their numerical calculations. Technical details of these experiments are shown in Fig. 5.23.

The experiments on the supersonic laser propulsion were made with a super-sonic nozzle that consists of a chamber connected by a tube with a cylindrical nozzle. There is an orifice in the end of that to inject argon gas from auxiliary high pressure vessel. Laser beam is focused inside the chamber also through this orifice in such a way that the laser-induced plasma originates in the beam focus region. The plasma originates as original chains following the high frequency laser pulses. The integrated shock wave, formed due to merging effect in front of the focus region, is

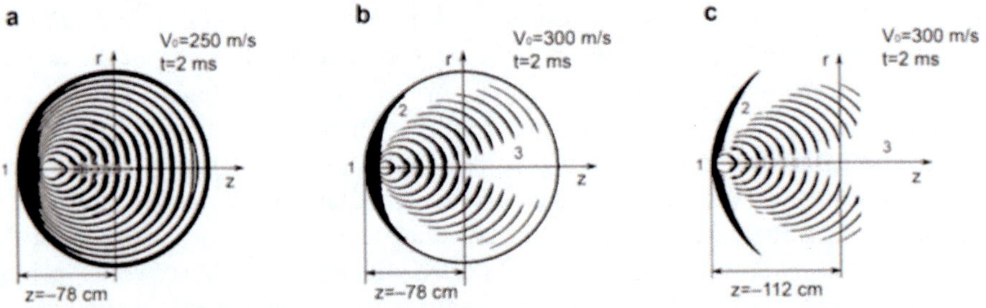

Fig. 5.22 Flow field patterns of merging of shock waves at various gas velocity and laser pulse parameters

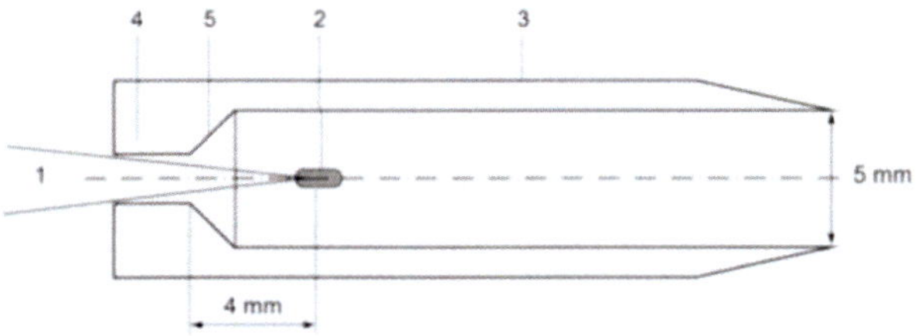

Fig. 5.23 Scheme of laser-propulsion engine tested[5]: (1) laser beam; (2) optical discharge in a gas; (3) reflector of shock waves; (4) orifice for incoming laser beam and injection of argon; (5) receiver of pressure impulse

Table 5.3 Principal characteristics and main results on a thrust produced due to merging of individual shock waves

	f, kHz	φ, kHz	D_r, H, mm	N	V, m/s	W, W	F_J, r	F_r, r	J_r, N/kW	Notice
1	45	RP	5–46		300	1300	80	40	0.61	1
2	45	RP	5–46		400	1300	141	69	1.06	1
3	100	RP	5–46		400	1200	155	54	1.08	1
4	100	1	15–50	30	300	720	49	4	0.085	1
5	45	1	15–50	15	50	720	0.9	2.1	0.042	1
6	45	1	15–50	15	300	720	49.1	4.5	0.09	1
7	45	1	15–50	15	50	720	1.2	1.4	0.028	2
8	45	1	15–50	15	100	720	6.3	5.6	0.11	2
9	45	1	15–50	15	300	720	62.7	4	0.08	2
10	45	1	15–50	5	170	500	17.7	3.5	0.1	2
11	45	2	15–50	5	100	600	6.3	4.8	0.11	2
12	45	2	15–50	5	164	600	18.5	7.5	0.18	2
13	45	2	15–50	5	300	600	70	–4	0.095	2
			15.12							
14	12.5	RP	25–35		60	430	2.4	4	0.13	3
15	12.5	RP	25–35		100	430	5	7	0.23	3
16	12.5	RP	25–35		150	430	11	11	0.37	3
17	12.5	RP	25–35		300	430	51	16	0.53	3
18	12.5	RP	25–35		50	430	6	1	0.033	4
19	12.5	RP	25–35		100	430	12	7	0.23	4
20	12.5	RP	25–35		300	430	195	−97	−3.2	4

Here, f, φ are the laser pulse repetition rate and pulse train regime; W average laser power; V argon gas velocity; D_r and H are the diameter and height of the shock wave reflector; F_r force produced by optical gas discharge; and *RP* repetitively pulsed

assumed to produce the thrust. In the experiments, the inner diameter of the nozzle was equal to 5 mm [5].

The main results of the thrust produced by the merging technique are listed in Table 5.3. The attention must be paid on such propulsion characteristics as coupling coefficient, J_r, and thrust, F_J. We also have an opportunity to compare these theoretical results of the supersonic laser propulsion with the experimental data obtained in Ref. [5].

It must be emphasized that all propulsion characteristics are unsteady ones, and they differ from run to run. For example, the momentum coupling coefficient, J_r, changes from maximum of 100 dynes/W in a magnitude (experiments under 2 and 3 numbers) to a negative value (experiment under 13 and 20 numbers). In our opinion, the thrust instability is caused by such factors as mismatching of the laser pulse power and subsonic gas flow velocity. It is seen from Table 5.3, the high momentum coefficient was obtained only in the experiments where the argon flow velocity corresponds to the front propagation of the generated shock waves indeed.

5.5 Supersonic Laser Ablation Propulsion

In this section, we validate a new technique of a thrust production at supersonic speeds of vehicles in the atmosphere by using high-power lasers. The proposed supersonic laser-propulsion technique is based on generation of a stable shock wave in the ramjet engine nozzle via interaction of ablation jet with a supersonic gas flow when the ablation jet is produced by laser within the nozzle.

5.5.1 The Effects of Gas Jet Injection into Supersonic Gas Flows

It is known that a gas jet injected into a supersonic flow wrapping a solid surface results in strong transformation of the flow, which is accompanied by strong gas-dynamic effects [27]. One of these effects is development of a bow shock wave that starts from a point located upstream the jet injection and continues into the flow. The shock wave power (Mach number) depends on the gas pressure and velocity of the gas jet as well as on the velocity of main gas flow.

Following Refs. [27, 28], we can choose basic parameters of the gas jet which affect the flow, namely, M_∞ as Mach number, T_∞ as the gas temperature, P_∞ as the static gas pressure of the main gas flow, and M_0 as Mach number of the injected jet. Moreover, such characteristics as viscosity and thermal conductivity of the injected gas have to be taken into account because they define the Prandtl and Reynolds numbers of the flows in the region of injection, which characterize the gas-dynamic effects to be considered (as usual, $Pr \sim 1$ and $Re \sim 10^4$–10^7) [28]. The expansion ratio of the lateral jet is also one of the important parameters of this phenomenon, which is determined as $n = P_0/P_\infty$.

Therefore, let us review the theoretical results on the flow structure generated in the region by injecting a lateral gas jet into a supersonic flow close to a solid plane.

Some patterns of the flow structure are shown in Fig. 5.24 [27]. We see that a trajectory of a lateral jet is curved under the dynamic velocity pressure of incoming supersonic flow. This jet trajectory and height of gas-dynamic barrier that originates at the jet depend on the balance between the dynamic velocity pressures developed by the main and secondary flows in the region of injection and depend on the injection angle, α_{inj}. The jet produces a barrier for the incoming flow in a form of irregular shaped step. As a result, strong pressure gradients are developed in the main stream in the area located close to the barrier which results in development of shock-induced detachment of the boundary layer that was formed on the solid surface.

The length of the detached flow domain that is separated from the leading edge of the injection orifice is proportional to the jet penetration depth into the incoming flow, and it is dependent on the same parameters of the flows. But there are some additional parameters which have to be taken into account if we will analyze the flow structure in more detail. These parameters are the injection angle, α_{inj}, and expansion ratio of the jet, n, as well as the distance between the jet injection and conjugation points, l_0.

It is complicated to get an exact analytical solution for effects caused by the lateral jet on supersonic flow under realistic conditions. But, adequate consideration of the experimental data with the jet allows formulating semi-empirical relation between the flow and jet gas-dynamic parameters [28]. It follows from the experiments that the most important parameters of the phenomenon are equivalent height of the jet barrier, *1*, and length of the lower domain behind the jet, 6 (see Fig. 5.24).

These features of the jet and the incoming flows interaction will be considered as two high-temperature streams that are mixed. This type of the interaction will be observed by ablating a solid surface into a supersonic flow inside an engine nozzle. In this case, the injected jet will possess a higher temperature than incoming flow. Simultaneously, the domains of the flow separation, as well as shock wave power, will be increased. This fact is a most interesting one because we assume this effect is used to produce a thrust via the laser ablation of solid propellants inside a supersonic nozzle. It follows from the physical considerations presented above, ablation jet parameters will determine efficiency of thrust production.

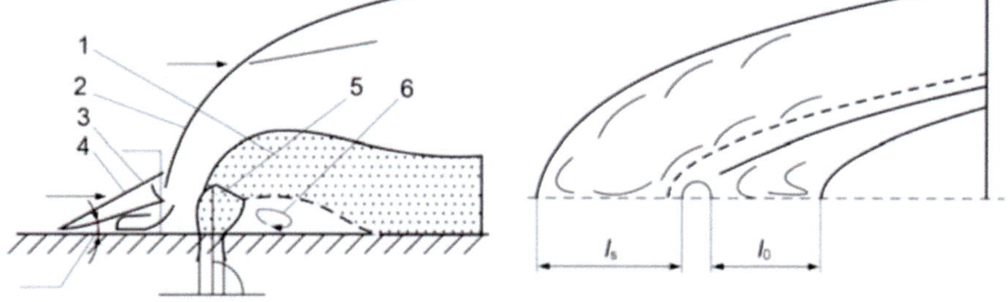

Fig. 5.24 The schemes of the lateral jet and supersonic flow interaction: (1) injected jet, (2) a curved shock wave, (3) detached flow domain, (4) oblique shock, (5) "Mach disk" formed in the jet, and (6) lower domain

5.5.2 *Theoretical Model of Supersonic Laser Ablation Propulsion*

As we discussed in Chap. 4, the laser ablation propulsion possesses certain advantages due to producing a thrust with a high specific impulse. The specific impulse can reach a few tens of thousands seconds if heat-resisting materials are ablated. Laser ablation of polymers allows using the internal energy stored in these materials to increase an efficiency of thrust production, expressed by an increase in the momentum coupling coefficient up to hundreds of dynes per Watt. Moreover, the laser ablation propulsion differs sufficiently from the pulsed laser propulsion because of a lack of strong shock waves which can choke a supersonic nozzle inlet if the laser pulse power is too high.

It follows from the previous section, the supersonic flow structure and its gas-dynamic characteristics can be changed by a vapor jet ablated from an inner surface of a supersonic nozzle. Let us consider a theoretical model of thrust production via the interaction of a laser-ablated jet with a supersonic gas flow in an engine nozzle as applied to the air-breathing laser propulsion. It is assumed that the gas-dynamic properties of the jet may be considered to be a supersonic gas jet injected into the supersonic flow. These properties are as follows: vapor pressure p_a, vapor velocity v_a in the jet, and ablated mass rate $\dot{m}$.

In accordance with the laser ablation theory [29], the momentum coupling coefficient C_m created by the ablation jet is defined as a ratio of specific recoil impulse $\dot{m}v_E$ produced by the jet to the laser radiation flux Φ in the case of pulsed lasers. In the case of *CW* lasers, this is a ratio of the ablated vapor pressure, p_a, to the laser radiation intensity, *I*.

To form a model of the process of the thrust production from interaction of the ablation jet with a supersonic flow, initial gas-dynamic parameters of the ablation jet are determined in accordance with the theory of a quasi-stationary evaporation of solid surfaces under the laser irradiance (see also Chap. 3). So, the ablated mass rate is determined as:

$$\dot{m} = \rho_a c_a \left[\mathrm{kg}/\left(\mathrm{m}^2\mathrm{s}\right)\right] \tag{5.1}$$

where ρ_a is the mass density of vapor jet and c_a is the adiabatic speed of sound in the vapor. The vapor pressure p_a can also be determined as:

$$p_a = \left[\left(1+\gamma M_a^2\right)\right]\rho_a c_a = \left[\left(1+\gamma M_a^2\right)\right]\dot{m}v_a/M_a \tag{5.2}$$

where M_a is Mach number of the jet, γ is the adiabatic exponent (usually $\gamma = 1.1$ if the vapor temperature exceeds 10^4 K), and v_a is the flow velocity of the vapor.

The vapor pressure in the jet depends on the modes of laser radiation interaction with plasma ignited by the laser in the jet [30]. We assume the plasma temperature is limited in magnitude by the effect of plasma screening of laser radiation (see

Table 5.4 Accepted ablation jet characteristics for the pulsed and CW laser radiation

Laser type	E, J	F, Hz	$P=F \times E$, W	Q^*, J/g	$\dot{m}$, g/s	v_E, m/s
Pulsed	0.18	5.6E+6	10^6	1.3E+4	77.0	5E+3
	0.18	5.6E+5	10^5	1.3E+4	7.7	5E+3
	0.18	5.6E+5	10^4	1.3E+4	0.77	5E+3
Special case	10.0	E+3	10^4	1.3E+3	7.7	
				1.3E+2	77.0	2E+3
CW	–	–	10^6	2.0E+3	500.0	2E+3
	–	–	10^5		50.0	2E+3

Here, E_i is the pulse energy; F is the pulse repetition rate; P is the average laser power; CW is continuous laser radiation. It is also assumed that the pulse laser length is 10 ns to define a peak power of laser pulse, which determine the laser ablation efficiency

Chap. 2). It is also assumed the plasma temperature depends on whether the radiation is pulsed or continuous (see Table 5.4).

In Table 5.4, the Q^* parameter is the pulse energy consumed to ablate one gram of the target material [29]; $Q^* = E_i/\Delta m$, where Δm is the material mass ablated by an E_i pulse energy. So, the total evaporated mass consumption will be determined as: $\dot{m} = P/Q^* = E_i f/Q^*$.

It is also required to fix the laser characteristics that will ablate solid materials. It is known (see Chap. 3), high-power laser irradiation of a solid surface is required to provide a continuous ablation jet. Therefore, a CW laser of megawatt power has to be applied to maintain the continuous ablation jet. But, such a high intensity of laser radiation can be easily achieved if a pulsed laser is used to ablate the material. To produce the quasi-continuous laser-ablated jet, the laser pulse repletion rate has to be sufficiently high and exceeds tens of kHz.

In Table 5.4, both the ablation jet and laser parameters are specified as the basis of the initial ablation data from Ref. [31] for various laser types. Also, the temperature of laser-ablated vapor is determined as an equilibrium temperature of the plasma produced at a specified radiation intensity.

It should be noted that the ablation jet is always directed perpendicular to a solid surface from which it is ablated, independently of the laser beam incident angle in contrast to the gas jet considered in the previous section. The flow velocity and mass rate of the vapor depend on the laser radiation intensity, absorption efficiency of the laser power, evaporation temperature of target material, and so on.

A numerical model of the supersonic laser ablation propulsion is proposed based on numerical simulation of semi-infinite gas flow by using the Navier-Stokes equations which are written in a vector form as follows:

$$\frac{\partial}{\partial t}\int_V W dV + \int_S (F - G)\cdot dS = 0 \tag{5.3}$$

In these equations, **W**, **F**, and **G** vectors have the following forms

$$W = \begin{bmatrix} \rho \\ \rho u \\ \rho v \\ \rho w \\ \rho E \end{bmatrix}, \quad F = \begin{bmatrix} \rho \mathrm{V} \\ \rho \mathrm{V} u + p\mathrm{i} \\ \rho \mathrm{V} v + p\mathrm{j} \\ \rho \mathrm{V} w + p\mathrm{k} \\ \rho \mathrm{V} E + p\mathrm{V} \end{bmatrix}, \quad G = \begin{bmatrix} 0 \\ \tau_{xi} \\ \tau_{yi} \\ \tau_{zi} \\ \tau_{ij} v_j + \mathrm{q} \end{bmatrix} \tag{5.4}$$

Here, the conventional symbols of Navier-Stokes equations are used. And the system of basic equations is completed by the equations of gas state, namely:

$$\begin{aligned} p &= p(\rho, T), \\ E &= e(\rho, p) + \frac{\rho V^2}{2} \end{aligned} \tag{5.5}$$

where $e(\rho, p)$ is the gas internal energy.

To improve the stability of the computational algorithm, a transition from the conservative to main variables is performed in the term containing the time derivative. The main difficulty in the numerical solution of the problem is associated with the computation of convective fluxes at the faces of the control volume. For this purpose, the counter flow scheme with separation of flows is used [32, 33]. This scheme makes it possible to obtain explicit expressions for convective fluxes at the faces of computational meshes without solving the Riemann problem. To determine the local time step, the Courant-Friedrichs-Lewy (CFL) criterion is used. In this case, the eigenvalues of the preconditioned system of equations are used as local velocity scales.

5.5.3 *Thrust Characteristics of Supersonic Laser Ablation Propulsion*

To test the proposed model of the supersonic laser ablation propulsion and to check the numerical technique for its accuracy of simulation, semi-infinite supersonic gas flow running close to an infinite solid plane are considered below. To do this, ablation jet characteristics were chosen to correspond to the jet described in Ref. [27], namely, ablation jet-flow velocity is of 429.2 m/s, mass rate is $\dot{m} = 40$ g/s. The velocity of main stream was $V = 592$ m/s, simulating a vehicle flight at the altitude of $H = 30$ km in the atmosphere and at Mach number of $M_0 = 2$.

Figure 5.25 illustrates the results of a numerical simulation of the "ablation jet-main stream" interaction with the initial conditions listed above [34]. As we can see, the bow shock wave is formed at a small distance upstream the ablation jet, and its wave front is a curved shape being initially perpendicular to the plane. We see that the flow velocity decreases to a subsonic value in the region downstream the wave, and then the velocity increases to a supersonic one right after the ablation jet

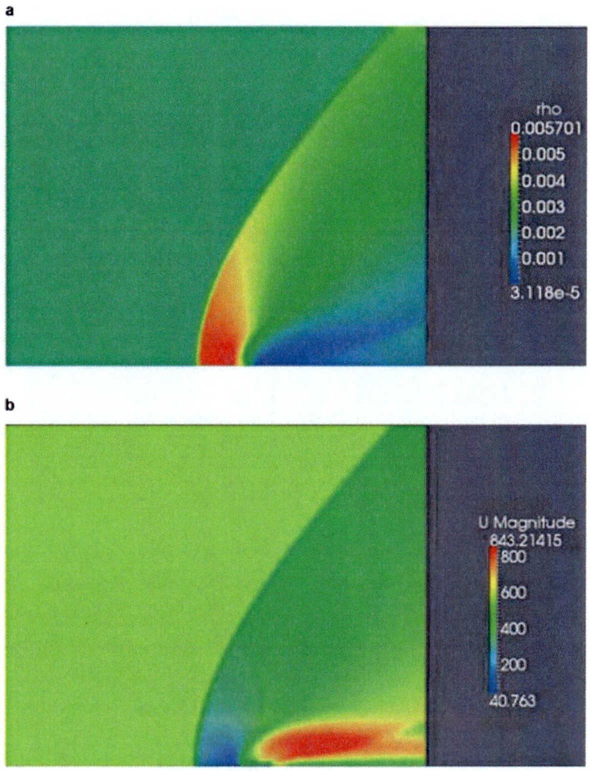

Fig. 5.25 (**a**) Iso-chores of the ablation jet, (**b**) lines of constant velocity in the vicinity of the ablation jet

injection point. Moreover, the flow velocity is twice larger as the velocity of the main stream flow. The latter fact means this additional acceleration of the main stream flow in a local region may be used to accelerate flows in supersonic nozzles.

Let us consider this effect as applied to the supersonic laser propulsion.

To prove the additional acceleration of a supersonic flow by a lateral ablation jet, we formulated two model cases of the jet-flow interaction inside two types of supersonic nozzles, including: an axial paraboloid and an off-axis paraboloid nozzle with a central body.

A parabolic nozzle consists of two parts, namely, a subsonic convergent inlet and an axial paraboloid (Fig. 5.26). It is assumed that the inner surface of the paraboloid can serve as a mirror to focus a laser beam into a focus plane of the paraboloid, where a solid propellant in the form of a cylinder is placed.

In Table 5.5, the main gas flow and the ablation jet initial characteristics, which were used to simulate the supersonic laser ablation propulsion production in the parabolic nozzle.

Detail patterns of the supersonic flow structures in the parabolic nozzle in the laser ablative mode of the thrust production are presented in Fig. 5.27. In Table 5.6, laser power (P) and ablation jet parameters for the case of 10^4 K plasma temperature, 5×10^3 m/s initial jet velocity at the specific ablation energy of $Q^* = 1.3 \times 10^4$ J/g

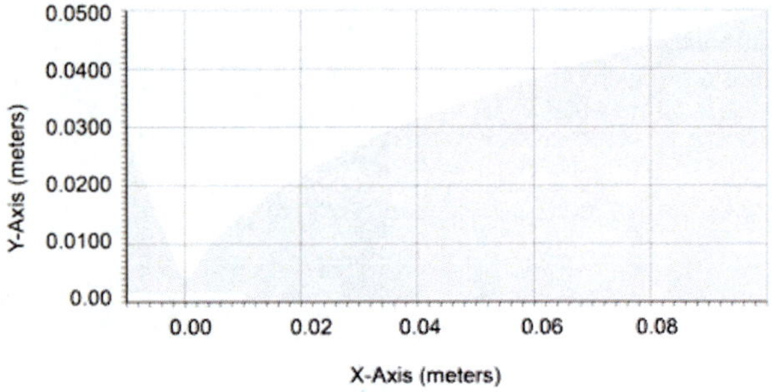

Fig. 5.26 Parabolic supersonic nozzle shapes

Table 5.5 Initial ablation jet and main gas flow characteristics in the parabolic nozzle

Inlet section	Outlet section	Ablation jet
$S = 2.82 \times 10^{-3}$ m	$S = 7.85 \times 10^{-3}$ m	$S = 2.45 \times 10^{-7}$ m
$P_0 = 6.0 \times 10^7$ Pa	$P_0 = 4.6 \times 10^7$ Pa	$P_0 = 1.9 \times 10^7$ Pa
$V = 23$ m/s	$V = 1707$ m/s	$V = 2291$ m/s
$Q = 2$ kg/s	$Q = 2$ kg/s	$\dot{m} = 7.7 \times 10^{-4}$ kg/s

Here, S is the square of section; Q is the initial mass rate of the gas flow in the nozzle; P_0 is the gas static pressure; V is the axial velocity of the flow; and $\dot{m}$ is the ablated mass rate

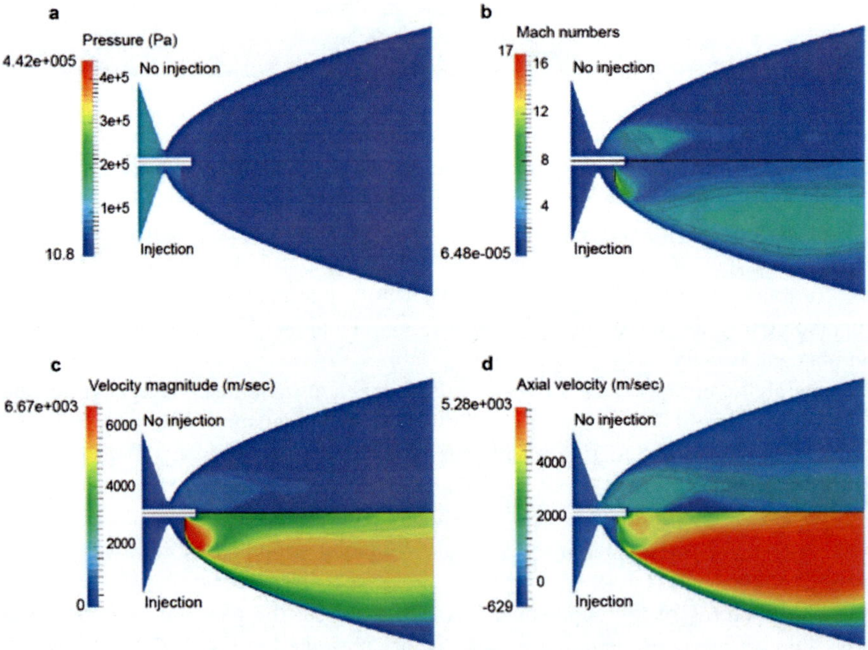

Fig. 5.27 Axial distributions of gas flow parameters inside the paraboloid nozzle with a solid propellant in the form of central rod installed in nozzle axis, including: (**a**) gas pressure, (**b**) Mach number, (**c**) velocity, and (**d**) axial velocity; upper part of every picture (no injection)—without ablation, and lower part (injection)—with ablation

Table 5.6 Laser-ablated jet and laser parameters for the parabolic nozzle flow

T_i s	E_i J cm^{-2}	F Hz	$\dot{m}$ g s^{-1}	$\dot{m}$ g s^{-1}	$\dot{m}$ g s^{-1}
1.0E-8	0.52	6E+4	0.77	7.7	77
1.0E-7	5.2	6E+3	0.77	7.7	77
1.0E-6	52	6E+2	0.77	7.7	77
P = 1E+4 W P = 1E+5 W P = 1E+6 W					

are summarized. Laser output power and laser pulse duration are chosen to be variable parameters.

Figure 5.27 illustrates the distributions of the flow Mach number and its gas pressure in an axial cross section of the nozzle. A stationary shock wave, formed via the interaction of the incoming gas flow with the ablation jet, results in a transformation of the gas flow initial structure so that the gas pressure increases at the nozzle wall. The shock wave also influences the distribution of flow Mach number in the nozzle itself, its maximum decreases slightly and shifts to the nozzle axis (see Fig. 5.27). Computations of the pressure distribution on the nozzle wall down the flow show that the interaction of the ablative jet with the incoming flow results in an increase of the axial component of the pressure, acting on the nozzle wall. And "this additional force" is equal to 310 N in the case under study.

Figures 5.28, 5.29, and 5.30 illustrate the parabolic exhaust jet characteristics for the case when the ablation mass rate is $\dot{m}$ = 77 g/s.

As it is seen from the figures, main gas flow pattern at the nozzle exit section corresponds to the similar pattern of an over-expanded exhaust jet in accordance with the initial gas-dynamic conditions.

To estimate the efficiency of thrust production by the supersonic laser ablation propulsion, we use a conventional definition of momentum coupling coefficient, C_m, namely as a ratio of the laser-induced thrust increment to laser power applied, $C_m = \Delta T/P$. In the case considered in the Fig. 5.27, numerical estimations show that C_m = 31 dyne/W. It must be noted that this result on thrust produced at supersonic laser ablation propulsion looks very attractive if we can optimize the laser-ablated jet characteristics in accordance with coming supersonic flow which is initially formed in the parabolic nozzle.

Numerical analysis of the supersonic laser ablation propulsion, as applied to the Lightcraft vehicle, shows similar results. The following analysis is based on the thrust production via a laser-ablated jet in the engine nozzle formed as a plug nozzle with an off-axial parabolic afterbody. The afterbody serves also as a laser beam concentrator (see Fig. 5.31) so that a ring ablation jet is injected from the nozzle wall simulating the Lightcraft engine cowl. Initial data for calculations are presented in Table 5.7.

It is assumed that these initial conditions correspond to flights of a laser-propelled vehicle of Lightcraft type at an altitude of 30 km above a ground in the Earth's atmosphere.

Figure 5.32 illustrates a pressure distribution in the nozzle for two conditions, namely, without ablation jet (a - top part of the figure) and with the jet (b - bottom of

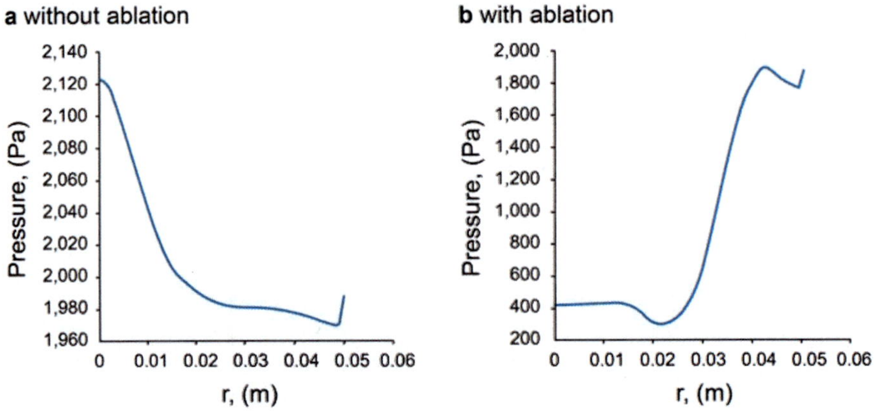

Fig. 5.28 Gas pressure distribution across the exit section of parabolic nozzle

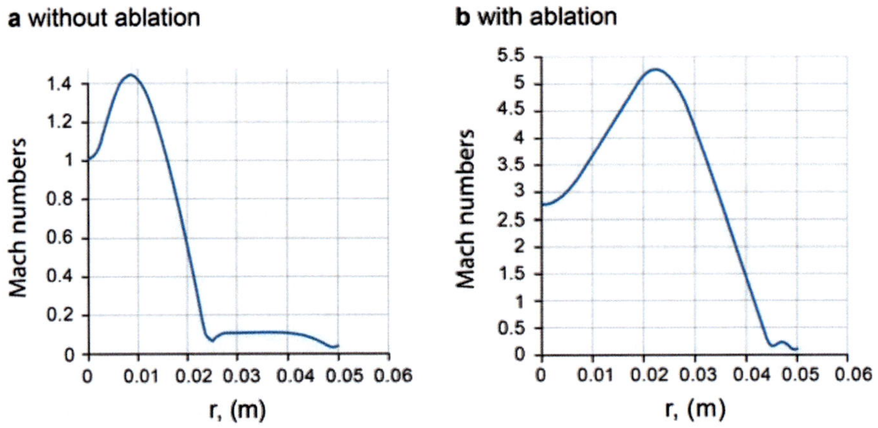

Fig. 5.29 Mach number distribution of the exhaust jet

the figure). Evidently, the ablation jet converts the initial supersonic flow significantly. A characteristic rarefaction zone located downstream of the nozzle inlet in the case (a) is disappeared in the case (b). In the case (b), there is a curved shock wave in the vicinity of the nozzle inlet via interaction of the flow with the ablative jet, which results both in a pressure increase in the vicinity of the afterbody surface and in producing additional thrust.

In this case, the thrust estimated as an axial component of the pressure force acting on the nozzle walls is equal to 1340 N. In this case, the coupling coefficient is 10^{-3} N/W (100 dynes/W), which looks very attractive for High-Power Laser Propulsion.

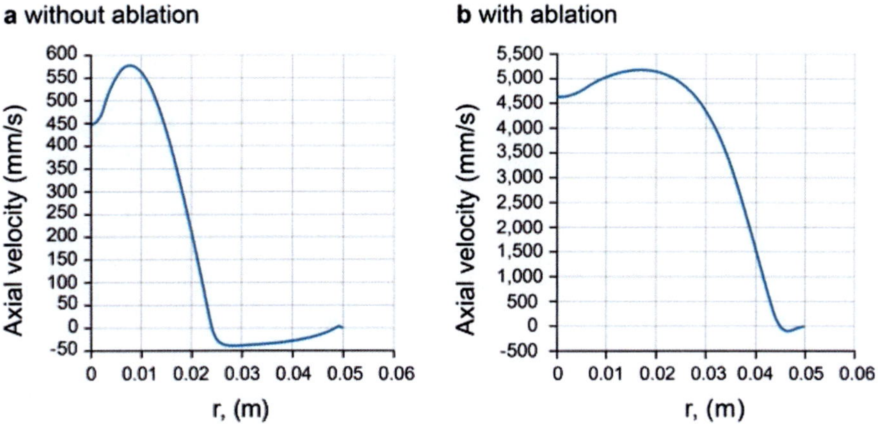

Fig. 5.30 Longitudinal component of gas flow velocity in the exhaust jet

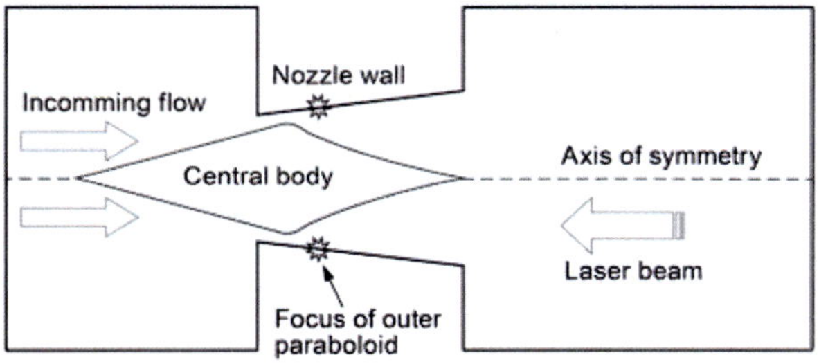

Fig. 5.31 Schematic diagram of the Lightcraft numerical model

Table 5.7 Initial data for the calculations of the supersonic laser ablation propulsion as applied to the Lightcraft vehicle

Initial characteristics of the main gas flow	$p = 2{,}000$ Pa $T = 220$ K
Mach number of incoming flow	$M = 3.2$
Laser power	150.0 kW
Absorbed laser power	60.8 kW (that corresponds to internal gas energy $U = c_v T$)
Square of focus region, S	2.45×10^{-7} m^2
Ablation jet characteristics	$T = 11{,}000$ K $V = 5000$ m/s

5.5.4 *Peculiar Properties of Thrust Production at the Supersonic Laser Ablation Propulsion*

Let us consider some effects of the ablation jet on a supersonic gas flow inside a parabolic nozzle of the Lightcraft type more detail to study the peculiar properties of thrust production at supersonic laser ablation propulsion. To be more precise, we

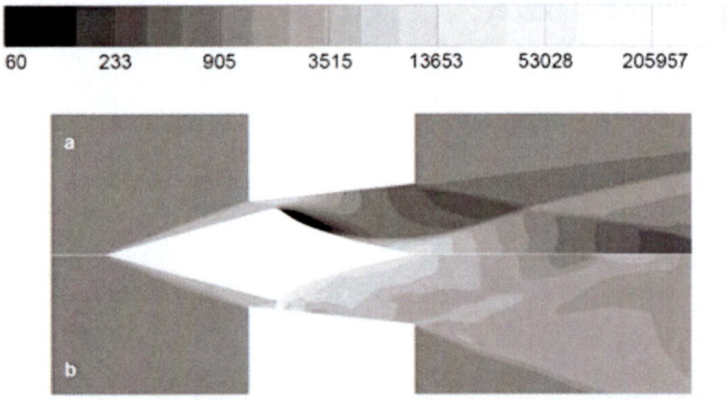

Fig. 5.32 Pressure distribution in the supersonic nozzle of the Lightcraft-type nozzle

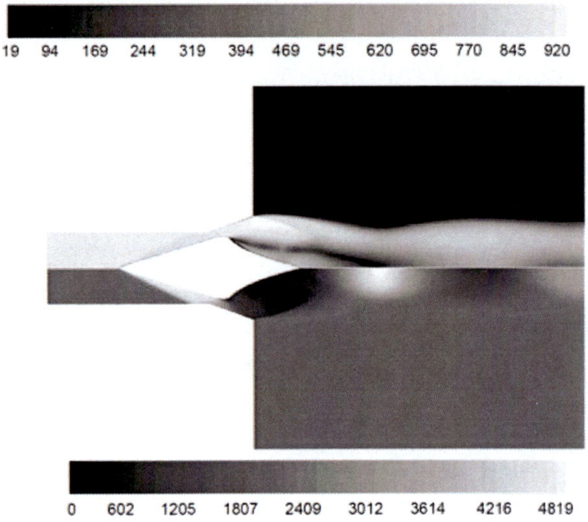

Fig. 5.33 Pressure distribution in the supersonic nozzle of the Lightcraft-type nozzle

assume that the incoming flow Mach number is equal to 5.0, and the gas temperature and gas density correspond to the atmospheric air characteristics at an altitude of 30 km. The engine nozzle is assumed to be formed by two paraboloids, so that the focus point of a central paraboloid (afterbody) is located on a surface of the second paraboloid, which is formed like a parabolic shroud of the Lightcraft engine. Moreover, the focus point of the parabolic shroud is located on the afterbody surface. We suppose such type of LPE nozzle shape would provide a more efficient utilization of the laser power delivered from a remote laser.

A few patterns of flow transformations are presented in Figs. 5.33 and 5.34, corresponding to the cases, when the ablative jet is formed inside the nozzle. It follows from comparison of the (a) and (b) parts of every figures, generation of a bow shock wave in the nozzle, due to the ablative jet, results in an acceleration of the flow and redistribution of gas pressure inside the engine nozzle. The gas pressure in the flow is transformed in such a way that the pressure peak is shifted into direction of the nozzle walls that leads to the production of additional thrust.

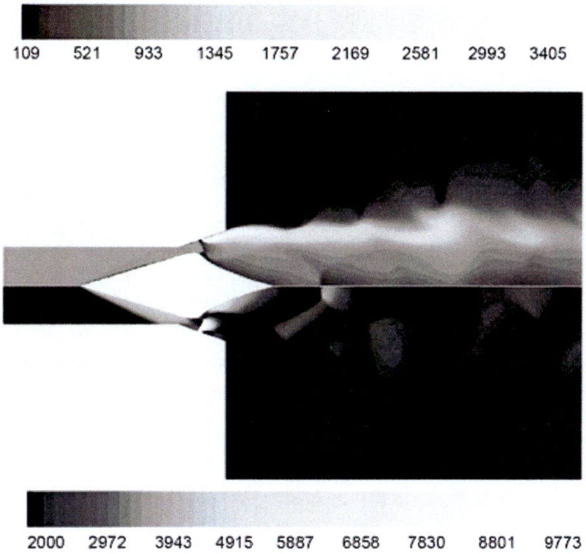

Fig. 5.34 Pressure patterns inside the parabolic nozzle: with ablative jet (bottom) and without ablative jet (top). Top and bottom scales denote the ranges of pressure change, Pa

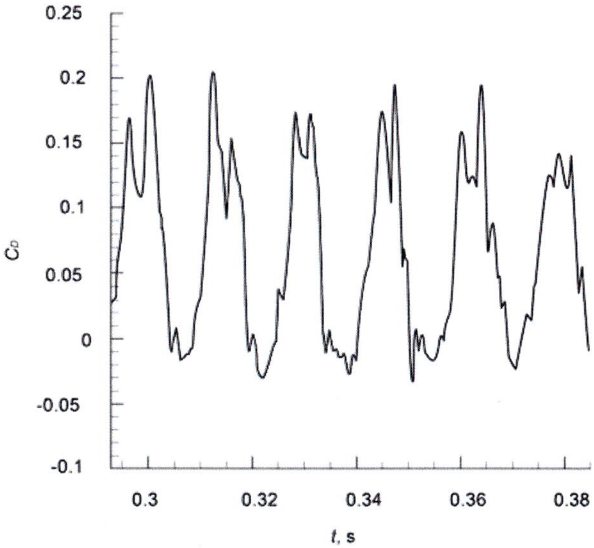

Fig. 5.35 Time fluctuations of the pressure coefficient for the parabolic nozzle

Another interesting fact that was discovered through the numerical calculations of the supersonic laser ablation propulsion [35] is the development of temporal instability of thrust in the nozzle when the mass flow rate in an ablated jet exceeds a certain value which depends on laser power. Figure 5.35 illustrates this instability in the form of a typical oscillation of the pressure coefficient when the laser power

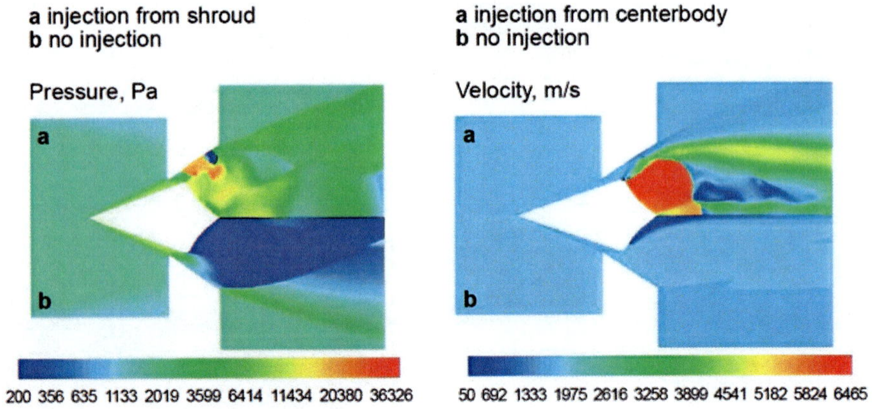

Fig. 5.36 Supersonic flow patterns in the Lightcraft nozzle under extremely high-laser power applied

exceeds certain maximum resulting in a generation of strong bow shock waves, which can reach the nozzle inlet section and chock it temporarily.

Let us consider some examples of the influence of the ablated jets on the supersonic flow patterns inside the Lightcraft nozzle, which are formed by the ablation jet at "right direction," are shown in Fig. 5.36. Here, the following partial cases are considered numerically, namely, (a) mass rate in the ablative jet is supposed to be equal to 0.25 g/s and (b) incoming mass flow rate is assumed to be 0.87 kg/s. The ablative jet characteristics correspond to a laser radiation power of 3 MW.

It must be underlined that there is an area of recirculating gas stream in the initial flow in the vicinity of the nozzle axis where the velocity vectors and axial velocity profile cross the nozzle exit. However, the ablative jet changes the flow pattern strongly. In the case of focusing the laser beam on the afterbody surface and generating a jet from its surface (Fig. 5.36(a)), the flow pattern is characterized by multiple shock waves and by zones of high velocity in a vicinity of the nozzle axis. Moreover, when the jet is injected from the afterbody in the direction of the shroud, the flow possesses more regular structure.

Comparing the pressure and velocity distributions in the nozzle for both cases, it can be concluded that the flow decelerates in the vicinity of the shroud where pressure increases via the ablation jet. This effect results in production of additional thrust, namely, axial projection of the pressure force acting on the nozzle structures changes from −275.3 N, in the case without the ablative jet, up to +802.0 N when the jet is generated ("minus" sign corresponds to deceleration, "plus" sign corresponds to acceleration of the flow). In particular, the coupling coefficient is equal to $C_m \sim 26$ dyne/W in last case.

The consequent effects of double ablative jets on a supersonic flow in the Lightcraft-type engine nozzle are observed with the assumption that mass flow

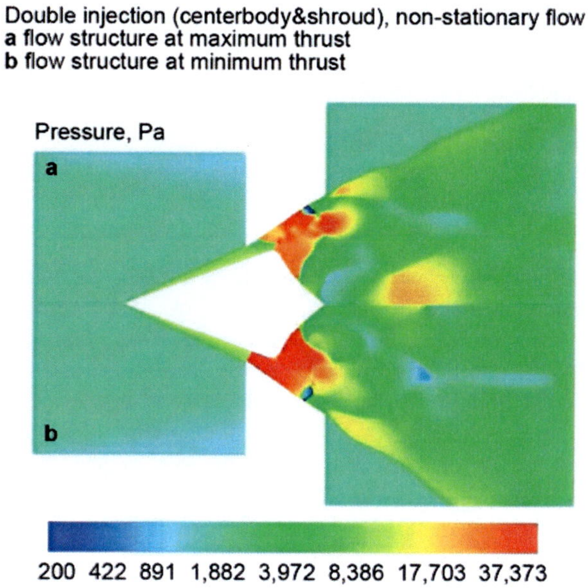

Fig. 5.37 Instability of a supersonic flow structure caused by the laser ablation jet of high rate of ablative mass

rates of double ablative jets equal to 2·0.2 kg/s that correspond to the total laser beam power approximately 5 MW (see Fig. 5.37).

It is seen that the supersonic flow structure becomes non-stationary and non-uniform in the case of laser extreme power. The reason is that the shock wave moving upstream at the direction of the engine inlet and chocking the inlet. This negative effect also results in both a decrease in thrust and its instability. Nevertheless, double jets produced in opposite phases of the non-stationary process provide positive integrated thrust, such a way that it is equal to +2122.0 N at a minimum, and +3295.3 N at a maximum that corresponds to $C_m \sim 42$ dyne/W and $C_m \sim 66$ dyne/W, respectively.

It should be noted that actual mass flow rates for ablative jets from the shroud and the afterbody will not be the same because of specifics of radiation distribution across the beam aperture. To simulate this effect, two sets of ablative jet parameters are considered below. The first set includes the mass flow rate of a shroud jet being 140 g/s that corresponds to the laser power of 2 MW, and the mass flow rate of an afterbody jet of 80 g/s that corresponds to a laser power of 1 MW. For the second set, the mass flow rate of a shroud jet is equal to 250 g/s at the laser power of 3 MW, and the mass flow rate of the afterbody jet is equal to 50 g/s at a laser power of 0.6 MW.

The simulation of thrust development for these two cases shows that the thrust produced would be 750.1 N, that is equal to $C_m \sim 28$ dyne/W at the first case, and of $T = 1102.6$ N that is $C_m \sim 30.6$ dyne/W in the second case. In both cases, the flow remains stationary with a structure of regularly interacting shock waves, described above, and with multiple detached shocks downstream.

Thus, we can conclude that the ablative jets enable not only to add energy into a supersonic flow but to produce additional thrust. Moreover, the jets provide total control of the flow structure by eliminating of the areas recirculation in the flow pattern. Implementation of a double paraboloid nozzle in the Lightcraft-type engine enhances these effects.

5.6 Conclusion

In the chapter, we considered the basic effects of laser propulsion in a supersonic mode. It is shown that those laser techniques which are used at the present time possess principal limitations caused by (1) instability of supersonic gas flow as a response to pulsed breakdown of the gas flow, and (2) generation of strong shock waves choking an inlet of a supersonic nozzle of Lightcraft-type vehicles. To eliminate these effects, we propose applying a laser-ablated jet to provide efficient laser power deposition in supersonic flows and as a consequence to produce a thrust. In this case, the thrust augmentation is a result of transformation of the supersonic flow via interaction of the ablative jet with the flow.

Numerical analysis of the supersonic laser ablation propulsion demonstrates the operational stability of the thrust production if the ablated mass rate does not exceed a threshold value. In that case, the efficient thrust production is observed. The momentum coupling coefficient of $C_m \sim 10^{-3}$ N/W can be obtained in this case. In our opinion, the proposed technique of using the laser-ablated jet interaction in supersonic flows could also be used to increase the efficiency of thrust production in hypersonic ramjet engines because the technique does not require a flow forced stagnation in the supersonic flow.

References

1. Kantrowitz, A.: Propulsion to orbit by ground-based lasers. Astron. Aeron. **10**, 74–76 (1972)
2. Simons, G.A., Pirri, A.N.: The fluid mechanics of pulsed laser propulsion. AIAA J. **18**, 835–842 (1977)
3. Harrland, A., et al.: Hypersonic inlet for a laser powered propulsion system. AIP Conf. Proc. **1402**, 145–157 (2011). https://doi.org/10.1063/1.3657023
4. Tichschenko, V.N., et al.: Gas-dynamic effects of interaction of optical discharge with a gas. J. Quant. Electr. (Rus.). **37**, 1–6 (2007)
5. Apollonov, V.V., Tischenko, V.N.: Laser propulsion on the basis of the effect of merching shock waves. J. Quant. Electr. (Rus.). **36**, 673–683 (2006)
6. Myrabo, L.N., Raizer, Yu.P.: Laser-induced air spike for advanced transatmospheric vehicles. AIAA Pap. 94-2451 (1994)
7. Rezunkov Yu, A., Schmid, A.A.: Supersonic Laser Propulsion. Appl. Opt. **53**, 155–165 (2014). https://doi.org/10.1364/AO.53.000I55
8. Langener, T., Myrabo, L., Rusak, Z.: Inlet aerodynamics and ram drag of laser-propelled lightcraft vehicles. AIAA Pap. 2009–4806 (2009). https://doi.org/10.1063/1.3435458
9. Richard, J.C., Myrabo, L.N.: Analysis of laser-generated impulse in an air-breathing pulsed detonation engine. Part 1. Beam. Ener.Prop. **766**, 265–278 (2004). https://doi.org/10.1063/1.1925150

10. Richard, J.C., Myrabo, L.N.: Analysis of laser-generated impulse in an airbreathing pulsed detonation engine. Part 2. AIP Conf. Proc. **766**, 279–291 (2004). https://doi.org/10.1063/1.1925150
11. Libeau, M.A., et al.: Combined theoretical and experimental flight dynamics investigations of a laser-propelled vehicle. AIP Conf. Proc. **664**, 125–137 (2002). https://doi.org/10.2514/6.2002-3781
12. Trizzini, F., et al.: Design of a Lightcraft model for experimentation on laser propulsion. In: Proceedings of COBEM 2011, 21st Brazilian Congress of Mechanical Engineering October 24–28, Natal, RN, Brazil (2011)
13. Katsurayama, H., Komurasakij, K., Arakawa, Y.: Numerical analyses on pressure wave propagation in repetitive pulse laser propulsion. AIAA-2001-3665 (2001). https://doi.org/10.2514/6.2001-3665
14. de Oliveira, A.C., et al.: Laboratory facilities and measurement techniques for beamed-energy propulsion experiments in Brazil. AIP Conf. Proc. **1402**, 31–46 (2011). https://doi.org/10.1063/1.3657014
15. Kenoyer, D.A., Salvador, I.I., Myrabo, L.N.: Axial impulse Generation of Lightcraft Engines with ~1 μs Pulsed TEA CO_2 laser. AIP Conf. Proc. **1402**, 82–92 (2011). https://doi.org/10.1063/1.3657018
16. Kenoyer, D.A., Salvador, I.I., Myrabo, L.N.: Beam-riding behavior of Lightcraft Engines with ~1 μs Pulsed TEA CO_2 laser. AIP Conf. Proc. **1402**, 93–105 (2011). https://doi.org/10.1063/1.3657019
17. Kenoyer, D.A., et al.: Flow Visualization of Thrust-Vectoring Lightcraft Engines with ~ 1μs Pulsed TEA CO_2 Laser. AIP Conf. Proc. **1402**, 106–114 (2011). https://doi.org/10.1063/1.3657020
18. Salvador, I.I., et al.: 2-D airbreathing lightcraft engine experiments in quiescent conditions. AIP Conf. Proc. **1402**, 158–173 (2011). https://doi.org/10.1063/1.3657024
19. Salvador, I.I., et al.: 2-D Air-breathing lightcraft engine experiments in hypersonic conditions. AIP Conf. Proc. **1402**, 174–186 (2011). https://doi.org/10.1063/1.3657025
20. Harrland, A.: Hypersonic inlet for a laser powered propulsion system. Master of Philosophy (Engineering). School of Engineering, Computer and Mathematical Sciences. The University of Adelaide 160 (2012). https://doi.org/10.1063/1.3657023
21. Tischenko, V.N.: Generation of a low-frequency wave by optical discharge moving in a gas with a subsonic velocity. J. Quan. Electr. (Rus.). **33**(1–8) (2003)
22. Zeldovich, Ya.B., Raizer, Yu.P.: Physics of shock waves and of high-temperature gas-dynamic phenomena. Physmatgiz, Moscow (1963)
23. Korotaev, T.A., Foment, V.M., Yakovlev, V.I.: Regimes of laser power deposition in gas flow. Proc. NGY. Ser.: Phys. **2**, 19–35 (2007)
24. Grachov, G.N., et al.: Stationary forces developed by optical pulsating discharge as applied to laser propulsion. J. Quant. Electr. (Rus.). **37**, 669–673 (2007)
25. Ghosh, S., Mahesh, K.: Numerical simulation of the gas-dynamic effects of laser energy deposition in air. J. Flu. Mech. **605**, 329–354 (2008). https://doi.org/10.2514/6.2017-1009
26. Lunev, V.V.: Hypersonic gas-dynamics. Mashinostroenie, Moscow (1975)
27. Tynibekov, A.K.: Gas jet in a supersonic flow. Herald KRSU. **8**, 128–133 (2008)
28. Beketova, A.O., Naimanova, A.G.: Numerical study of supersonic gas flow with lateral gas jet. Appl. Mech. Theor. Phys. **45**, 72–80 (2004). https://doi.org/10.1007/BF02466865
29. Phipps, C.N.: Review laser-ablation propulsion. J. Prop. Pow. **26**, 609–637 (2010)
30. Raizer, Y.P.: Physics of gas discharge. Nauka, Moscow (1987)
31. Phipps, C.R., Reilly, J.R., Campbell, J.M.: Optimum parameters for laser launching objects into low Earth orbits. Las. Part. Beam. **18**, 661–695 (2000). https://doi.org/10.1017/S0263034600184101

32. Smirnov, E.M., Zaitsev, D.K.: Method of finite volumes as applied to gas-hydro-dynamics and heat exchange in regions of complex geometry. Sci. Techn. Bull. (Rus.) (2004)
33. Roe, P.L.: Characteristic-based schemes for the euler equations. Ann. Review Flu. Mech. **18**, 337 (1986). https://doi.org/10.1146/annurev.fl.18.010186.002005
34. Rezunkov, Y.A., Schmidt, A.A.: Development of laser propulsion at supersonic modes. J. Tech. Phys. (Rus.). **83**, 33–41 (2013)
35. Schmidt, A.A., Rezunkov, Yu.A.: Nonstationary interaction of incoming flow with ablative jet in supersonic laser propulsion. In: 30-th International Symposium on Shock Waves, vol. 1, pp. 527–529 (2017). https://doi.org/10.1007/978-3-319-46213-4_90

Chapter 6
Space Mini-vehicles with Laser Propulsion

Abstract Application of high-power laser propulsion to explore a near-Earth space is discussed for a long time, starting from the pioneer investigations by Arthur Kantrowitz (1972) and Prof. A.M. Prokhorov (1976). A number of theoretical and experimental works on the laser propulsion were carried out in Russia, the USA, Germany, Japan, China, Brazil, and Australia since this time. Various models of spacecrafts with a laser propulsion were proposed with all this going on. One of these vehicles is the Lightcraft Technology Demonstrator developed by Prof. Leik Myrabo to be applied to launching of satellites into near-Earth space. In the chapter, we present a space mini-vehicle with laser propulsion produced by the aerospace laser-propulsion engine. The mini-vehicle optical system is designed to satisfy the principal conditions defining the vehicle application in a space. Particularly, one of these conditions is an independence of the vehicle orbital maneuver on a mutual orientation of the vehicle and laser power beam. This is achieved by using auxiliary onboard optical units such as receiver telescope, optical turret and hinges.

In the chapter, the applications of the mini-vehicle for removing space debris from the low and geostationary Earth orbits are considered.

Keywords Satellite clusters · Lightsat vehicle · Space debris · Electrodynamics tethers · Space-tug · Trash-collector · Orbital maneuver · Circular orbits · Hohmann transfer orbit · Geocentric equatorial coordinate system · Propellant mass consumption · Specific impulse · Thrust · Propulsion efficiency · Space mini-vehicle · Receiver telescope · Optical turret · Optical switch · Electric-mechanical deflector

6.1 Introduction to the Problem

Space mini-vehicles (SMV) possess some operational advantages when compared with the large and heavy spacecraft, namely [1]:

Y. A. Rezunkov, *High Power Laser Propulsion*, Springer Series on Atomic, Optical, and Plasma Physics 116, https://doi.org/10.1007/978-3-030-79693-8_6

- cost-performance operation on low Earth orbits of 200–1000 km altitudes, where heavy vehicles cannot survive for long without special techniques of the orbit correction
- reduction of expenses and risks by using special technologies at designing SMV
- fast renovation of the SMV design to solve a wide range of space tasks as applied to specific applications
- reduced intrinsic gravitational, electrical, and gaseous backgrounds, decreasing the influence of the satellite's platform on operation capability of target electronic equipment
- short-term period of development of the special SMV devices, demanding only 3–5 years from start point up to flight tests
- reduced cost of SMV operation.

Operation experience of hundreds of SMV confirms its high operational efficiency both for the civil and military applications [2]. SMV allows solving single-application tasks at least as well as large space vehicles. Moreover, it should be noted that SMV can be exploited in the form of a satellite-inspector to monitor large space vehicles after its long-term exploitation period.

In accordance with the hands-on experience of the space vehicle classification [2], we separate the SMV as follows: (a) mini-vehicles with a mass of (100–500) kg, (b) micro-vehicles with a mass of (10–100) kg, and (c) pico-vehicles with a (0.1–1.0) kg mass. At the present time, many advanced nations take an active part in the development of "cluster constellation" on the basis of micro- and nano-vehicles with distributed operation goals. Hence, main advantages of the clusters are assumed to be:

- global monitoring of the Earth surface
- high operability from the orbital vehicle cluster
- SMV series manufacturing, and so on

Although the development of the new class of space vehicles such as SMV is a positive achievement of new space exploration capabilities, it is necessary to meet some specific requirements as the operation continuity and global monitoring of the Earth ground at developing SMV devices. Moreover, if we assume orbital satellites cluster to be on the basis of SMV then every vehicle has to fulfill certain operational functions within the frames of a common program. Such a type of the satellite cluster can operate in the form of both federated and individual space vehicles [3, 4].

These satellite clusters can serve as the communication or radio navigation systems which provide:

- data acquisition by using automatic ground equipment to detect atmosphere characteristics, technological states of oil pipe line, electrical transmission facilities, and so on
- banking corporate networks to control distributed production complexes and aircraft flights
- global computer networks
- detection of geographic coordinates of ground movable vehicles, and so on

Moreover, the low orbit satellite cluster with a lot of satellites possesses an increased survivability because destruction of one or a few of the satellites does not affect its operability. The satellite cluster can be exploited for development of mobile telephone communication lines for the nations and areas where the ground telecommunication networks haven't been developed. A two-way exchange by phone communication can be achieved between vehicle drivers, ships, and rescue services by using these mini-vehicles systems. These advantages of the low Earth orbit satellite clusters as applied to these applications listed are especially important in Russia with its long distances between the residential provinces with poorly developed telecommunication networks. Estimates show that to achieve this goal, the satellite cluster has to include about 200 space mini-vehicles.

To achieve an effective development of state-of-the-art space communication networks, new technologies for designing the vehicles have to be developed. For this case, every vehicle in the satellite cluster must have an optimized payload and satellite number in the cluster, depending on the key problems to be solved. Optimization of the cluster total mass means that every vehicle is developed by using appropriate technologies for thrust production both for launching and orbital maneuver of the vehicle.

Satellites are usually launched by using carrier rockets of various types. Nevertheless, aerospace systems are being considered for launching satellites, including special carriers being based on exotic launch systems. For example, such Russian rocket systems as "Calm-21" or "Iskander" are among them. The US carrier-rocket named "Pegasus" is used for atmospheric launching of satellites made by DAPRA and NASA. The point is that the aerospace systems possess certain advantages as compared with ground-based rocket launchers [5].

Thus, high-power laser propulsion is assumed to be required both for launching of vehicles into low Earth orbits and for orbital maneuvers of mini-vehicles. One of the examples of such type of vehicles is the Lightcraft vehicle that has been considered in previous chapters in detail. Lightcraft is a laser-propelled single-stage vehicle without the stages which must be detached and returned to Earth.

To arrange space missions by using the HPLP launchers, following basic problems must be solved:

- development of space vehicles of specific design, including an onboard optical receiver system to collect a laser radiation from a remote laser and to produce a thrust by using an onboard laser-propulsion engine
- thrust has to be produced independently on a mutual spatial orientation of the vehicle and laser
- precise illumination of the vehicle by a remote laser

The first step in this direction was made by Prof. Leik Myrabo [6–8], considering the Lightsat system developed within the SDIO Laser-Propulsion Program. The Lightsat system focused on the concept of a Lightcraft drone boosted by a 100 MW-class laser. The Lightsat orbital cluster had to monitor the Earth atmosphere and to serve as the radio and optical booster converters.

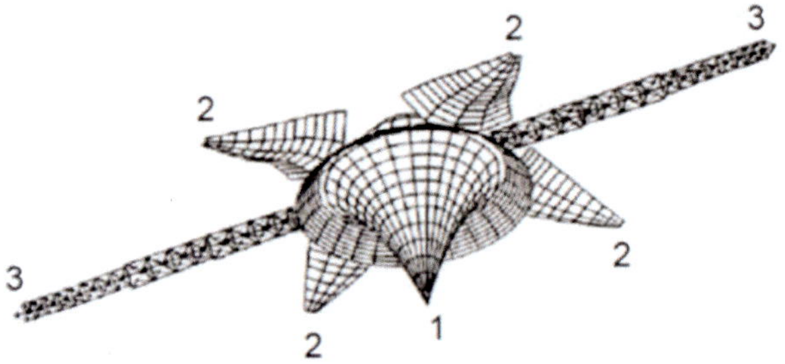

Fig. 6.1 Conceptual design of the Lightsat vehicle: (1) optical beam concentrator, (2) solar cell batteries, (3) radio antennas arranged on astro-masts

Every Lightsat vehicle (Fig. 6.1) was assumed to include an onboard computer controlling orbit location of the vehicle, orientation detectors, and radio booster convertors transferring space data to the Earth, solar antennas collecting a solar power and feeding onboard electronic equipment.

But, one of the technical issues of the vehicle was onboard gaseous thrusters which have to control its orbits, limiting space lifetime of the vehicle because of restricted mass of stored onboard propellant.

It is evident that application of HPLP system to support SMV space missions will be attractive if the vehicle motion in orbits will not be dependent on a mutual orientation of high-power laser and the vehicle. To fulfill this condition, aerospace laser-propulsion engine is developed as the engine integrated with the onboard optical system of specific design (see Chap. 4). Here, we consider the basic principles of designing the onboard laser-propulsion system that can be developed by using present-day optical technologies.

One possible use of laser-propelled SMV is considered to be space debris removal away both from low Earth orbits and from geostationary one [9], including the objects of large mass. One should mention that a number of projects on removal of the debris are under development at the present time [10]. The projects are usually based on both passive and active engineering techniques. As the examples of the passive techniques, we can mention the following:

(a) electrodynamics tethers with a length of a few hundred meters, which are attached to debris, which allows removal of the object back to the Earth due to interaction of the tether with the Earth's electric-magnetic field
(b) solar sails which are unfolded on the object, allowing removal of some debris objects far from near-Earth orbits

The use of space-tugs which are used to dock large space objects and transfer it into storage or burial orbits is an example of active engineering tools to remove space debris [10]. The tugs or trash-collectors are usually considered as a part of the space transportation system that has to be developed specially. It is estimated that the use of one trash-collector will allow removal of 750 objects from GEO to the Earth in the course of 10 years of its operation. Taking into account the fact that about 1500 used objects are in GEO, about 20 years will be needed to make the GEO free of these objects by using only one trash-collector.

The trash-collector system has a great attraction used to deal with the space debris problem, but more specific requirements are needed to design the collector system cope with the debris problem. For example, to remove non-operational satellites from GEO, the trash-collector must approach every satellite at a short distance to capture it, to dock the satellite and return to its operational orbit. In this case, every rendezvous will require a large mass of propellant increasing the inertial mass of the tug. In this regard, MDA Canadian Company has developed a Space Infrastructure Servicing station arranged in GEO, which could allow fueling the trash-collector if necessary.

In our opinion, space mini-vehicle with laser propulsion, which is prearranged on a trash-collector board and is used to capture and to dock the used satellites on the trash-collector board, is one of the perspective ways of solving the GEO debris problem [11]. In this case, the laser that propels the SMV would be arranged on the trash-collector board too. In this chapter, we consider the design and application of space mini-vehicles with onboard laser-propulsion system as a part of the space transportation system that would be used to remove the space debris both from low Earth orbits and from GEO.

6.2 Scenario of the SMV Orbital Maneuvers

It is known that a lifetime of a low-orbital satellite is limited by a variety of reasons [12]; for example, a satellite leaves its orbit because of non-homogeneity of the Earth's gravitational field as well as aerodynamic drag of the upper atmosphere, and others reasons. To prevent the satellite leaving its orbit, it is necessary to carry out correction of the vehicle's orbit altitude at a precisely time. Hence, the altitude correction must be done at a minimal consumption of onboard fuel (propellant).

To bring the satellite back to its operational orbit, its orbital maneuver in a near-Earth space is usually performed through intermediate transfer orbits, namely (see Fig. 6.2) [13]: (a) spiral transfer trajectory that is achieved by a maneuver under a small thrust or (b) Hohmann transfer corresponding the satellite motion along an elliptical trajectory under a large thrust. Despite the fact that both transfer orbits are used in practice of space missions, it is necessary to take into account some peculiar properties of SMV with laser propulsion when considering the satellite orbit correction.

Both of these transfer orbits possess the individual peculiar properties, and in the regime of a small thrust, we have the following:

- Orbit correction period takes a long temporal period so that the satellite executes a number of orbits around the Earth especially near the low orbits.
- The later means that laser-propulsion system has to operate all of the time which the satellite is maneuvering, resulting in a continuous motion of the satellite by a laser.

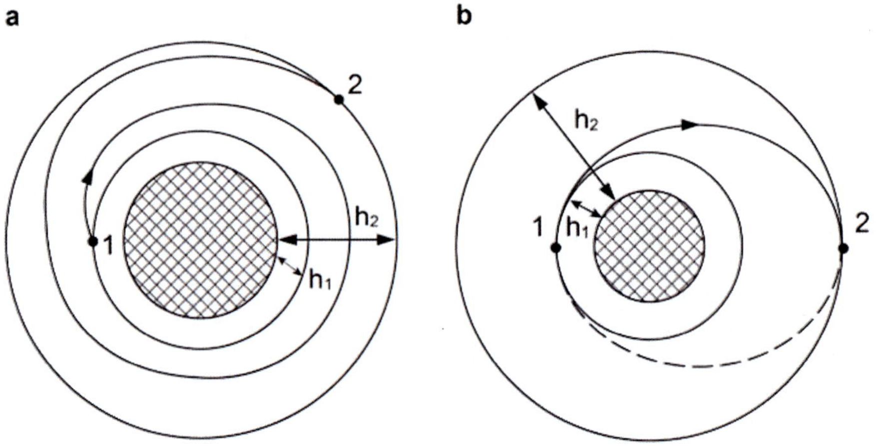

Fig. 6.2 Transfer trajectories of satellites at the mode of (**a**) small thrust or (**b**) large thrust; h_1 and h_2 are the altitudes of the satellite initial and operational orbits correspondently, 1 and 2 are the satellite initial and final positions

In the regime of a large thrust, that is in the Hohmann transfer orbit, we have the following:

- The satellite orbit correction is achieved in a short time due to the satellite motion by the shortest space trajectory due to a high increment of the satellite initial velocity.
- Only two short periods of switching the laser propulsion are required depending on the velocity increment required and its decrement which have to be imparted to the satellite.

So, the low Earth orbit satellite trajectory correction is considered by taking into account these modes of thrust production. To simplify the general statement of the consideration, the satellite is assumed to be as a material point with a mass of M_s, moving under the action of two forces, namely, Earth's gravitational force and drag force of the Earth's upper atmosphere [13, 14]. It is also assumed that the Earth's shape is spherical.

In accordance with a geocentric equatorial coordinate system, OXYZ, the origin of these coordinates coincides with the center of the Earth, that is M, and Z-axis coincides with the Earth axis of rotation. Hence, the X-axis is directed to a point of the vernal equinox, and Y-axis completes the coordinate system up to a right coordinate system (see Ref. [15]).

The satellite orbiting takes place under the F_z gravitational force that can be expressed by following formula for a mass point:

$$\vec{F}_z = -\gamma \frac{M_E \cdot m_s}{r^2} \vec{r}_0, \quad \vec{r}_0 = \frac{\vec{r}}{r} \tag{6.1}$$

where γ is the gravitation constant, 6.67 × 10^{-11} $m^3/(kg \times s^2)$, M_E is the Earth mass that is of 59,736 × 10^{24} kg, m_s is the satellite mass, r_o is a unit vector, $\vec{r}$ is a radius-vector originated from the center of the Earth up to the satellite, and r is a radius from the center of the Earth to the satellite.

The system of motion equations describing relative orbiting of the two mass points, M and m, is determined in accordance with the Newton's second law in a common form as follows:

$$\begin{aligned} \frac{d^2\vec{p}_1}{dt^2} &= -\gamma\frac{M}{r^2}\vec{r}_0, \\ \frac{d^2\vec{p}_2}{dt^2} &= \gamma\frac{m_s}{r^2}\vec{r}_0, \\ \vec{p}_1 &= \vec{p}_2 + r\cdot\vec{r}_0 \end{aligned} \tag{6.2}$$

where $\vec{p}_1$ is the radius-vector fixed from the inertial coordinates origin to the point m, $\vec{p}_2$ is the radius-vector fixed from the inertial coordinates origin into the point M (see Fig. 6.3), and t is a time.

The motion equation for the mass point m in respect of attractive center M is deduced in the inertial coordinate system by using the system (6.2) and by deducting the second equation from the first one, namely:

$$\frac{d^2\vec{r}}{dt^2} = -\gamma\frac{M_E + m_s}{r^2}\vec{r}_0 \tag{6.3}$$

Since $m_s << M_E$, then we can neglect the acceleration that is imparted to the Earth by the satellite, and we can arrange the inertia coordinate system origin in the center of the Earth M. Then, we have $\vec{p}_1 = \vec{r}, \vec{p}_2 = 0$, and the equation of satellite orbital

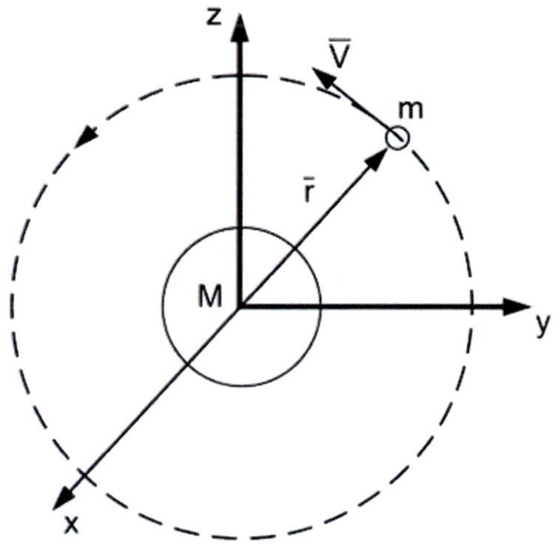

Fig. 6.3 Satellite (m) orbiting in a geocentric equatorial coordinate system, OXYZ

motion with respect to the gravitational center, M taking account of the disturbing forces that may be written in the following form:

$$\frac{d^2\vec{r}}{dt^2} = -\frac{\mu}{r^2}\vec{r}_0 + \frac{\vec{F}}{m} \tag{6.4}$$

where $\mu = \gamma M$ is the Earth gravitation constant, and $\vec{F}$ is a total vector of the disturbing forces.

The vector Eq. (6.4) can be written in the X, Y, and Z projections as follows:

$$\begin{aligned} \frac{d^2x}{dt^2} &= -\frac{\mu}{r^3}x + \frac{\vec{F}_{\mathrm{X}}}{m} \quad \text{or} \quad \frac{d^2x}{dt^2} = -\frac{\mu}{r^3}x + a_{\mathrm{X}}, \\ \frac{d^2y}{dt^2} &= -\frac{\mu}{r^3}y + \frac{\vec{F}_{\mathrm{y}}}{m} \quad \text{or} \quad \frac{d^2y}{dt^2} = -\frac{\mu}{r^3}y + a_{\mathrm{y}}, \\ \frac{d^2z}{dt^2} &= -\frac{\mu}{r^3}z + \frac{\vec{F}_{\mathrm{Z}}}{m} \quad \text{or} \quad \frac{d^2z}{dt^2} = -\frac{\mu}{r^3}z + a_{\mathrm{Z}} \end{aligned} \tag{6.5}$$

where a_{x}, a_{y}, a_{z} are the components of disturbing acceleration.

The force of aerodynamic drag affecting the satellite in low Earth orbit, R_{x}, can be expressed in the following form:

$$R_{\mathrm{x}} = C_{\mathrm{x}}\frac{\rho(h)\cdot V^2}{2}S_{\mathrm{m}} \tag{6.6}$$

where $\rho(h) = \rho_0 \times \exp\,(-0.023758 \times h[\mathrm{km}] - 17.2332627)$ is the atmosphere density at the altitude of $h = (200\text{–}300\ \mathrm{km})$, and C_{x} is the aerodynamic drag coefficient of the atmosphere. The drag force produces a disturbance acceleration that is directed tangentially to the satellite orbit. The X, Y, and Z projections of the acceleration a_{X}, a_{B}, can be determined from the following.

The acceleration vector that is $\vec{a}$ is opposite to the satellite's velocity vector, namely:

$$\vec{a} = -C_{\mathrm{x}}\frac{\rho(h)}{2m}S_{\mathrm{m}}|V|^2\vec{e} \tag{6.7}$$

where $\vec{e} = \left[V_{\mathrm{x}}/|V|, V_{\mathrm{y}}/|V|, V_{\mathrm{z}}/|V|\right], |V| = \sqrt{V_{\mathrm{x}}^2 + V_{\mathrm{y}}^2 + V_{\mathrm{z}}^2}$ is a unit vector of the acceleration, and S_{m} is the area of the satellite cross-section. And consequently:

$$a_{\mathrm{X}} = -C_{\mathrm{X}}\frac{\rho|V|}{2m}S_{\mathrm{m}}\cdot V_{\mathrm{x}},\ \ a_{\mathrm{y}} = -C_{\mathrm{X}}\frac{\rho|V|}{2m}S_{\mathrm{m}}\cdot V_{\mathrm{y}},\ \ a_{\mathrm{Z}} = -C_{x}\frac{\rho|V|}{2m}S_{\mathrm{m}}\cdot V_{\mathrm{z}} \tag{6.8}$$

Substituting the expressions (6.8) into (6.5), we finally obtain the system of equations which describe the satellite orbiting under the Earth gravitation field and under the action of the Earth atmosphere drag, including under thrust produced by a rocket engine [13, 14]:

$$\begin{aligned}\frac{d^2x}{dt^2} &= -\frac{\mu}{r^3}x - C_x\frac{\rho|V|}{2m}S_m \cdot V_x + \delta(t) \cdot a_{xt},\\ \frac{d^2y}{dt^2} &= -\frac{\mu}{r^3}y - C_x\frac{\rho|V|}{2m}S_m \cdot V_y + \delta(t) \cdot a_{yt},\\ \frac{d^2z}{dt^2} &= -\frac{\mu}{r^3}z - C_x\frac{\rho|V|}{2m}S_m \cdot V_z + \delta(t) \cdot a_{zt}\end{aligned} \tag{6.9}$$

where $\rho = \rho(h)$, a_{xt}, a_{yt}, a_{zt} are the components of the satellite acceleration on X, Y, and Z axis, and $\delta(t)$ is a function of switching on laser-propulsion engine:

$$\delta(t) = \begin{vmatrix} 1, t_0 = t_1 = 0 \leq t \leq t_2 \\ 0, t_2 \leq t \leq t_3 \\ 1, t_3 \leq t \leq t_4 \\ 0, t_4 \leq t \leq \infty \end{vmatrix} \tag{6.10}$$

Here, $t_0 = 0$ is first moment of switching on the propulsion, [t_1, t_2] and [t_3, t_4] are temporal intervals during which the thrust is produced.

The system of Eq. (6.9) is applied to determine the basic characteristics of the satellites motion. For example, some parameters of the equation's solution are listed in Table 6.1, which were obtained for a satellite with a mass of 100 kg, orbiting at low Earth orbit with an altitude of 200 km or at operational orbit with an altitude of 300 km. It was also assumed that the atmospheric drag coefficient C_x is equal to 0.1, and $S_m = 1.0$ m^2, $\delta(t) = 0$. In Table 6.1, parameter t_i illustrates the period during of which the satellite stays at every orbit.

Some details of the satellite orbital flights are important ones for considering the satellite orbit correction. By definition, the satellite orbital velocity corresponding a 280 km altitude is equal to $V = 7740$ m/s, including $V_x = -7721$ m/s, $V_y = -542.7$ m/s, and $V_z = 0$, respectively, as X, Y, and Z components of the velocity. The velocity characteristics corresponding a 200 km orbit will be $V = 7786.5$ m/s, when $V_x = 3556.38$ m/s, $V_y = -6926.88$ m/s, $V_z = 0$, including the radial and tangential components of the velocity equal to $V_r = 0.0296$ m/s and $V_n = 7786.5$ m/s. By comparing the velocity parameters with similar ones listed in Table 6.1, it is seen that the satellite's velocity on both orbits correspond closely to a

Table 6.1 Orbital parameters of the satellite orbital motion

h_i, km	R_i, km	V_i, km/s	T_i, h	n_i	t_i, h
200	6578.39	7.79	1.47	6.51	9.6
280	6658.39	7.75	1.50	–	–
300	6678.39	7.73	1.51	318.3	480

Here, n_i is a general number of orbits passed by the satellite during its life in orbit circular velocity for every orbit. But nevertheless, every velocity has a radial component! And the angle between the velocity vector and tangential component of the velocity is equal to 2.14×10^{-6} rad for a 280 km orbit and to 3.8×10^{-6} rad for a 200 km orbit.

Consequently, we can propose the following scenario of satellite orbit maneuver to perform corrections to a satellite's orbit by using a Hohmann trajectory. When a satellite reaches orbit with a minimal attained altitude, the laser is switched on a time period of t_1 to produce a thrust by using a laser-propulsion engine. During this period, the satellite velocity increases by a ΔV_1 increment, and it passes into an elliptical orbit. Then the laser is switched off, and the satellite keeps its motion of the elliptical trajectory in a passive regime up to the moment when it reaches a 300 km orbit. At that moment of time, the laser is switched on once again to produce a thrust during a t_2 time period, during which the satellite's velocity is decreased by a ΔV_2 decrement that is required to hold the satellite in this operation orbit. It is assumed that the velocity magnitude and its vector will reach the 300 km orbit parameters in a t_2 time period. The correction procedure can be repeated when the satellite needs to come down again to the 200 km or 280 km orbits again.

Figure 6.4 illustrates the Hohmann orbit, transferring the satellite between two circular orbits. The transfer orbit is calculated in accordance with some preliminary assumptions, including:

- initial and final orbits are circular
- the planes of the initial, transfer, and final orbits coincide with the Earth's equatorial plane, that means the orbits inclination angle is equal to zero
- thrust vector lies in the satellite orbit plane

Numerical results of calculations of the transfer elliptical orbit parameters for two initial satellite orbits are listed in Table 6.2.

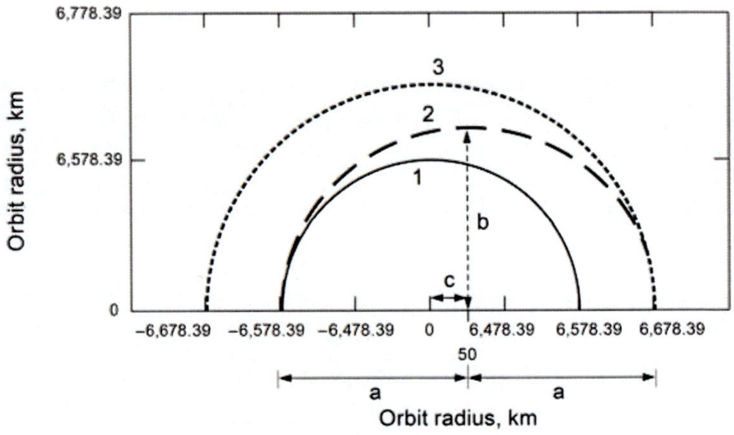

Fig. 6.4 Elliptical transfer trajectory (2) of SMV

Table 6.2 Transfer elliptical orbit parameters for two initial SMV orbits

Initial orbit altitude	*a*, km	*b*, km	*c*, km	*p*, km	*e*, eccentricity
280 km	6668.39	6668.25	10	6668.12	0.0063997
200 km	6628.39	6628.20	50	6628.01	0.0075433

Here, *p* is a focal parameter

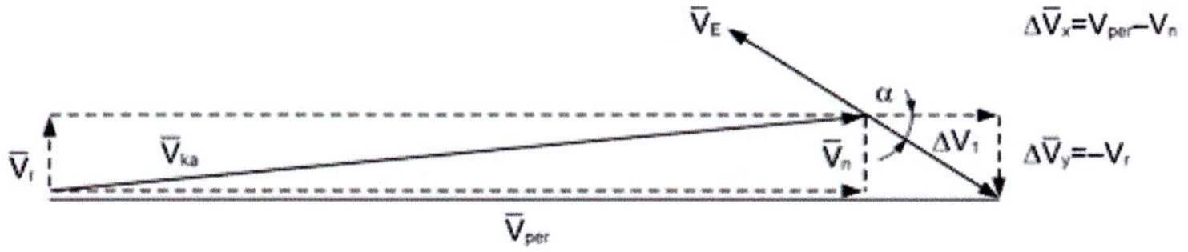

Fig. 6.5 Vector components of the satellite's velocity at a perigee point of its transfer orbit

In the figure, starting from orbit with 200 km or 280 km (1) altitudes, (3) is operational orbit of the satellite. The a, b, c are large, small semi-axis and focal distance of the elliptical orbit. The Earth's geometric center is located in (0, 0) coordinate point.

The satellite's velocities at the apogee and perigee points of the orbit are equal to:

$$V_{\text{per}} = \sqrt{\frac{G \cdot M_3}{a} \cdot \frac{1+e}{1-e}}, \quad V_{\text{apo}} = \sqrt{\frac{G \cdot M}{a} \cdot \frac{1-e}{1+e}} \tag{6.11}$$

where $G = 6.673 \times 10^{-11}$ m^3/(kg $\times$ s^2), $M_{\text{E}} = 5.977 \times 10^{24}$ kg.

The satellite's velocity vectors in the perigee point are shown in Fig. 6.5. Similar vector components in the apogee point take place.

Here, V_{s}, V_{r}, V_{n} are vectors of the satellite's velocity and the velocity radial and tangential components in respect of the circular orbit; V_{per}, V_{apo} are the satellite's velocity vectors in the perigee and apogee points, respectively, and V_{E} is an exhaust jet velocity vector, α is an angle between the V_{E} and V_{n} vectors.

Taking into account the vector diagrams (Fig. 6.5), we can write:

$$\Delta V_{\text{x}} = V_{1n} = V_{\text{per}} - V_{\text{n}}, \quad \Delta V_{\text{y}} = V_{1r} \tag{6.12}$$

and the velocity increment in a perigee as well as the velocity vector angle α will be:

$$\Delta V_1 = \sqrt{V_{1n}^2 + V_{1r}^2}, \text{ and } \alpha = \text{arctg}\left(\frac{V_{1r}}{V_{1n}}\right) \tag{6.13}$$

Radial $\Delta V_{2\text{r}}$ and tangential $\Delta V_{2\text{n}}$ components of the satellite's velocity decrease, ΔV_2, at the 300 km orbit, where satellite reaches the orbit, can be calculated by assuming the velocity radial component is directed out from the center of the Earth, that is:

$$\Delta V_x = V_{2n} = V_1 - V_n, \ \ \Delta V_y = V_{2r} \tag{6.14}$$

Consequently, the decrease in the satellite's velocity that is needed to achieve the transfer orbit apogee, as well as the angle α between the velocity tangential component and the velocity decrement vector ΔV_2 are:

$$\Delta V_2 = \sqrt{V_{2n}^2 + V_{2r}^2}, \ \ \alpha = \mathrm{arctg}\left(\frac{V_{2r}}{V_{2n}}\right) \tag{6.15}$$

The results of numerical calculations of the system of Eq. (6.9), which are obtained with the use of (6.12)–(6.15) formula, are listed in Table 6.3.

Similar results on orbital transfer of the satellite with laser propulsion propelled by an airborne laser are listed in Table 6.4.

These data can be used to estimate the laser-propulsion characteristics which are to meet the listed parameters of the satellite's orbit. Hence, some auxiliary assumptions are attracted, which follows from the specific properties of the laser propulsion. It is assumed that the efficiency of laser propulsion is of 40% (see Chap. 2).

Table 6.3 Vector components of the satellite's velocity at the elliptical transfer trajectory

		V, m/s	V_n, V_r, m/s	ΔV, m/s	α, radn
Start orbit 200 km	Perigee	7815.81	$V_n = 7786.50$ $V_r = 0.0296$	$V_{1r} = -0.0296$ $V_{1n} = 29.308$ $\Delta V_1 = 29.308$	0.0010
	Apogee	7698.78	$V_n = 7698.76$ $V_r = 7.22$	$V_{2r} = -7.22$ $V_{2n} = 29.23$ $\Delta V_2 = 30.107$	0.2422
Start orbit 280 km	Perigee	7761.91	$V_n = 7740.056$ $V_r = 0.0166$	$V_{1r} = -0.01656$ $V_{1n} = 21.858$ $\Delta V_1 = 21.858$	7.6×10^{-4}
	Apogee	7663.20	$V_n = 7708.75$ $V_r = 0.00142$	$V_{2r} = -0.00142$ $V_{2n} = 19.234$ $\Delta V_2 = 19.234$	7×10^{-5}

Table 6.4 Time scenarios of the SMV orbit transfer

Transfer stages	Parameter	ΔV, m/s	Day's duration of the correction
200 km	Orbit decrease 200 km	–	158,684 days
	First correction	29.308	≤49.78 s
	Passive flight	–	2464.9 s
	Second correction	30.107	≤76.07 s
280 km	Orbit decrease 280 km	–	65 days
	First correction	21.858	≤69.546 s
	Passive flight	–	1537.6 s
	Second correction	19.234	≤76.07 s

Table 6.5 LPE characteristics to correct the SMV orbit altitudes

h, km	F_m, N	C_m, dyne/W	P, kW	η, %	I_{sp}, s	$\dot{m}$, g/s	m_{t1}, kg	t_1, s	m_{t2}, kg	t_2, s	m_Σ, kg
280	31.43	31.43	100	40	259.5	12.35	0.859	69.546	0.749	60.672	1.608
200	58.87	58.87			138.65	43.3	2.156	49.78	2.166	50.04	4.322

Here, h is initial orbit altitude of the satellite; $F_m = \frac{\Delta V_1 \cdot m_0}{t_{el}}$ is the thrust produced; C_m is the momentum coupling coefficient; P is laser power; I_{sp} is the specific impulse of the exhaust jet; $\dot{m}$ is the propellant's specific mass rate in the jet; m_{t1}, m_{t2}, m_Σ are the propellant masses required for every transfer stage as well as the total mass of propellant for one transferring cycle; $t_1, t_2 = \frac{\Delta V_2 \cdot (m_0 - m_{t1})}{F_m}$ are the LPE operation time periods, and $m_0 = 100$ kg is the satellite mass

There are the onboard laser-propulsion engines characteristics given in Table 6.5, which satisfy the orbital maneuvers of SMV.

It follows from Table 6.5, assuming a low efficiency of the laser propulsion leads to a small specific impulse of the exhaust jet (about 200 s only), and it results in a large consumption of the onboard propellant mass. To improve the situation with enhanced laser-propulsion efficiency, we propose using the CHO-polymers as the onboard propellant, efficiency of which was discussed in Chap. 3. For example, Polyphormaldegid (Polyoxymethylene) is considered to be the propellant candidate which possesses specific heat of combustion of $Q_{cm} = 18{,}787$ J/g as well as specific energy of evaporation of $Q^* = 2692$ J/g [16].

During the satellite orbit maneuver, the propellant mass consumptions and the satellite's velocity at different orbit points can be expressed as follows:

$$\begin{aligned} & m_{t1} = m_0 - \dot{m} \cdot t_1, \quad m_{t2} = m_{t1} - \dot{m} \cdot t_2 = m_0 - \dot{m} \cdot (t_1 + t_2), \quad \text{and} \\ & \Delta V_1 = -I_{sp} \cdot g \cdot \ln\left(\frac{m_{t1}}{m_0}\right), \quad \Delta V_2 = -I_{sp} \cdot g \cdot \ln\left(\frac{m_{t2}}{m_{t1}}\right) \end{aligned} \tag{6.16}$$

Substituting the first pair equations in to the second pair equations in (6.16) correspodingly, and then subtracting first new equation from secon new equation, we get the following equality:

$$\exp\left(\frac{\Delta V_1 - \Delta V_2}{I_{sp} \cdot g}\right) = \frac{m_0 \cdot (m_0 - \dot{m} \cdot (t_1 + t_2))}{(m_0 - \dot{m} \cdot t_1)^2} \tag{6.17}$$

After transformation, this expression has the following form:

$$\frac{A \cdot t_1^2}{m_0} \cdot \dot{m}^2 - (2 \cdot A \cdot t_1 - (t_1 + t_2)) \cdot \dot{m} + (A - 1) \cdot m_0 = 0 \tag{6.18}$$

Finally, the equation describing mass consumption of onboard propellant is:

$$f(\dot{m}) = a \cdot \dot{m}^2 - b \cdot \dot{m} + c = 0,$$

$$\text{where } a = A \cdot \frac{t_1^2}{m_0}, \quad b = 2 \cdot t_1 \cdot A - (t_1 + t_2), \quad c = m_0 \cdot (A - 1), \quad A = \exp\left(\frac{\Delta V_1 - \Delta V_2}{I_{sp} \cdot g}\right) \tag{6.19}$$

It is evident that Eq. (6.19) cannot have negative solutions, then the following relations take place if $a > 0$, $b > 0$ и $2a > b$:

$$\frac{1}{2 \cdot A - 1} \cdot t_2 < t_1 \leq 69.546 \text{ s} \tag{6.20}$$

At this assumption, we have the following dependencies of laser-propulsion characteristics and laser power on the specific impulse of a jet engine to achieve orbital correction of the SMV—see Fig. 6.6.

It can be seen from Fig. 6.6(a) that (6.20) inequality is satisfied if the specific impulse is varied from 100 s to 3000 s throughout the period of the second switching on the laser-propulsion engine, which is equal to $t_2 = 60$ s. Then the duration of the first switching on laser propulsion must not exceed a time limit arising from this inequality.

Now, we can examine the data obtained from increasing the laser-propulsion efficiency and using energetic propellants such as Delrin and similar ones. Using data from Chap. 3, the propulsion efficiency, η, of two energy propellants as functions of specific impulse are shown in Fig. 6.7.

One of the propellants is Delrin, and the second one is a hypothetic polymer (HP) with a 57,000 J/g specific energy of combustion, that exceeds the specific energy of Delrin by a factor of 3. As is seen from the figure, the specific impulse of thrust produced by Delrin cannot exceed 662 s because of the propulsion efficiency must not exceed unity by definition. But using the hypothetic polymer, the propulsion efficiency $\eta = 70\%$ at a specific impulse of 1000 s could be achieved!

Based on the obtained results, we can also deduce that the general requirements of the laser propulsion as applied to correction of the satellite orbit altitude being performed by a Hohmann trajectory—see Table 6.6.

It follows from the table that a laser power of 8–22 kW is required, varying depending on the altitude of the satellite's initial orbit and laser-propulsion engine characteristics, to correct the orbit of SMV with a 100 kg inertial mass. It should be noted that the specific impulse is considered as an independent parameter of the correction in this analysis. This follows because the specific impulse defines propellant mass consumption during the satellite's maneuver. At the same time, there are a few limitations on the efficiency of laser-propulsion production, being considered in previous chapters.

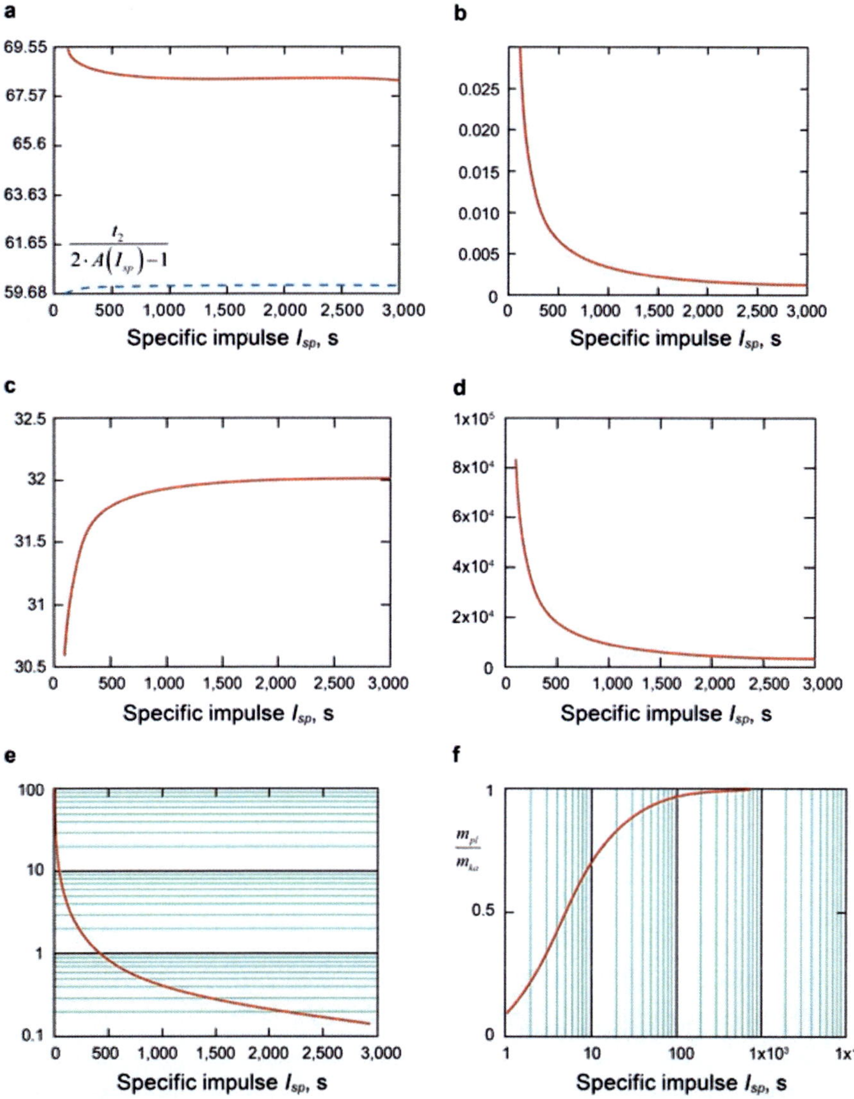

Fig. 6.6 (**a**) LPE operation time, (**b**) propellant mass consumption, (**c**) required laser-produced thrust, (**d**) required laser power, (**e**) total propellant storage mass for one transfer cycle, and (**f**) a part of SMV payload mass as a ratio of total satellite inertia mass

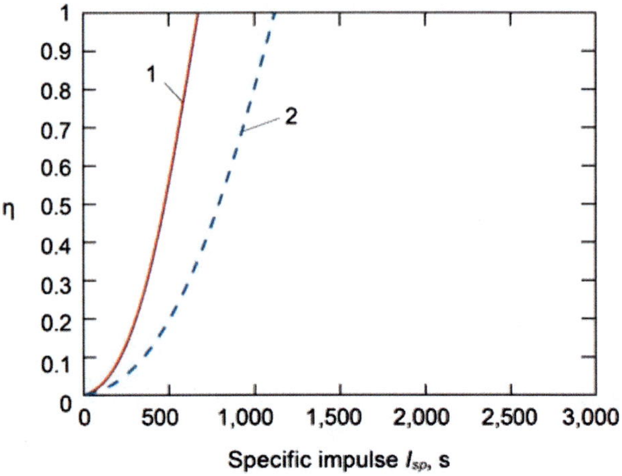

Fig. 6.7 LPE efficiency as functions of specific impulse for two types of energetic propellants. (1) Delrin, (2) hypothetic propellant (see Chap. 3)

Table 6.6 General requirements to laser propulsion characteristics to correct an orbit of SMV

	η, %	I_{sp}, s	C_m, dynes/W	F_t, N	$\dot{m}$, g/s	P, kW	m_{pl}, m	Q_{cm}, J/g	C_e, J/g
Start orbit 280 km	70	Delrin 598	217.8	31.82	5.4	14.6	0.993	18,787	15.1
		HP 1000	364.0	31.9	3.2	8.8	0.9958	57,000	9.1
Start orbit 200 km		Delrin 600	219.0	49.6	8.4	22.7	0.989	18,787	22.0
		HP 1000	364.0	49.8	5.1	13.7	0.994	57,000	13.1

m_{pl} is a mass of the vehicle payload; C_e is specific energy consumption for the satellite maneuver

6.3 Space Debris Removal Out of Geosynchronous Earth Orbit (GEO) by Using Laser-Propelled Space Mini-vehicles

Let us consider a removal scenario for defunct satellites which are arranged in GEO by using satellites with laser propulsion by the analogy with the previous section. We assume that a space-tug with the SMV and power laser onboard moves in an orbit altitude that is $L = 100$ km low than the GEO altitud (see Fig. 6.8) [15]. As we mention in Sect. 6.1, the concept of space-tugs used for docking big debris and transferring the objects into the storage orbits is considered by various countries. The tugs or trash-collectors are usually assumed to be a part of a space transportation system based on new propulsion technologies.

Then possible scenario of removal of the satellites out from the GEO may be as follows. When the space-tug moves to a rendezvous orbit and approaches to the GEO satellite at some distance of $(L + \Delta)$, SMV leaves the tug and starts moving by a

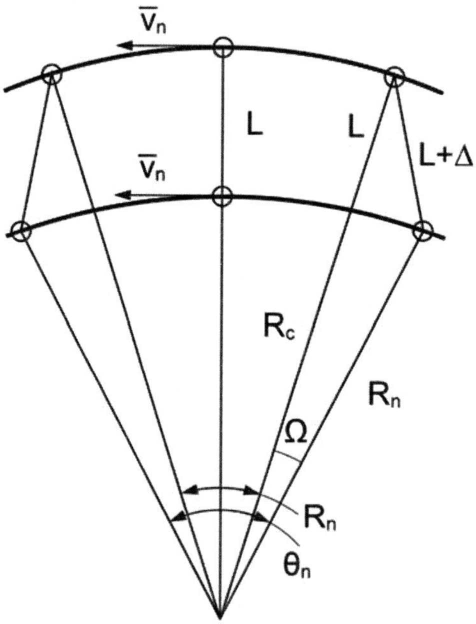

Fig. 6.8 Relative location of the defunct satellite and space-tug. Δ is the initial extra length between the satellite and space-tug; R_n and R_c are the orbit radii which are with respect to the center of the Earth; V_n and V_c are the satellites orbital velocities, θ_n and θ_c are the satellite angular positions varying during the tug mission

special orbit towards the satellite. At a short distance from the satellite, the SMV starts regulating its orbital velocity to catch the defined satellite. After catching this satellite, the SMV moves back to the space-tug with the satellite to load it on the tug. If necessary, the SMV can maneuver to another GEO satellite repeat the action. When the space-tug is fully loaded with satellites, the tug transfers the space debris to a storage orbit, and then it comes back to the rendezvous orbit to collect new debris objects abandoned on GEO.

To follow the scenario, the SMV space mission can be achieved by using:

- only the laser-propulsion technique with a laser arranged onboard the tug board
- a hybrid rocket engine system that includes both the laser-propulsion and electrojet engines.

In both cases, the SMV has to be provided with a receiver telescope designed to collect the laser power and to direct it to onboard laser-propulsion engine to produce a thrust.

Let us estimate the general characteristics of the SMS orbital maneuver in order to determine the requirements of the laser propulsion, including propellant mass consumption during the mission. We know that the GEO satellite's period of rotation around the Earth (T) is equal to one-star day, namely, 23 h 56 min and 4.06 s. Moreover, the satellites move in a circular orbit with a zero inclination angle of its orbit plane with respect to the Earth's equatorial plane. The GEO altitude over the Earth ground is of 35,786 km.

The initial distance $L + \Delta$ between the SMV and defunct satellite is determined by a maximum time period t during which the SMV has to reach the satellite, to capture

Table 6.7 "Space tug-satellite" contact time interval

Δ, km	1	2	3	4	5	10	14	16	20	100
t, h	0.72	1.02	1.25	1.45	1.63	2.33	2.78	2.99	3.37	8.78

it, and to transfer this satellite to a space-tug that is moving to a rendezvous orbit. This interval may be considered as a tug-satellite contact period. In Table 6.7, the contact period is presented by using the following formula [15]:

$$t(\Delta) = 2 \times a\cos\left(\frac{R_c^2 + R_n^2 - (L+\Delta)^2}{2R_cR_n}\right) / \left|\frac{V_n}{R_n} - \frac{V_c}{R_c}\right| \tag{6.21}$$

This time interval of the tug-satellite operation contact determines the initial requirements for the velocity and propellant mass consumption of the SMV orbit maneuver. Let us apply it to the geocentric equatorial coordinate system shown in Fig. 6.8 with the origin arranged in the center of the Earth. Here, OX-axis (horizontal one) is directed to the first point of Aries, OZ-axis is directed along the Earth's rotation angle, and Y-axis forms a right coordinate system and it is perpendicular to the OXZ-plane so that the equator and geostationary planes coincide with the OXZ-plane.

The choice of the optimum trajectory of the SMV orbit mission, as well as laser-propulsion characteristics, will be defined by a balance between the propellant mass consumption and the SMV maneuver interval. The propellant mass is assumed to be of 10% of the SMV total mass during a single mission. In this case, the SMV orbital flight to the GEO satellite and back to the space-tug is assumed to be fulfilled by an elliptical orbit allowing a saving in the use of onboard propellant [17]. But, a minimal mission period between coplanar orbits is provided if the tug and GEO orbits are close to each other when the propellant consumption will be moderate in mass.

To transfer the SMV from a circular orbit to an elliptical one, it is necessary to impart a velocity V_e to the SMV, which is directed at an angle δ with respect to the satellite's circular orbit at a certain point of the orbit (see Fig. 6.9). Hence, it is necessary to increase the vehicle orbital velocity, V_k, by a vector increment, ΔV, which is directed in respect of vector V_k under an angle β.

The SMV intercept trajectory lasts until the moment of time when its elliptical orbit crosses the GEO satellite orbit. Then, the SMV approaches the GEO satellite at a short distance that can be considered as a distance of capturing.

Detailed results of consideration of the SMV maneuver to a GEO satellite and back to the space-tug are listed in Tables 6.8 and 6.9 for various initial increments of the SMV velocity, ΔV. The time interval required the SMV onboard engine to operate under the laser power is determined by the following formula [13]:

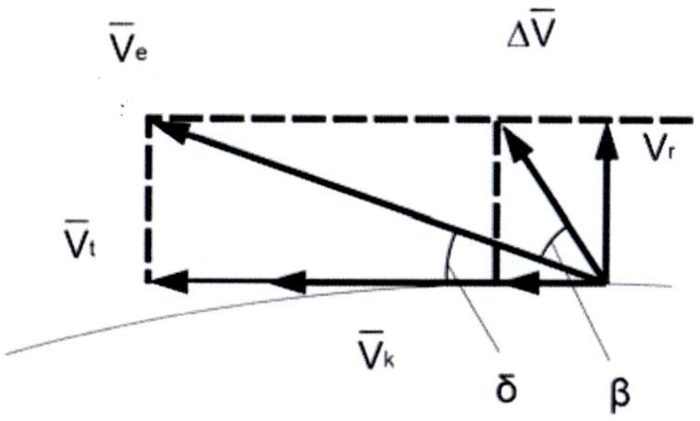

Fig. 6.9 Vector diagram of SMV transferring to elliptical orbit

$$V_e = \Delta V / \ln\left(\frac{m}{m - F_t/(I_{sp}g) \times t_d}\right) \tag{6.22}$$

where: $V_e = I_{sp} \times g$, $m = 50$ kg is the SMV mass, $g = \mu/R_n^{\ 2}$ is the acceleration of a gravity at the space-tug orbit, and μ is the gravity parameter.

The onboard laser-propulsion engine characteristics are assumed to satisfy following: thrust: $F_t = 1.5$–3.0 N, specific impulse $I_{sp} = 2300$ s–1200 s, consequently, $C_m = (1.5$–$3.0) \times 10^{-3}$ N/W if the general efficiency of the laser propulsion is of 40%.

In Tables 6.8 and 6.9, the SMV mission parameters are presented, including: β an angle between the local horizon and thrust vectors; t_n a total time interval of the SMV mission, lasting from space-tug departure to GEO satellite intercept, and $\Sigma\, t_{pr}$ is the engine operation time; υ is the space angle of the SMV total mission; ΔV_g, ΔV_u are velocity losses caused by the Earth gravity and trajectory control; $\Sigma\, \Delta V_p$ is a general velocity factor of the mission with the velocity losses account; $\pi/2$–δ_k is an angle of the SMS elliptical orbit crossing the GEO orbit; $\Sigma\, M_{P.B.}$, $\Sigma\, M_{PB.p}$ is propellant mass required for the mission.

These data show that the period of the SMV mission to GEO is decreased by increasing the SMV velocity increment and angle β between the velocity and its increment vectors. But, to reach the velocity increment, the laser-propulsion engine has to operate for longer time that leads to greater consumption of an onboard propellant. For example, if $\Delta V = 50$ m/s, then the propellant mass consumption is 16 kg when a 1.5 N thrust produced, and it is 26 kg for a 3 N thrust. At the same time, the propellant consumptions will be only 1 kg if $\Delta V = 2$ m/s, but it will take about 10 h for the SMV to move it to GEO.

One has to pay an attention on the following fact. The contact mission period decreases by approximately 10% by increasing thrust from 1.5 N up to 3 N. But, the propellant mass consumptions increase twice at percentage. It follows that there is some optimal scenario of the contact mission by varying the SMV velocity increment. The table's data also show that angle β also can be optimized.

The flight of the SMV with a GEO satellite onboard back to the space-tug orbit, achieved by the same technique, is illustrated by Tables 6.10 and 6.11. The mission

Table 6.8 SMV mission at: $F_t = 1.5$ N, $C_m = 1.5 \times 10–3$ N/W, and $I_{sp} = 2367$ s

β, °	ΔV_1, m/s	t_{1n}, s	$\Sigma\, t_{pr}/t_{d\ p}$, s		ΔV_2, m/s	$\Delta V_g + \Delta V_u$, m/s	υ_1, grad	$\Sigma\, \Delta V_p$, m/s	$\pi/2–\delta_k$, grad	$\Sigma M_{P.B.}$, kg	$\Sigma M_{P.B.p}$, kg
0	2	34,977	159	197	2.79	1.146	145.8	5.94	0.042	0.45	0.56
10		34,398	155	191	2.67	1.027	143.4	5.69	0.039	0.44	0.53
20		35,483	142	163	2.27	0.626	147.9	4.89	0.026	0.40	0.46
0	10	12,928	868	1610	16.7	23.14	50.7	49.84	0.287	2.44	4.53
60		9674	751	1395	13.0	19.73	37.4	42.74	0.241	2.11	3.92
90		11,793	582	1057	7.75	14.46	46.9	32.2	0.127	1.64	2.97
0	50	8545	3303	5812	59.6	101.6	22.3	211.17	0.691	9.29	16.4

Table 6.9 SMV mission parameters at: $F_t = 3$ N, $C_m = 3.0 \times 10^{-3}$ N/W, and $I_{sp} = 1183$ s

β, °	ΔV_1, m/s	t_{1n}, s	$\Sigma\, t_d/t_{d\ p}$, s		ΔV_2, m/s	$\Delta V_g + \Delta V_u$, m/s	υ_1, grad	$\Sigma\, \Delta V_p$, m/s	$\pi/2\text{-}\delta_\kappa$, grad	$\Sigma M_{P.B.}$, kg	$\Sigma M_{P.B.p}$, kg
0	2	34,995	79	98	2.79	1.135	145.8	5.93	0.042	0.89	1.10
10		34,414	77	94	2.67	1.015	143.4	5.68	0.039	0.87	1.05
20		35,458	71	46	2.27	0.618	147.9	4.89	0.026	0.79	0.91
0	10	13,231	424	747	16.7	22.38	50.7	49.07	0.287	4.76	8.40
60		9940	368	649	13.0	19.09	37.4	42.11	0.241	4.13	7.30
90		11,999	286	501	7.75	14.14	46.9	31.88	0.128	3.22	5.64
0	50	9065	1498	2325	59.6	87.91	22.3	197.53	0.691	16.85	26.2

Table 6.10 SMV transferring back to the space-tug at: $F_t = 1.5$ N, $C_m = 1.5 \times 10^{-3}$ N/W, and $I_{sp} = 2367$ s

β, °	ΔV_1, m/s	t_{2n}, s	$\Sigma\, t_{pr}$, s	ΔV_2, m/s	υ_2, grad	$\pi/2-\delta_k$, grad	$\Sigma M_{P.B.}$, kg
160	2	36,294.6	987.8	2.28	147.757	−0.026	2.78
180		35,919.8	1107.7	2.8	145.647	−0.042	3.12
200		36,294.6	987.8	2.28	147.757	−0.026	2.78
180	10	18,145.6	6043.0	16.7	50.483	−0.288	17.0
225		14,627.5	5590.4	14.7	37.750	−0.267	15.7
270		15,260.0	4047.2	7.74	46.966	−0.127	11.4
278		18,615.4	3525.7	5.42	63.255	−0.037	9.92
270	50	23,077.8	21,070.6	49.6	8.406	−0.919	59.3

Table 6.11 SMV transferring back to a space-tug at: $F_t = 3$ N, $C_m = 3.0 \times 10^{-3}$ N/W, and $I_{sp} = 1183$ s

β, °	ΔV_1, m/s	t_{2n}, s	$\Sigma\, t_{pr}$, s	ΔV_2, m/s	υ_2, grad	$\pi/2-\delta_k$, grad	$\Sigma M_{P.B.}$, kg
160	2	35,796.0	489.0	2.28	147.757	−0.027	5.5
180		35,360.0	548.2	2.8	145.647	−0.042	6.17
200		35,796.0	489.0	2.28	147.757	−0.027	5.5
180	10	15,033.1	2930.6	16.7	50.483	−0.288	33.0
225		11,753.2	2716.2	14.7	37.750	−0.266	30.5
270		13,191.7	1978.9	7.74	46.966	−0.127	22.3
278		16,817.3	1727.6	5.42	63.255	−0.037	19.4
270	50	11,589.0	9581.0	49.6	8.406	−0.919	107.8

parameters of the SMV transfer from GEO to the tug orbit are calculated for different velocity decrements. It is assumed that the GEO satellite mass is 300 kg.

One of the principal parameters of the contact mission is the angular distance between the SMV initial and final orbit points in which the SMV meets with the space-tug. This angle depends on the total time period of the mission, which is determined as a sum of (a) t_1 the time interval of the SMV flights from the space-tug orbit to GEO, (b) t_{pr} the time to catch the GEO satellite, (c) t_2 the time of flight of the SMV, with the GEO satellite onboard, back to the tug orbit.

The angular size of the space-tug contact mission is determined by the following formula:

$$\upsilon_H = (V_n/R_n) \times (t_{1n} + t_c + t_{2n}) \tag{6.23}$$

And the SMV mission angular size follows:

$$\upsilon_{SMV} = \upsilon_1 + (V_c/R_c) \times t_c + \upsilon_2 \tag{6.24}$$

Equating these formulae to each other, find the time interval that can be used to catch the GEO satellite, namely:

$$t_c = \left(\upsilon_1 + \upsilon_2 + \Delta\upsilon_{pr} - (V_n/R_n) \times (t_{1n} + t_{2n})\right) \times (V_n/R_n - V_c/R_c)^{-1} \quad (6.25)$$

where $\Delta\upsilon_{pr}$ is the angular distance that is turned through during the SMV active flight.

It follows from Tables 6.8, 6.9, 6.10, and 6.11, the SMV positioned onboard a space-tug could remove the defunct satellites of 300 kg mass out from the GEO with a propellant mass consumption of less than 10% of the initial SMV mass for every contact mission. This usage of propellant mass can be decreased by optimal controlling of the SMS contact maneuvers, choosing the optimal thrust during the orbital flight, increasing the general efficiency of laser propulsion by using the CHO-polymers as propellant, and so on.

6.4 Onboard Laser-Propulsion System as Applied to SMV

In this section, we suggest an onboard propulsion system for an SMV to control both low orbit and GEO missions by using remotely located lasers. The onboard propulsion system consists of an optical system to collect the laser power and a laser-propulsion engine. A low orbit SMV with laser propulsion is distinguished by the type of maneuver characterized by variable thrust and its vector. The laser is assumed to be arranged on the surface of the Earth, on an airplane or in space. But in any case, the SMV orbital maneuver must not depend on its spatial orientation with respect to the laser beam angle.

SMV orbital maneuvers lead to the need to develop a special onboard optical system with the following principles:

- independence of the satellite orbit maneuver on the laser beam position by using a laser-propulsion engine of special design that is based on ASLPE (see Chap. 4) and a receiver telescope wide angle field of regard, which is optically coupled by using onboard optical elements
- minimal laser power losses of the laser radiation through this optical system and matching the transmitter and receiver telescope apertures by spatial coupling of the telescopes axes
- fixing a general optical axis of the onboard propulsion system as the design axis of the SMV device, with respect to which all other optical units are to be oriented

Using these principles, we present the total schematic diagram of the SMV shown in Fig. 6.10 [18]. In this scheme, such optical units as optical switchers and turrets are used, which allow guiding the spatial orientation of the telescope axes collecting the laser radiation during any maneuver of the vehicle.

In the figure, X-axis is the design axis of the SMV device.

The SMV optical system incorporates two telescopes with an individual turret, which are arranged symmetrically with respect to the device geometric center. The telescopes are united by a common optical axis (i.e., the design axis), and the turrets can rotate about this axis. Every turret provides a 170° field of regard in an X-Y-

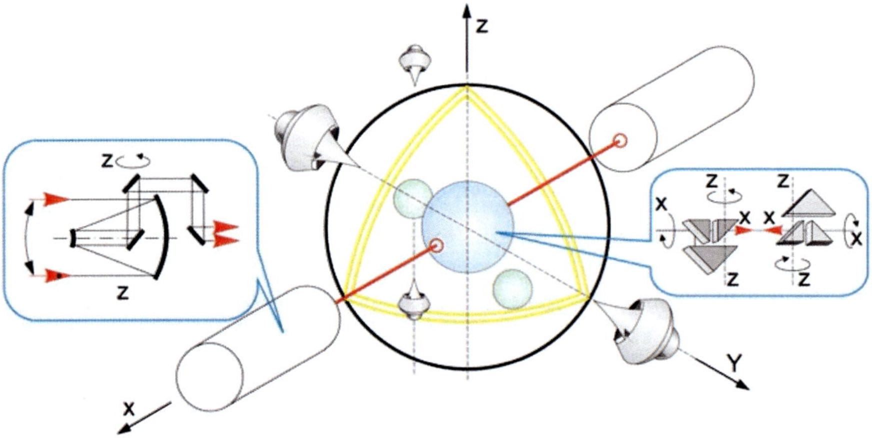

Fig. 6.10 Schematic diagram of the SMV laser-propulsion system

plane to receive laser radiation. To change the plane of the angular domain, the telescope can rotate about the X-axis over 360°.

Spatial orientation of the SMV is based on the use of a conventional navigation system operating on the basis of star trackers [19]. Star trackers provide the satellite orientation accuracy of 10^{-3} rad with respect to stars. The satellite guidance system allows stabilizing the telescope axis orientation in the direction of a remote laser with an accuracy of 10^{-6} rad. Every telescope collects the laser radiation and redirects it to the laser-propulsion engines by using optical devices to direct the laser beam to every engine that has to operate at any specified moment.

The onboard propulsion system incorporates three engine units, which are arranged so that the thrust vector of every engine crosses the satellite center of mass. Moreover, the device incorporates six engines operating as angular orientation engines, which means that there are two engines for each spatial axis. Doubling of engines amount is used to take into account possible extreme conditions which could take place during a long-term period of the satellite exploration [20].

One of the optical channels of the onboard propulsion system is shown in Fig. 6.11 as a functional scheme, illustrating couplings between the system's optical units. The channel includes a receiver telescope (M1, M2), plane mirrors (M3–M6) transmitting the laser radiation to optical switcher (M7–M10) as well as laser-propulsion engine.

Let us consider design features of each unit of the onboard propulsion system as applied to the low Earth orbit and GEO missions.

6.4.1 Receiver Telescope

The receiver telescope is used to collect laser radiation transmitted from a remote laser. The telescope is made of two parabolic mirrors, M1 and M2, off-axis optical

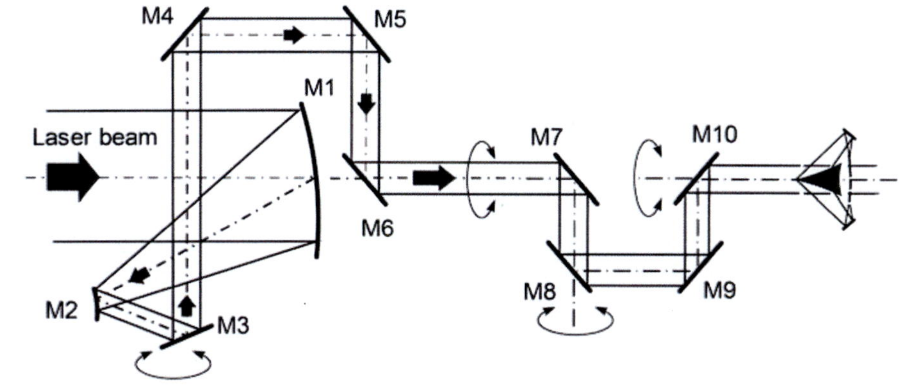

Fig. 6.11 Optical schematic diagram of one of the optical channels of the onboard propulsion system

scheme eliminate the laser power loss caused by screening of the laser beam aperture as happens in centered telescopes. The output beam of the telescope is a parallel one without any intermediate focus points which could initiate an optical breakdown of the medium inside the telescope.

Here, we consider the SMV that could be used to remove GEO satellites during its maneuvers away from a space-tug to GEO at distances of 200–300 km. In this case, the laser transmitter telescope with an aperture of D_0 arranged on the tug and the SMV receiver telescope with a diameter D are matched because of a diffraction divergence of the laser beam if the laser is on the tug board. But, the receiver telescope aperture D is restricted by limitations imposed on the SMV mass. The telescope mass can be decreased if modern optical technologies of space mirror production is used. For example, the optical mirrors made of CO-115M glass-ceramics possess a weight-reduction factor of 0.72 [21].

Similar reason, all mirrors of minimal sizes in the laser-propulsion system are assumed to be used. In this case, the mirrors have to be made of materials which could exclude possible breakdown damage of its surfaces under high laser power. This condition also imposes limitation on the minimum size of the mirrors. Special attention should be paid to the telescope's second mirror M2 whose diameter D_B is determined by the general magnification of the telescope. At present, the optical technologies allow production of mirrors with an intensity resistance of 5 J/cm^2 under the laser radiation with a 1.06μm wavelength if the mirror reflection coefficient of the radiation is of 0.998.

6.4.2 Optical Turret

Turning the receiver telescope through a wide field of regard is achieved by the use of a mirror system to stabilize the axis of a laser beam in space in spite of arbitrary

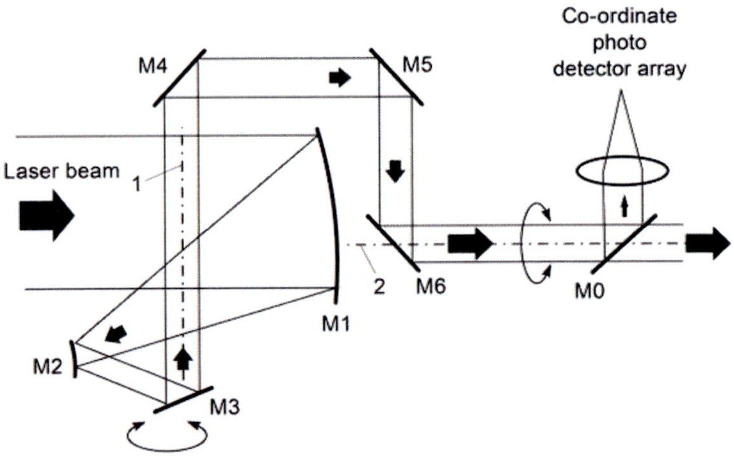

Fig. 6.12 Receiver telescope schematic diagram. M1–M6—mirrors of the system, M0—is auxiliary mirror

direction of arrival of the incoming laser beam (Fig. 6.12). The optical turret consists of M3–M6 plane mirrors.

We examine a case when the receiver telescope collects a laser beam that comes from a wide range of the space angles within 170° around the axis (1) passing through M3–M4 mirrors and within azimuth angle of 360° coinciding with the SMS design axis (2).

The procedure of adjusting the receiver telescope with respect to the laser transmitter telescope positioned on the tug can be achieved by the following technique. The transmitter telescope is turned on to the direction of SMV spatial location, which can be determined by an active radio system on the SMV, and then the SMV is irradiated by auxiliary laser rangefinder arranged on the tug. Corner reflectors on the receive telescope primary aperture reflect the rangefinder signal. When the reflected signal is received, SMV space position is defined exactly by using a coordinate photo detector array. Then, the adjustment procedure is repeated until the positions of irradiated and received signals on the detector coincide to designed or given accuracy. The receiver telescope is also equipped with detectors registering the laser radiation, signals from which will allow turning the telescope in the direction of the tug by a similar procedure.

The laser rangefinder radiation is intercepted by the SMV receiver telescope and is then directed by the M3–M0 mirror system into a photo detector array to control the solid angular position of the receiver telescope axis. The electronic guidance system of the telescope turns it onto the direction of the transmitter telescope until the signal from the rangefinder is centered on the detector. The same electronic guidance system is used to hold both telescopes to be co-aligned.

Capture of the laser radiation characteristics including radiation intensity, power, and so on, as well as checking the telescopes, including the adjusting accuracy are carried out by special electronic equipment arranged on the SMV. For that purpose, a part of the laser radiation reflected from the optical crystal, M0, is used.

6.4.3 Optical Switch

One of the principal units of the SMV laser-propulsion system is an optical switch shown schematically in Fig. 6.13. The optical switch is designed to provide optical coupling between the receiver telescope and the laser-propulsion engines arranged in different parts of the SMV. Every engine operates individually independent on the maneuver to be executed. The optical switch is made of conventional optical and electric-mechanical deflectors [22].

The optical switch operates as follows. A laser beam comes to an input element of the switch, made of a semi-transparent optical plate (see a left side of Fig. 6.13). The reflected part of the laser radiation comes into upper part of the switch, which is a fixed unit, and then this radiation is directed to optical exit 1. Out of the exit 1, the radiation is transmitted to one of three engine blocks (see Fig. 6.11). The laser radiation that passed through the optical plate is reflected by a mirror installed behind the plate into exit 2 or exit 3 which are optically coupled with the other two blocks of laser-propulsion engines. The optical switch distributes the laser power between different engine blocks depending on the SMS maneuver that reads to be executed.

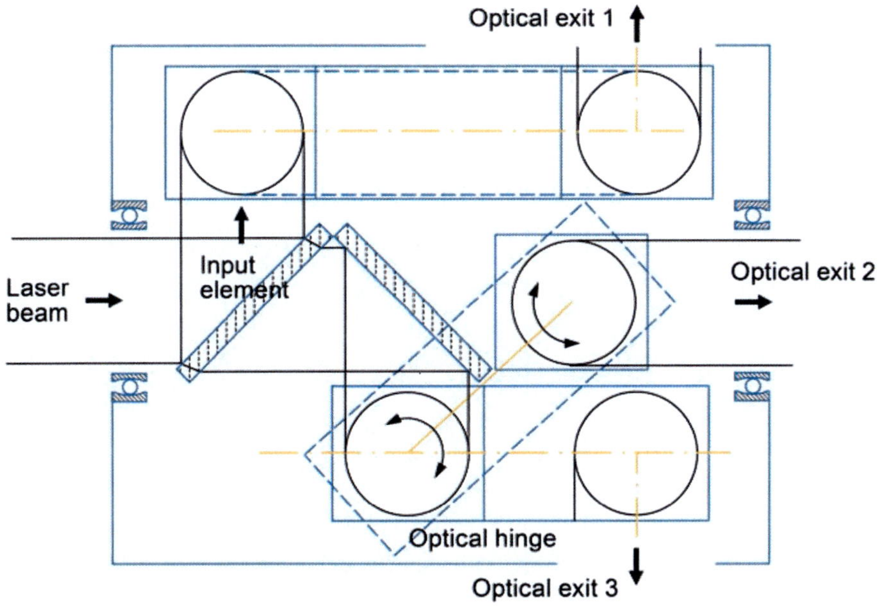

Fig. 6.13 Schematic diagram of the optical switch

6.4.4 The Unit of Laser-Propulsion Engines

We propose to use a block construction of the laser-propulsion engines onboard of the SMV. One of the possible versions of the unit is shown in Fig. 6.14 [23, 24]. The unit construction includes four engines of the ASLPE type which are arranged symmetrically on each side of a square frame. In the center of the frame, a four-way mirror pyramid is installed. Each reflecting surface of the pyramid is assumed to direct the incoming laser beam on to the first parabolic reflector of each engine. This block construction is named like a correcting propulsion unit—CPU.

The basic modes of the correcting propulsion unit operation are shown schematically in Fig. 6.15. In this, the principle of a pair of propulsion engines operating on one of the three spatial axes is used to orientate and stabilize the SMS in a space. A reference laser beam is used to adjust the high-power laser beam channel and to direct the laser power into each propulsion engine precisely. The procedure of adjusting the laser beam is achieved by using the reference beam reflected from a corner reflector installed on a bevel tip of the pyramid.

Solid propellants for laser propulsion are assumed to be used in the form of a long wire that is fed into the engine nozzle axis point corresponding to the laser beam focusing point. Solid propellant can be stored easily in a compact device with minimal mass and volume. In this case, laser ablation propulsion is one of the prospective methods of thrust production under space conditions. Moreover, to increase efficiency of the laser propulsion, various polymers can be used as space propellants.

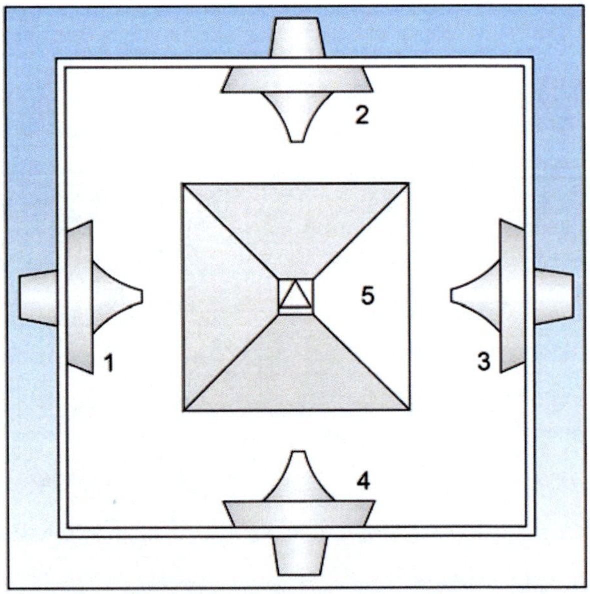

Fig. 6.14 Block construction of onboard laser-propulsion engines. (1–4) are ASLPE, and (5) is a four-way mirror pyramid

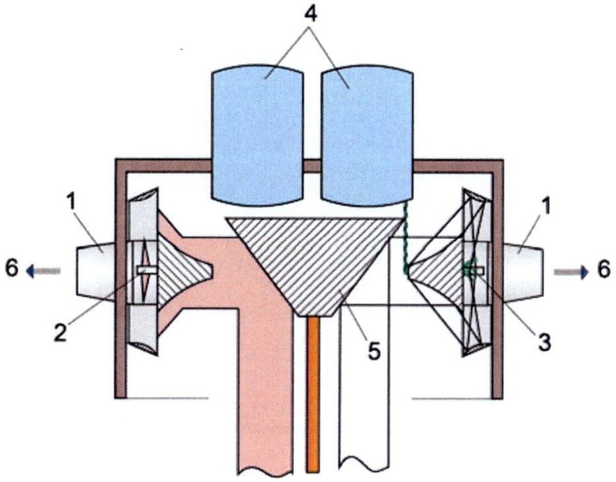

Fig. 6.15 Principal schematic diagram of the CPU operation modes (two versions of onboard propellant storage are shown in the figure). (1) ASLPE, (2, 3) propellants, (4) propellant store devices, (5) mirror pyramid, (6) exhaust jet

6.4.5 Requirements to Optical Elements of the Onboard Laser-Propulsion System

Let us consider the basic requirements of an onboard laser-propulsion system which arise as "optical point of view." The requirements follow from the design considerations which should be considered choosing the development of the system. For example, the conventional interconnection between the diameters of the transmitter and receiver telescopes, providing collection of 84% of the transmitted laser power collected by the receiver telescope, has to satisfy the following relations, namely [25]:

$$D \cdot D_0 = 4\lambda L_d \tag{6.26}$$

$$\eta_{ef} = 1 - J_0^2\left(\frac{\pi D D_0}{2\lambda}\right) - J_1^2\left(\frac{\pi D D_0}{2\lambda}\right) \tag{6.27}$$

Here, L_d is Fresnel distance of the beam propagation; J_0 and J_1 are Bessel functions.

Equation (6.26) determines the correlation between the transmitter and receiver telescopes diameters if the distance between them is less than the Fresnel distance. Diffraction losses of the laser power at distances exceeding L_d by introducing the delivery efficiency coefficient determined by formula (6.27) has to be taken into account. For the example, the results of calculations of correlation between collected power and receiver diameter are shown in Fig. 6.16.

In the figure, 1–4 lines correspond to the laser power collected by the receiver telescope with a 0.25 m diameter. The dependence of the receiver diameter as function of the distance between the laser and SMS is presented by 9-th line for this case. The 2 and 4–6 lines correspond to the collected power for the receiver mirror diameter of $D_0 = 0.5$ m and the dependence of the receiver diameter on the

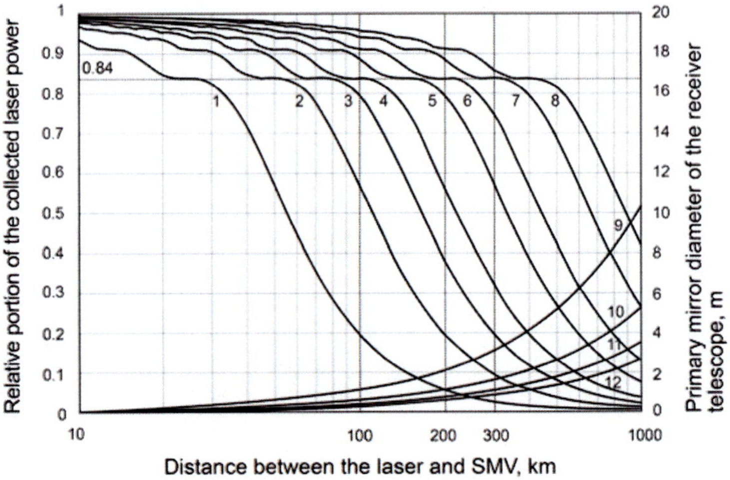

Fig. 6.16 Laser power delivered (coefficient, lines 1–8) and receiver telescope primary diameter (lines 9–12) as functions of distance between the space-tug and SMS for laser radiation with a 1.06μm wavelength

distance between the laser and SMS is presented by 10-th line for this case. The 4 and 6–8 lines are the results of calculations when $D_0 = 1$ m. It is seen that a portion of the collected power is less than 84% at distances of 200–300 km if the telescope diameter is less than 1 m.

In that case, the most acceptable size of the telescope diameter of the onboard laser-propulsion system would be equal to $D_0 = 1$ m and/or 0.5 m at a distance of 200 km, which corresponds to the lines under number 6 and 12 in the figure; and it is equal to $D_0 = 1$ m and/or $D = 0.75$ m for a distance of 300 km (lines 7 and 12).

Now, let us estimate parameters of other mirrors in the receiver telescope system. Taking into account the requirements following from the radiation resistance of the mirrors, the minimum diameter of these elements would be of 0.055 m if the laser pulse energy of 100 J is used. Usually, the mirror diameter is chosen by a criterion of a factor of two above the radiation resistance threshold. So, the mirrors with 0.11 m in diameter have to be installed into the system. In this case, the telescope magnification coefficient will be equal to:

$$M = \frac{F_{\mathrm{M1}}}{F_{\mathrm{M2}}} = 4.5$$

And the light aperture of the M3–M10 plane mirrors should be more than 0.155 m. Here, F_{M1} and F_{M2} are areas of the M1 and M2 mirrors.

Principal geometric parameters of the telescope mirrors are presented in Tables 6.12 and 6.13.

What is more important is the technology requirements imposed onto optical surface quality of the mirrors and to the spatial accuracy of the mirrors adjustment in

Table 6.12 Geometry parameters of the SMV receiver telescope mirrors

Mirror	Curvature radius, m	Aperture diameter, m	Optical axis displacement	Conical constant, e^2	Distance, m
M1	−1.50	0.50	0.50	−1	−0.63
M2	−0.24	0.12	0.08	−1	0.35
M3	Plane	0.20	0.42	0	−0.50
M4	Plane	0.20	0.30	0	0.20
M5	Plane	0.20	0.21	0	−0.50
M6	Plane	0.20	0.15	0	0.50
M7	Plane	0.20	0.10	0	−0.30
M8	Plane	0.20	0.08	0	0.30
M9	Plane	0.20	0.05	0	−0.20
M10	Infinity	0.20	0.04	0	0.50

Table 6.13 Total telescope parameters

M6-LPE distance, m	1	Aperture ratio for M1	1:1.5
Telescope magnification, M	4.5	Aperture ratio for M2	1:1.2
Telescope general field of view	6’		

Table 6.14 Optical technology aberrations of the receiver telescope system

Aberration type	Δz	Δy	Δe_{M1}	Δe_{M2}	$\Delta \alpha$
Magnitude	943 λ	92 λ	4×10^{-5}	6×10^{-4}	0.01 rad

the telescope—see Table 6.14. All requirements to the parameters listed in Table 6.14 are indicated for a 1.06μm wavelength. It is also assumed that every parameter inaccuracy leads to the laser power losses less than 10% in the plane of LPE beam concentrator. These technology parameters include themselves:

Δz—change of axial distance between the M1 and M2 mirrors of the telescope
Δy—axial misalignment of the M1 and M2 mirrors of the telescope
Δe—technology inaccuracy of the telescope mirrors M1 and M2 caused by variation in its eccentricity
$\Delta \alpha$—axial inclination of the M1 and M2 mirrors in respect of the telescope axis

As it is seen from Table 6.14, the requirements for the receiver telescope mirrors are high. But, the other plane mirrors of the laser-propulsion system can be made by relatively simple conventional optical technologies.

6.5 Brief Outcome

We assume that the SMV-s with the onboard laser-propulsion system possess good perspectives for application in space exploitation in the near future, which are determined by the development of both space optical technology [26] and of high-power lasers of various deployments including near-Earth space [27, 28]. Among them, solar-pumped space-based lasers are being developed to be used to transmit power at space distances to prolong satellites lifetime and so on [29, 30].

As for ground-based lasers, their prospects will depend on solving the problems of efficient delivery of the laser power through the Earth atmosphere to space vehicles.

References

1. McRidenko, L.A., Volkov, S.N., Khodenko, V.P.: Conceptual consideration of development and application of small satellites. Iss. Electrom. (Rus.). **114**, 15–26 (2010)
2. Small satellites. Regulatory challenges and chances. Marboe, I., Nijhoff, B. (ed.) Ser.: Studies in Space Law 11 (2016)
3. Sevastianov, N.N., et al.: Analysis of state-of-the-art development of small satellites for the earth remote observation. Proc. MFTI (Rus.). **1**, 14–20 (2009)
4. Zinchenko, O.N.: Small optical satellites for remote sensing. Rakurs, Moscow (2010)
5. Ketsdever, A.D., et al.: An overview of advanced concepts for space access. J. Spac. Rock. **47**, 238–250 (2010). https://doi.org/10.2514/1.46148
6. Myrabo, L.: Lightcraft Technology Demonstrator, Transatmospheric Laser Propulsion. Fin. Tech. Rep. Contract N. 2073803 (1989)
7. Richard, J.C., et al.: Earth-to-Orbit laser launch simulation for a Lightcraft Technology Demonstrator. AIP Conf. Proc. **830**, 564–575 (2005). https://doi.org/10.1063/1.2203298
8. Myrabo, L.: Advanced beamed-energy and field propulsion concepts. Fin. Rep. bdh/h-83-225-tr. (1983)
9. Orbital Debris Modeling.: NASA Orbital Debris Program Office. National Aeronautics and Space Administration (2011). http://orbitaldebris.jsc.nasa.gov/
10. Kaplan, M. H.: Space debris realities and removal. https://info.aiaa.org/tac/SMG/SOSTC/Workshop
11. Contzen, J.P., Muylaert, J.: Scientific aspects of space debris reentry. ISTC Workshop on mitigation of space debris, Moscow (2010)
12. Sevastianov, N.N., et al.: Analysis of technical abilities of development of small satellites for earth observation. Proc. of MFTI. **1**, 14–22 (2009)
13. Levantovsky, V.I.: Mechanics of space flights. Nauka, Moscow (1980)
14. Burdakov, V.P., Zigel, F.Y.: Physical basis of astronautics. Atomizdat, Moscow (1975)
15. Egorov, M.S.: Investigation and development of the onboard optical system for small satellites with laser propulsion. A Thesis Submitted to the Graduate Faculty of Saint-Petersburg ITMO University in Partial Fulfillment of the Requirements for the degree of doctor of philosophy. S.-Petersburg (2016)
16. Ageichik, A.A., et al.: Combustion of propellants of CHO chemical composition as applied to laser propulsion. J. Tech. Phys. **79**, 76–83 (2009)
17. Galabova, K.K., Dereck, O.L.: Economic case for the retirement of geosynchronous communication satellites via space tugs. Act. Astron. **58**, 485–498 (2006). https://doi.org/10.1016/j.actaastro.2005.12.014
18. Egorov, M.S., et al.: Corrective laser propulsion system for space vehicles. J. Opt. Tech. **77**, 8–15 (2010)
19. Tappe, J.A.: Development of star tracker system for accurate estimation of spacecraft attitude. Naval Postgraduate School Monterey, CA 93943–5000 December (2009)
20. Egorov, M.S., Nosatenko, P.Y., Rezunkov, Y.A.: Optical system of mini-satellite with laser propulsion. J. Opt. Tech. (Rus.). **81**, 16–27 (2014)

21. Smith, W.J.: Modern optical engineering. The design of optical systems. Kaiser Electro-Optics Inc. Carisbad, California and Consultant in Optics and Design (2000)
22. Liou, J.-C.: Orbital debris modeling. NASA Orbital Debris Program Office, Johnson Space Center, Houston, TX (2012)
23. Rezunkov, Y.A.: Laser propulsion for LOTV space missions. AIP Conf. Proc. **702**, 228–241 (2004). https://doi.org/10.1063/1.1721003
24. Rezunkov, Y.A., et al.: Laser fine-adjustment thruster for space vehicles. AIP Conf. Proc. **1230**, 309–318 (2009). https://doi.org/10.1063/1.3435446
25. Artukhina, N.K.: Circuit analysis of principles for simulation of mirror systems with variable characteristics. Metrol. Instr. Eng. (Rus.). **2**, 22–28 (2013)
26. Savitzky, A.M., Sokolov, I.M.: Design feature of light-weight mirrors for space telescopes. J. Opt. Tech. (Rus.). **76**, 94–98 (2009)
27. Avdeev, A. V., et al.: Construction arrangement of space HF-DF-laser facility based on space vehicle. Proc. MAI **71** (2013). http://www.mai.ru/science/trudy/published.php
28. Carroll, D.L.: Overview of high energy lasers: past, present, and future? AIAA, 2011–3102 (2011). https://doi.org/10.2514/6.2011-3102
29. Summerer, L., Purcell, O.: Concepts for wireless energy transmission via laser. Int. Conf. Sp. Opt. Sys. Appl. (ICSOS) (2009)
30. Danilov, O.B., Gevlakov, A.P., Yuriev, M.S.: Oxygen-iodine lasers with optical (solar) pumping. Opt. Spectr. (Rus.). **117**, 151–158 (2014)

Chapter 7
Laser Power Transfer to Space Vehicles with Laser Propulsion

Abstract Creation of the HPLP system depends on solving such technical problems as: (a) generation of a laser beam with extremely small beam divergence at high power, (b) compensation for the laser beam distortions caused by atmosphere and by optical elements of the system. The technical solution of these problems defines the ultimate HPLP system design.

To compensate for beam phase distortions, dynamic holography techniques are considered, including phase conjugation. State-of-the-art laser systems with PC possess wide field of view, high responsibility, and high speed of response. In the chapter, the experimental data on phase conjugation of beam distortions of CO_2 lasers by using such nonlinear medium as a sulfur hexafluoride are analyzed as applied to laser power transfer to space vehicle with onboard ASLPE.

The proposed laser adaptive optical system is based on the algorithm of compensation for the atmospheric optical distortion, which includes the following principal operation procedures:

– generation of a reference coherent radiation source in an upper atmosphere to record the atmospheric distortions in the laser beam propagation path
– generation of the high-power laser beam with a conjugated beam wave front by using a dynamic hologram-corrector that is recorded with the use of the reference source beam

Keywords Adaptive laser system · Adaptive two-circuit system · Space missions · Solid-state lasers · Optical technology · Aerosols extinction and scattering of laser radiation · HITRAN databank · Kolmogorov's isotropic turbulence · Inner gravitational waves · Transmitter telescope magnification · Freed radius · Isoplanatic angle · Rytov parameter

Y. A. Rezunkov, *High Power Laser Propulsion*, Springer Series on Atomic, Optical, and Plasma Physics 116, https://doi.org/10.1007/978-3-030-79693-8_7

7.1 Introduction into the Problem

In the chapter, we consider the atmospheric effects which influence laser radiation during its propagation through upper layers of the Earth's atmosphere up to a laser-propelled vehicle. The atmosphere effects distort the phase and intensity of a laser beam, which depend not only on the atmosphere but also on laser radiation parameters. The models of the upper atmosphere are applied to study the laser beam propagation in the atmosphere and to evaluate the influence of the atmosphere on spatial profile of the irradiation at the vehicle's receiver. First of all, much attention is paid to the solid and gaseous aerosols models of the atmosphere and on the laws of its distribution by the atmospheric altitude.

At present, a number of models of the Earth's atmosphere have been developed on the basis of experiments of laser beam propagation through the atmosphere. Using the results of these experiments, various techniques have been developed to compensate for the laser beam distortions caused by the atmosphere [1–3]. We consider only a few models of the atmosphere and some phase conjugating techniques which can be applied to the problem of high-power laser power deliver through upper atmosphere to a laser-propelled vehicle. Hence, airborne lasers are assumed to be used to develop HPLP. We think of AirBorne Laser (ABL) developed in the USA and airborne lasers developed in Russia that are the prototypes for the laser-propulsion systems [4, 5].

At present, more fantastic ideas concern the assumption of using space-based lasers arranged in GEO (or even on the Moon) for space missions of laser-propelled vehicles. Such types of ideas have been developed during projects for space power transmission from space and the Moon on to the Earth. The space-based solar pumped lasers and HF/DF lasers were considered in these projects [6, 7]. Similar ideas are considered by Prof. Leik N. Myrabo in his *Lightcraft. Flight handbook* [8].

Here, we try to restrict the considerations of the high-power laser systems as applied to laser propulsion by the lasers with average power of 100 kW or less which are widely developed around the world to the present time. We see from previous chapters of this book, a 100 kW laser power is sufficient to be used for launching small payloads into low Earth orbits and for space vehicles orbiting in the near-Earth space.

It should be noted also that there are certain requirements to the laser radiation characteristics, which define an efficiency of laser propulsion and have to be taken into the consideration of the radiation propagation through the Earth's atmosphere.

Therefore, some part of this chapter is devoted to analysis of the adaptive optical systems to compensate for the laser beam distortions caused by upper layers of the Earth's atmosphere as applied to lasers arranged on an aircraft. Airborne lasers are most likely to be used to propel space satellites with laser propulsion for various space missions (see Chap. 6). We consider the technical solutions of this problem as applied to delivery of high-power laser radiation to the satellites.

As is known, the basic limitations of the development of high-power solid-state lasers (SSL) is heat removal out from optical active elements of the laser which effect on the laser beam divergence significantly. One of the possible solutions of this heat problem is the use of the active elements in the form of "active mirrors" in which radiation propagates in the direction transverse to the active layer and it leaves the layer after reflection from the mirror rear side. As it is shown in Refs. [9, 10] the scaling of Yb:YAG laser up to 200 kW output power for operation both in CW and pulse repetition modes is possible on a basis of "master oscillator—power amplifier" (MOPA) configuration with phase conjugation mirror (PCM) to compensate for the amplifier optical distortions. Pumping of the "active mirrors" is performed by matrix of laser diodes arranged between the "mirrors." The laser can operate in quasi-CW mode.

Usually, the radiation of high-power laser, operating in near-IR spectral range, is collimated by a transmitter telescope with a primary mirror of 1 m in a diameter and is focused at a specified distance. The beam guidance is usually realized by using auxiliary communication subsystems which are based on extra lasers of less power. All operation procedures, including scanning remote target, telescope-rider guidance, transmitting and receiving the radiation of the low and high-power lasers, are carried out by using one-and-the-same receiver-transmitter telescope. Brining in and brining out of radiation of the low power laser into the telescope aperture is fulfilled by using auxiliary optical path containing spectral selective optical mirrors with a high reflection coefficient for the extra laser radiation. Such type of the adaptive laser system is developed by using the state-of-the-art optical technologies [6].

7.2 Models of the Aerosols and Gases Attenuation, Absorption, and Scattering of Laser Radiation in the Upper Atmosphere

7.2.1 Models of the Atmospheric Aerosols and Gases

It should be mentioned that there are very few experimental data on the propagation of high-power laser beams in the upper atmosphere at altitudes higher than 10 km, which could be considered to apply to launching vehicles by using lasers. Therefore, some attention should be given to those models of the atmosphere aerosols and gases distributions with the atmosphere altitude, which include optical characteristics of these atmospheric components.

Currently, a number of models of the atmospheric aerosols have been developed around the world [11–14]. But, the main criteria for choosing the aerosol model as applied to high-power laser propulsion is its statistical validity concerning optical characteristics of the aerosol particles.

Most broad-scale studies of the aerosols, and its components distributions with an altitude in a stratosphere, have been carried out by using such equipment as SAM-2, SAGE-I, and SAGE-II [12, 13]. These experiments collected vertical distribution of the aerosol's extinction of laser radiation for different wavelengths (0.385, 0.453,

0.525, and 1.02μm) in the troposphere and stratosphere [14]. Similar experiments were made by means of the "MIR" space station and SPOT satellite. The experimental data were used to develop and to correct methodically the upper atmosphere aerosol models which were included into such numerical software as Lowtran, Fascod, and Modtran [15]. Comparison of the numerical data on the atmosphere transmittance of laser radiation shows that the Lowtran software is most effective as its results coincide with experimental data most exactly.

At present, there are a few models of the atmospheric air composition and aerosol components which are used in scientific and applied studies. These models present averaged experimental data on the altitude distributions of temperature, air pressure, air density, and molecules concentration of the air components. In Russia, there are models of the State Optical Institute (GOI) and the Russian Institute of Atmosphere Optics (IOA) [16]. The IOA model is referred to a regional model with its basic contents and presents great interest, as the model includes statistical data on variability of the Earth's atmosphere.

The so-called AFGL model was developed at the US AirForce Phillips Laboratory (Geophysics Directorate) [17, 18], and the model have been used widely, and reported in scientific literature for solving some problems of transferring the laser radiation in the atmosphere. The AFGL model is corrected periodically on the basis of new experimental data as it is performed using the Lowtran aerosol models. That is why, the AFGL model is usually applied to consider the laser radiation absorption and scattering in the atmosphere (see Fig. 7.1). In this figure, VCAE is the volume coefficient of aerosol extinction.

Further, we consider the atmospheric effects only for a limited number of the laser radiation wavelengths because the number of high-power lasers is also limited, that is, the radiation wavelengths of 0.53, 1.06, 1.315, 3.0, 5.5, and 10.6μm.

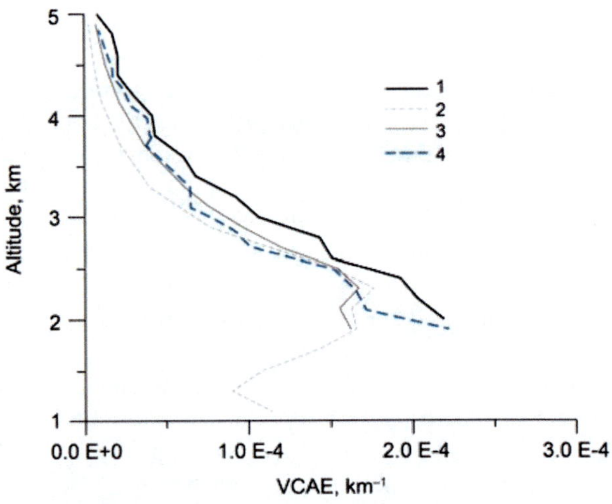

Fig. 7.1 Data of various models of aerosol extinction of radiation with a 1μm wavelength (1) Ozon-MIR data; (2, 3, 4) SAGE data averaged through a month (1997)

Numerical calculations of the atmospheric extinction and absorption of laser radiation are carried out by using the models mentioned above. The following expression for monochromatic absorption coefficient is used:

$$A_{\mathrm{h}}(v) = \int_{-\infty}^{\infty} \{[1 - \exp(-D_{\mathrm{h}}(v'))]\} \mathrm{d}v \tag{7.1}$$

where $D_{\mathrm{h}}(v')$ is optical thickness of the laser radiation path, l, in the atmosphere, namely:

$$D = \int_{l} K(\lambda, l)\, \mathrm{d}l$$

where the extinction coefficient, K, is a sum of the absorption coefficient K_{a} and scattering coefficient K_{R} both of which are also determined as a sum of the gaseous K^{M} and aerosol's K^{a} components of the Earth's atmosphere extinction, namely:

$$K = K^{\mathrm{M}} + K^{\mathrm{a}} = K_{\mathrm{a}}^{\mathrm{M}} + K_{\mathrm{a}}^{\mathrm{a}} + K_{\mathrm{R}}^{\mathrm{M}} + K_{\mathrm{R}}^{\mathrm{a}} \tag{7.2}$$

To determine the aerosol absorption and scattering coefficients, the Lowtran model is used. Determination of the coefficient of molecular scattering of radiation K^{M} is straightforward. But, the problem is an accurate determination of the molecular absorption because of the complicated spectrum of laser radiation absorption with different wavelengths by air molecules, which consists of the diffusion and structural absorption bands. To estimate the molecular absorption of laser radiation, it is necessary to use three bases of data, namely:

- composition and parameters of the gas and aerosol components of the atmospheric air for different seasons and geographic regions
- parameters of the spectral lines of the various gas component of the atmosphere which include rotational components
- absorption cross sections of the gas components with unresolved structure of rotation spectral lines

Data catalog of the spectroscopic parameters of atmosphere gases spectral absorption lines is presented in the HITRAN databank [19]. To calculate absorption of a 0.53 wavelength radiation, the databank of gas components with unsolved rotational line structure are used, for example. At the same time, one can force with the problem of estimation of continuum component of the radiation absorption because of limited data on corresponding spectroscopic information. To solve the problem, we propose applying the UVACS databank on the absorption cross sections of combustion products of rocket engines [20].

More detailed information on the proposed algorithm of calculation of the atmospheric absorption is presented in Ref. [21]. Here, we present some results of

Fig. 7.2 Altitude distributions of the components of extinction coefficients of the atmosphere, namely: ....K^a_____K^M____K

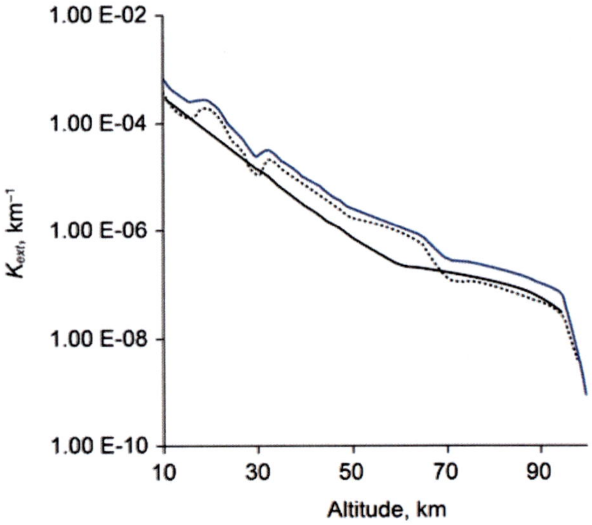

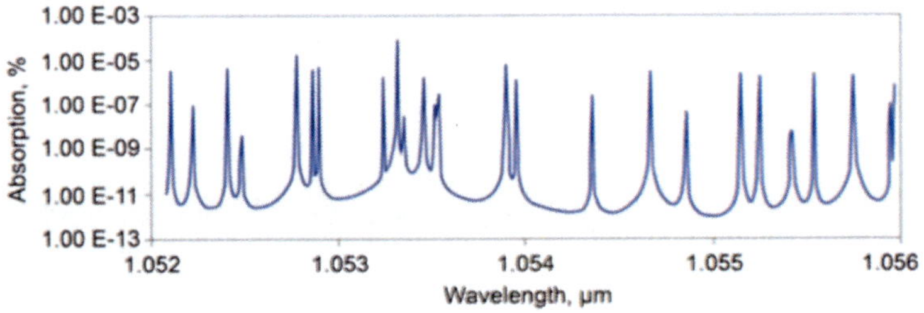

Fig. 7.3 Atmosphere spectral absorption of the radiation at 10 km altitude

the aerosol K^a and molecular K^M extinction coefficients of the laser radiation with a 1.06μm wavelength as examples (see Fig. 7.2).

The spectral distribution of the molecular absorption of radiation with a wavelength that lies within the range of 1.052–1.056μm is shown in Fig. 7.3 for a 10 km altitude atmosphere, which was calculated in accordance with Eq. (7.1) [20].

As can be seen from the figure, the radiation with a 1.053μm wavelength is in a spectral range that is located between two lines of strong absorption of radiation. The later means that a small shift of the radiation spectral line can increase the radiation absorption greatly. Nevertheless, even in this case, the molecular absorption isn't high and it reaches only a few thousandth of a percent.

Another situation is with the laser radiation of $\lambda_1 = 0.5265\mu m$, which is a second harmonic of Nd:YAG laser radiation. In this case, the molecular absorption is determined mainly by the ozone which is characterized by a smoothed spectral dependence for the altitude of 10 km. It should be mentioned that ozone gas has a maximum concentration at 35 km altitude.

7.2.2 Nonlinear Effects Developed During Propagation of High-Power Laser Radiation in the Upper Atmosphere

The aerosol and molecular absorption of laser power can cause various nonlinear atmospheric effects developing under high-power laser radiation. One of these effects which have a low threshold of its development is the so-called thermal blooming [22] that results in wandering and defocusing of the laser beam. To be more precise, we consider possible nonlinear effects as applied to propagation of high-power Yb:YAG laser radiation in the upper atmosphere. The laser characteristics are listed in Table 7.1, and the laser is considered as most real candidate for high-power laser propulsion [9, 10] at present.

Here, to estimate atmosphere nonlinear effects, the numerical technique of phase screens distributed along the laser beam propagation path is used [23]. The nonlinear effects [22, 24] are determined as numerical solutions of following system of the equations describing the atmospheric characteristics:

$$n_2(x, y, z) = \frac{\partial n}{\partial \rho} \Delta \rho \tag{7.3}$$

$$\frac{\partial}{\partial x}\left[\left(M^2 - 1\right)\frac{\partial^2}{\partial x^2} - \frac{\partial^2}{\partial y^2}\right]\Delta\rho = \frac{\alpha}{VC_{\mathrm{p}}T}\left(\frac{\partial^2}{\partial x^2} + \frac{\partial^2}{\partial y^2}\right) < I(x, y, z) > \tag{7.4}$$

where $n_2\ (x, y, z)$ is index of refraction of the atmospheric air nonlinear disturbances, and $\Delta\rho$ is disturbance of the air density which are caused by the laser radiation intensity, $I\ (x, y, z)$. This system can be solved for specified distribution of the laser intensity, $<I\ (x, y, z)>$, averaged across the beam aperture, aircraft velocity, V (it is assumed the laser is arranged within the fuselage of a plane), and initial air temperature, T. Here, $M = V/c_{\mathrm{s}}$ is Mach number, c_{s} is the local speed of a sound, α—absorption coefficient, and C_{p}—thermal capacity coefficient at a constant pressure.

If the aircraft velocity is subsonic that is $M < 1$, then the solution of Eq. (7.3) may be written in the following form:

Table 7.1 Basic parameters of diode-pumped solid-state laser (Yb:YAG-laser) assumed to be used for laser propulsion

Laser pulse parameters[a]	
Pulse energy E, J	40.0
Pulse repetition rate f_{i}, Hz	5000
Average power P, kW	200
Pulse length τ_{i}, ns	10

[a]Listed characteristics of the solid-state laser are most preferable to produce the supersonic laser ablation propulsion, being considered in Chap. 5

$$\Delta\rho = \rho_1 + \rho_2,$$
$$\text{where } \rho_1 = \frac{\alpha}{VC_pT} \int_{-\infty}^{x} < I(x', y, z) > \mathrm{d}x' \tag{7.5}$$

$$\rho_2 = \frac{A}{2} \int_{-\infty}^{\infty} \int_{0}^{\infty} \times \left[\widetilde{I}(x + x', p, z) - \widetilde{I}(x - x', p, z)\right] \exp\left(ipy - \frac{|p|x'}{\sqrt{1 - M^2}}\right) \mathrm{d}x'\mathrm{d}p \tag{7.6}$$

$$A = \frac{\alpha}{VC_pT} \frac{M^2}{(1 - M^2)} \tag{7.7}$$

$$\widetilde{I}(x, p, z) = \frac{1}{2\pi} \int_{-\infty}^{\infty} < I(x, y, z) > \exp(-ipy)\mathrm{d}y \tag{7.8}$$

Here, z is a spatial coordinate along which the beam propagates and (7.8) is a Fourier spectrum of the laser radiation distribution along the y-coordinate. To simplify the analysis, let's consider the laser beam as a Gaussian one with the following intensity distribution at the aperture:

$$< I(x, y, z) >= I_0(z) \exp\left(-\frac{x^2 + y^2}{a^2}\right) \tag{7.9}$$

In this case, we have:

$$< \widetilde{I}(x, p, z) > \; = I_0(z) \frac{a}{2\sqrt{\pi}} \exp\left(-\frac{x^2}{a^2} + \frac{p^2a^2}{4}\right) \tag{7.10}$$

$$\rho_2(x, y, z) = -\frac{Aa}{\sqrt{\pi}} I_0(z) \int_{0}^{\infty} \times \int_{0}^{\infty} sh\left(\frac{2xx'}{a^2}\right) \cos(py) \exp\left(-\frac{x^2 + (x')^2}{a^2} - \frac{p^2a^2}{4} \frac{|p|x'}{\sqrt{1 - M^2}}\right) \mathrm{d}x'\mathrm{d}p \tag{7.11}$$

The function ρ_2 and its partial derivatives $\partial\rho_2/\partial y$, $\partial^2\rho_2/\partial x^2$, $\partial^2\rho_2/\partial y^2$, $\partial^2\rho_2/\partial x \partial y$ are equal to zero on the beam axis ($x = 0$, $y = 0$). Therefore, the ρ_2 function can be expressed in the following form:

$$\rho_2(x, y, z) = -\frac{1}{2}\frac{\alpha}{VC_\mathrm{p}T}\frac{I_0(z)}{a}g(M)\frac{x}{a}, \tag{7.12}$$

$$\text{where}: g(M) = \frac{4}{\sqrt{\pi}}\frac{M^2}{1-M^2}\int_0^\infty\int_0^\infty x\exp\left[-x^2 - \frac{p^2}{4} - \frac{|p|x}{\sqrt{1-M^2}}\right]\mathrm{d}x\mathrm{d}p \tag{7.13}$$

Then, the function ρ_1 can be expressed as

$$\rho_1(x, y, z) = -\frac{1}{2}\frac{\alpha a}{VC_\mathrm{p}T}I_0(z)\left[\sqrt{\pi}\left(1 - \frac{y^2}{a^2}\right) + 2\frac{x}{a}\right] \tag{7.14}$$

Then, the total disturbance of the atmosphere air density will be:

$$\rho_1(x, y, z) + \rho_2(x, y, z)_{x,y\to 0} = -\frac{1}{2\pi}\frac{\alpha}{VC_\mathrm{p}T}\frac{P(z)}{a} \times\left[\sqrt{\pi}\left(1 - \frac{y^2}{a^2}\right) + [2 + g(M)]\frac{x}{a}\right] \tag{7.15}$$

where $P(z) = \pi a^2 I_0(z)$ is the laser radiation power.

The $g(M)$ function is presented in Table 7.2 as the example.

In what follows, the influence of the atmosphere effects is considered in the approach of a thin lens with nonlinear phase φ_non:

$$\varphi_\mathrm{non}(x, y) = k\int_0^L n_2(x, y, z)\mathrm{d}z = \frac{\partial n}{\partial \rho}k\int_0^L [\rho_1(x, y, z) + \rho_2(x, y, z)]\mathrm{d}z \tag{7.16}$$

Here, L is beam path length in the atmosphere. And for a part of the beam located closely to the beam axis where ($x << a$, $y << a$)

$$\varphi_\mathrm{non}(x, y) = k\left[S_0 - \beta_\mathrm{x}x + \frac{y^2}{2F_\mathrm{y}}\right] \tag{7.17}$$

where S_0 is a constant, and the beam front inclination β_x, its radius of curvature F_y are defined by the following formula

$$\beta_\mathrm{x} = \frac{1}{VC_\mathrm{p}T}\frac{P_0}{\pi a^2}\frac{\partial n}{\partial \rho}[1 + g(M)/2]T_\mathrm{abs} \tag{7.18}$$

Table 7.2 Numerical example of $g(M)$ function

M	0.1	0.2	0.3	0.4	0.5	0.6	0.7	0.8	0.9
$g(M)$	0.010	0.041	0.097	0.182	0.309	0.500	0.800	1.333	2.588

$$F_{\mathrm{y}} = \left[\frac{\sqrt{\pi}}{VC_{\mathrm{p}}T}\frac{P_0}{\pi a^3}\frac{\partial n}{\partial \rho}T_{\mathrm{abs}}\right]^{-1} \quad (7.19)$$

Here, $T_{\mathrm{abs}} = \int_0^L \alpha(z)\mathrm{d}z$ and the following parameters as P, V, C, T, a are assumed to be constant. If we denote the zenith angle as ϑ and the aircraft altitude as H, assuming $\vartheta < 70°$, then we obtain the following expression:

$$T_{\mathrm{abs}}(\lambda, H, \vartheta) = T_{\mathrm{abs}}(\lambda, H, 0)/\cos(\vartheta) \quad (7.20)$$

T_{abs} $(\lambda, H, 0)$ is the total coefficient of the laser power absorption. The data on T_{abs} is calculated with the use of the assumed models of the atmospheric components (see previous section) for two possible altitudes of the aircraft flight and are listed in Table 7.3.

To estimate the thermal blooming effect caused by absorption of the laser power, let us estimate the nonlinear divergence of the laser beam due to its defocusing in accordance with the following formula [22]:

$$\beta_{\mathrm{non,y}} = \frac{a}{F_{\mathrm{y}}} = \frac{\sqrt{\pi}}{VC_{\mathrm{p}}T}\frac{P_0}{\pi a^2}\frac{\partial n}{\partial \rho}T_{\mathrm{abs}} \quad (7.21)$$

This parameter coincides with the beam wave front phase inclination, accurate within a factor of $\sqrt{\pi}/(1 + g(M)/2)$. Therefore, the relative influence of the beam thermal defocusing may be characterized by a ratio of this parameter $\beta_{\mathrm{non,y}}$ to the diffractive divergence of the beam that is $\beta_{\mathrm{d}} = \lambda/a$, where a is the transmitter telescope radius, namely:

$$Q(\lambda, H, P, \vartheta) = \frac{\beta_{\mathrm{non,y}}}{\beta_{\mathrm{d}}} = \frac{\sqrt{\pi}}{VC_{\mathrm{p}}T}\frac{P_0}{\pi a\lambda}\frac{\partial n}{\partial \rho}\frac{T_{\mathrm{abs}}(\lambda, H, 0)}{\cos(\vartheta)} \quad (7.22)$$

The effect of the beam's thermal defocusing will be negligible if $Q(\lambda_1, P, \vartheta) < 0.1$. If $P_0 = 200$ kW, then for the vertical beam path in the atmosphere we have:

$$Q(\lambda_2, P_2, \vartheta = 0) = 0.007 \quad (7.23)$$

Here, it is assumed that:

Table 7.3 The total coefficient of the laser power absorption for two possible altitudes of the aircraft flight

	Winter season	Summer season
T_{abs} ($\lambda = 1029$μm, $H = 10$ km)	2.86×10^{-5}	3.98×10^{-5}
T_{abs} ($\lambda = 1029$μm, $H = 15$ km)	3.77×10^{-6}	3.77×10^{-6}

$$a = 0.5\ \text{m}, C_p = 10^3\ \text{J/kg} \cdot \text{grad}, T = 293\ \text{K}, \frac{\partial n}{\partial \rho} = 2.3 \cdot 10^{-4} \text{m}^3/\text{kg}, V = 100\,\text{m/s}$$

$$\lambda_1 = 0.5265\ \mu\text{m}, \lambda_2 = 1.053\ \mu\text{m}.$$

These examples show that we can neglect the thermal nonlinear effect on propagation of a 200 kW laser radiation in the upper atmosphere, starting from a 10 km altitude [25].

7.3 Self-Empirical Models of the Upper Atmosphere Turbulence

One of the characteristic effects of atmosphere turbulence on a laser beam profile which is displayed during the beam propagation through the upper atmosphere are random spatial fluctuation of the beam energy center as well as broadening of the beam [26, 27]. These effects depend on the laser radiation wavelength, turbulent index of refraction, beam propagation path length in the atmosphere, and so on. The turbulence induces phase aberrations of the beam wave front, which can be transformed into the amplitude distortions of the beam profile during a certain distance.

Unfortunately, there is no clearness of experimental data on the turbulence located higher than 10 km above a ground in literature. It is known that investigations have been carried out on this subject particularly in Ref. [26]. In these experiments, propagation of the laser radiation of the visible and near-IR spectral ranges with a 1 m diameter beam was studied along horizontal atmosphere paths of a 149 km distance at the altitude of 3000–3500 m between two observatories located on the Big Island of Hawaii and Maui Island. The experiments were carried out during darkness when the atmosphere is usually calm and the turbulence level characterizing by the refractive index structure constant C_n^2, was not recessively. It is known that other experiments [27] were carried out at the same navy observatory located in the Maui Island by using an adaptive optical technique to correct the star image located near the horizon and recorded by a telescope with a 363.0 cm diameter. It was demonstrated that the possibility image correction of the star but the quality of star images was not sufficient.

Two observatory optical systems exist also in the Russian Federation, namely, first one is the Altai telescope [28] with a 600 m diameter to observe space vehicles and a second one is the unique adaptive optical telescope with a 760 m diameter of a primary mirror to look at the Sun [29].

But, to estimate turbulence effects theoretically, detail information on the turbulence nature is required. It is known that inhomogeneous of the atmospheric air index of refraction, δn, are determined mainly by the relative temporal and spatial fluctuations of the air temperature, that is $\delta n \approx -N\,\delta T/T$ [30].

Nevertheless, among the models of atmosphere turbulence, the Kolmogorov's isotropic turbulence [30] and model of inner gravitational waves (IGW) [32–34] are recognized around the world. Theoretical predictions from Kolmogorov's turbulence model have been validated reasonably well by various experiments on radiation propagation through near-ground layers of the atmosphere.

As for the upper atmosphere, one of the interesting effects is a stable temperature stratification of the atmospheric air, which creates favorable conditions to generate anisotropic layers being elongated horizontally with respect to the earth's surface. This effect is named as inner gravitational waves (IGW) generated in the stratosphere. The IGW were observed from the MIR space station for the first time by radar monitoring of the stratosphere. One should expect a strong influence of the anisotropic IGW on a laser beam at its propagation as inclined to these layers as it travels to a space vehicle.

So as applied to high-power laser propulsion, the general spectrum model of turbulent fluctuations of the atmospheric air index of refraction, Φ_n, may be written in a common form as follows [31]:

$$\Phi_{\mathrm{n}}(\kappa_1,\kappa_2,\kappa_3;h) = C^2(h)\eta^2(\kappa_1^2+\eta^2(\kappa_2^2+\kappa_3^2)+\kappa_0^2)^{-\mu/2}\exp(-\frac{\kappa_1^2+\eta^2(\kappa_2^2+\kappa_3^2)}{\kappa_{\mathrm{m}}^2}) \tag{7.24}$$

with the assumption of statistical symmetry of the fluctuations. Here, κ_1 is the number of vertical mode of fluctuations, κ_2 and κ_3 are numbers of horizontal modes of fluctuations, h is the current altitude of the atmosphere layer as a parameter, and C^2 (h) is structural constant of the turbulence that characterizes altitude dependence of fluctuations of the air index of refraction. Here, η is a parameter introduced as anisotropy coefficient determined as a ratio of air horizontal fluctuations to vertical ones, and μ as a parameter of the spectrum exponential dependence.

In accordance with the Karman model [32] of the atmospheric turbulence, there is the turbulence outer scale $L_0 = 2\pi/\kappa_0$ that is introduced to define its spatial spectrum more accurately jointly with the turbulence inner scale $l_0 = 2\pi/\kappa_{\mathrm{m}}$ introduced in accordance with the Novikov's model, which defines spectrum damping (molecule or vortex one) in the small spatial scale of the spectrum. Thus, there are four parameters which characterize the atmospheric turbulence in the upper atmosphere, namely, C^2, η, L_0, and l_0, which have to consider every case when we try simulating laser beam propagation in the Earth's atmosphere.

Unfortunately, the experimental data on the turbulence structure in the troposphere and the stratosphere are not complete enough to develop detailed turbulence models. Nevertheless, not all of these parameters are fundamental for studying the turbulence induced wandering of a laser beam. One of the principal parameters affecting the laser beam quality is a spatial intensity spectrum of every type of the turbulence. Moreover, the introduced anisotropy coefficient and outer scale of the turbulence are important parameters if we consider the inner gravitational waves.

The Kolmogorov model of isotropic turbulence is stated when $\mu = 11/3$, $\eta = 1$ in (7.24), and $C^2 = 0.033 \times C_{\mathrm{n}}{}^2$, where $C_{\mathrm{n}}{}^2$ is the refractive index structure constant of turbulent fluctuations of the air index of refraction. Then, Φ_n spectrum of the Kolmogorov turbulence can be written in the form:

$$\Phi_n^k(\kappa;h) = 0.033\, C_n^2(h)\left(\kappa^2+\kappa_0^2\right)^{-11/6}\exp\left(-\frac{\kappa^2}{\kappa_m^2}\right) \tag{7.25}$$

where $\kappa = \sqrt{\kappa_1^2+\kappa_2^2+\kappa_3^2}$ is an amplitude of the turbulence spatial wave number estimated under the assumption of statistical isotropy of the air fluctuations. It should be mentioned that there is a more precise relation between the inner turbulence scale l_0 and wave number κ_m as $l_0 = 5.92/\kappa_m$ in this case.

These relations for the Kolmogorov model of the turbulence were developed for the atmosphere of 10–20 km altitudes on the basis of the experimental data published in Ref. [33] as well as auxiliary data on radar monitoring of the upper atmosphere. To develop turbulence models for the atmosphere altitudes of 20–50 km, the data on stellar scintillation were used which were obtained by the MIR space station [34]. The numerical data on the turbulence characteristics are presented in Table 7.4 in accordance with the Kolmogorov model.

As for the IGW model, vertical distribution of temperature fluctuations corresponding to the power law can be written in the following form [31]:

$$V_T^g(\kappa_1) = AT^2\frac{\varpi_{B.V.}^{\;4}}{g^2}\kappa_1^{-3} \tag{7.26}$$

where κ_1 is the vertical wave number, A is a numerical coefficient, $\varpi_{B.}$ is the Brenta frequency so that $\varpi_B = 0.022$ rad/s for regular model of the atmosphere. In accordance with the theory of saturated IGW, recommended coefficient A is equal to 0.1 when $\kappa_1 > 0$ [31, 35, 36].

Table 7.4 Altitude profiles of the Kolmogorov turbulence parameters

H, km	$(C_n^2)_{min.}$, m$^{-2/3}$	$(C_n^2)_{avr}$, m$^{-2/3}$	$(C_n^2)_{max}$, m$^{-2/3}$	$l_{0,}$ m	$L_{0,}$ m
	weak	middle	strong	turbulence	
10	2.60E-18	3.95E-17	6.00E-16	0.017	300
12	2.08E-18	2.14E-17	2.20E-16	0.023	300
14	1.23E-18	8.37E-18	6.70E-17	0.031	300
16	6.10E-19	4.80E-18	1.99E-17	0.042	300
18	4.00E-19	2.50E-18	8.50E-18	0.057	300
20	1.81E-19	9.03E-19	5.20E-18	0.077	300
22	1.42E-19	7.09E-19	3.54E-18	0.103	300
24	7.20E-20	3.60E-19	1.80E-18	0.139	300
26	4.13E-20	2.07E-19	1.03E-18	0.186	300
28	2.48E-20	1.24E-19	6.21E-19	0.249	300
30	1.65E-20	8.27E-20	4.14E-19	0.334	300
32	8.49E-21	4.25E-20	2.12E-19	0.449	300
34	2.94E-21	1.47E-20	7.35E-20	0.602	300

Then, 1-D single-sided spectrum of the IGW fluctuations ($\kappa_1 > 0$) can be written in the form:

$$V_{\rm n}^{\rm g}(\kappa_1) = AN^2 \frac{\varpi_{\rm B.V.}{}^4}{g^2} \kappa_1^{-3} = {\rm Cl}_{\rm g}^2 \kappa_1^{-3} \tag{7.27}$$

where N is the index of refraction ${\rm Cl}_{\rm g}^2 = AN^2 \frac{\varpi_{\rm B.V.}{}^4}{g^2}$. Introducing formally both the outer and inner scales of these fluctuations, one can deduce 3-D spectrum $\Phi_{\rm n}{}^{\rm g}$ of the IGW, namely:

$$C_{\rm g}^2 = \frac{(\mu - 2)}{4\pi} {\rm Cl}_{\rm g}^2 = AN^2 \frac{3\varpi_{\rm B.V.}{}^4}{4\pi g^2} \tag{7.28}$$

where $C_{\rm g}{}^2$ is an analog of the structure constant of the atmospheric turbulence, and

$$\Phi_{\rm n}^{\rm g}(\kappa_1, \kappa_2, \kappa_3; h) = C_{\rm g}^2\,(h)\eta^2 \left(\kappa_1^2 + \eta^2\left(\kappa_2^2 + \kappa_3^2\right) + K_0^2\right)^{-5/2} \exp\left(-\frac{\kappa_1^2 + \eta^2\left(\kappa_2^2 + \kappa_3^2\right)}{K_{\rm m}^2}\right) \tag{7.29}$$

η is the coefficient of the IGW anisotropy, and $\eta >> 1$, K_0 and $K_{\rm m}$ are wave numbers corresponding to the outer scale $L_0{}^{\rm g} = 2\pi/K_0$ and to inner scale $l_0{}^{\rm g} = 2\pi/K_{\rm m}$.

All experimental data on the IGW published to the present time give the following values of the waves parameters, namely, $A = 0.1$; $\eta = 150$; $\varpi_{\rm B..V} = 0.022$ rad/s; $L_0{}^{\rm g} = 2.5 \times 10^3$ m; $l_0{}^{\rm g} = 50$ m, and altitude dependence of the air index of refraction N is given by a model of the standard atmosphere. It is also assumed that the internal scale of IGW does not depend on the altitude between 10 and 50 km.

Altitude dependencies of the IGW parameters are listed in Table 7.5 for three versions of the wave's power: low, medium, and strong. It should be emphasized that all the data in this table are based on the experimental investigations of IGW. But, these data are not normal ones, and it is necessary to make the data more exact in the near future.

Now, we can estimate how the upper atmospheric turbulence affects the laser beam passing through the upper atmosphere (z_0) and then in a vacuum from an airborne laser to the vehicle with a laser-propulsion engine. The following initial conditions are assumed to be fulfilled.

1. The laser beam is a collimated beam with a Gaussian profile of intensity distribution across the beam aperture of $a = 0.75$ m in radius. The radiation wavelength is $\lambda = 1.03$μm.
2. Initial coherent function of the laser beam is defined as

$$\Gamma_2^0\left(\vec{R}, \vec{\rho}\right) = I_0^0 \exp\left(-\frac{R^2}{a^2} - \frac{\rho^2}{4a^2}\right) \tag{7.30}$$

where R is a radius vector of the beam's energy peak.

Table 7.5 Altitude profiles of the IGW parameters

H, km	$(C_g^2)_{min}$, m^{-2}	$(C_g^2)_{avr}$, m^{-2}	$(C_g^2)_{max}$, m^{-2}	η	l_0, m	L_0, m
10	1.62E-19	5.05E-19	1.52E-18	150	50	2500
12	9.19E-20	2.87E-19	8.62E-19	150	50	2500
14	4.90E-20	1.53E-19	4.59E-19	150	50	2500
16	2.62E-20	8.18E-20	2.45E-19	150	50	2500
18	1.40E-20	4.37E-20	1.31E-19	150	50	2500
20	7.52E-21	2.35E-20	7.05E-20	150	50	2500
22	3.91E-21	1.22E-20	3.67E-20	150	50	2500
24	2.09E-21	6.53E-21	1.96E-20	150	50	2500
26	1.11E-21	3.47E-21	1.04E-20	150	50	2500
28	5.94E-22	1.86E-21	5.57E-21	150	50	2500
30	3.20E-22	1.00E-21	3.00E-21	150	50	2500
32	1.73E-22	5.41E-22	1.62E-21	150	50	2500
34	9.25E-23	2.89E-22	8.67E-22	150	50	2500
36	5.00E-23	1.56E-22	4.69E-22	150	50	2500
38	2.72E-23	8.51E-23	2.55E-22	150	50	2500
40	1.51E-23	4.72E-23	1.42E-22	150	50	2500
42	8.48E-24	2.65E-23	7.95E-23	150	50	2500
44	4.82E-24	1.51E-23	4.52E-23	150	50	2500
46	2.77E-24	8.66E-24	2.60E-23	150	50	2500
48	1.64E-24	5.13E-24	1.54E-23	150	50	2500
50	9.93E-25	3.10E-24	9.31E-24	150	50	2500

3. The altitude of the aircraft can change from 10 km through 15 km.
4. The aperture radius of the receiver telescope that is arranged on a space vehicle board is $R_s = 2.25$ m (see Chap. 1, section 1.4).
5. The space vehicle orbit changes from 300 km through 40,000.00 km.

As the calendared parameters of the estimations, the following are determined as functions of the z, λ, z_0, θ parameters, namely:

- $R_0(\lambda, z_0, \theta, z)$ is the diffractive radius of the beam that passed through homogeneous atmosphere.
- $R_{eff}(\lambda, z_0, \theta, z)$ is the effective radius of the beam that passed through the atmospheric turbulence characterized by the Kolmogorov model.
- $E_f(\lambda, z_0, \theta, z)$ is a portion of the laser power delivered to the receiver aperture of the space vehicle.

It is also assumed that the laser beam is completely coherent with a Gaussian intensity profile. In this case, the averaged intensity across the beam aperture is written as [36]:

$$< I\left(L,\vec{R}\right) >$$
$$= I_0(L)\int\limits_0^{\infty} J_0\left(\frac{2kaR}{Lg(L)}t^{1/2}\right)\exp\left\{-t-\frac{1}{2}D_\phi\left(L,\frac{2a}{g(L)}t^{1/2}\right)\right\}\mathrm{d}t \quad (7.31)$$

where J_0 is the Bessel function, $k = 2\pi/\lambda$, and $I_0(L)$ is the axial intensity of the diffracted beam, namely:

$$I_0(L) = I_0^0 a^2/R_0^2(L) \quad (7.32)$$

and

$$R_0(L) = \sqrt{a^2 + L^2/k^2a^2},\quad g(L) = \sqrt{1 + k^2a^4/L^2} \quad (7.33)$$

Here, $D_\Phi(L, \rho)$ is structural function of complex phase of a spherical wave.

We also use the assumption that $\kappa_m\, a/g(L) >> 1$ for inclined atmosphere paths, then [30]:

$$D_\phi(L,\rho) = 0.73C_\varepsilon^2(z_0)k^2X_{eq}\rho^{5/3} \quad (7.34)$$

$$X_{eq} = \frac{L}{C_\varepsilon^2(z_0)}\int\limits_0^1 C_\varepsilon^2(L\tau,\theta)(1-\tau)^{5/3}d\tau \quad (7.35)$$

and $C_\varepsilon^{\ 2} = 4\ C_n^{\ 2}$ is structure characteristic of the air fluctuations.

Then, the radiation intensity in the axis of the beam when $\vec{R} = 0$ may be written as:

$$< I(L,0) > = I_0(L)\int\limits_0^{\infty}\exp\left\{-t-\frac{1}{2}D_\phi\left(L,\frac{2a}{g(L)}\right)t^{5/6}\right\}\mathrm{d}t \quad (7.36)$$

To estimate the effective radius of the beam, R_{ef}, the law of energy conversation is applied, namely:

$$< I(L,0) > R_{ef}^2 = I_0(L)R_0^2 \quad (7.37)$$

and the portion of the laser power delivered to the vehicle is determined as:

$$E_f(\lambda, z_0, \theta; z) = R_s^2/R_{ef}^2 \ \text{ if } \ R_{ef} \geq R_s \quad (7.38)$$

Figure 7.4 illustrates one of the results of the analysis in view of the efficiency (or portion) of laser power collected by the vehicle receiver telescope of 5.0 m in

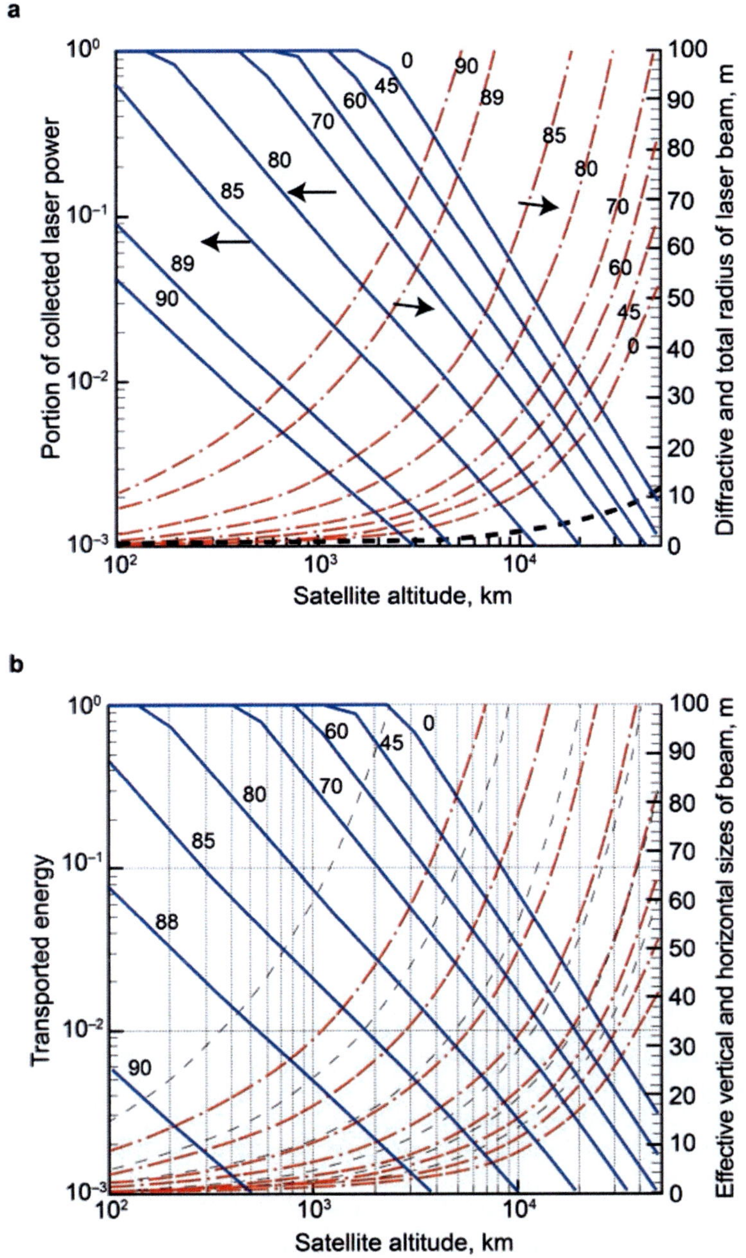

Fig. 7.4 Delivery efficiency of the laser power through the turbulent upper atmosphere. Numbers above curves is zenith angle. (**a**) only Kolmogorov's turbulence, (**b**) plus IGW are taken into account

diameter (solid lines) and beam radius (dashed lines) as functions of the vehicle orbit radius for various zenith angles of laser beam path in the turbulent atmosphere. The turbulence is assumed to be of middle intensity. Dashed lines correspond to the diffractive beam aperture.

As is seen from the figure, the whole laser power can be delivered into the receiver telescope aperture through an inclined atmospheric path at a distance of 1000 km if the zenith angle is less than 60°. If the zenith angle is $\theta \geq 80°$, then only 10% of the power is collected by the receiver telescope. If the vehicle reaches GEO, then only 3% of the laser power will be collected by the receiver.

Natural existence of the internal gravitational waves in upper atmosphere will cause additional fluctuations of the beam wave front as compared with the turbulence effects. Main part of these fluctuations will be observed through a vertical direction with respect to a horizontal one [32]. Assuming both types of the fluctuations to be independent, the total beam wandering effect can be defined as a root of a sum of squares of all these independent wave front fluctuations. These estimations of the laser beam parameters show that the IGW additive to the Kolmogorov's turbulence results in a sufficient decrease in the laser power delivery if zenith angles of the beam propagation exceed 80°.

7.4 Phase and Intensity Profiles of the Laser Beam That Passed Through a Turbulent Atmosphere

In previous Chaps. 1–5, we assume that the laser beam had a uniform intensity distribution across the beam aperture. But, we see from the previous section that the atmosphere turbulence can bring distortions into the beam's wave front, which could affect the intensity distribution across the beam aperture. To illustrate this effect, numerical modeling of the beam propagation in the turbulent atmosphere on the basis of stochastic phase screens is preferred [23].

To generate the isentropic stochastic phase screens, the spectral techniques are applied. These techniques allow creating spontaneous fields, in which the largest scales of the phase fluctuations which don't exceed half of a numerical field size. On the other hand, the minimum numerical steps have not to be less than one sixth of smallest scale of beam phase fluctuations.

As shown in the previous section, the upper atmosphere possesses some peculiar properties which can effect on simulating the spontaneous phase fields of the refractive index of the air. For the example, the inner scale of the turbulence-induced fluctuations of the refractive index is of $l_0 = 0.036$ m and outer scale $L_0 = 300$ m at 15 km altitude. These parameters mean that the numerical nets must include 100,000 by 100,000 grid points. It is evident that the calculation of the random fields by using such large nets will take a long operation time.

The method of phase screens that is applied here to calculate the propagation of the laser beam in the turbulent atmosphere is based on a combination of the spectral and modal approaches developed in Ref. [23]. But, this combination technique requires a special approach to solving the problem being considered. The technique is used to generate .stochastic phase screens characterized by a wide range of fluctuating scales. In this case, the stochastic field S is represented in the form of a sum of two (or more) phase fields being statistically independent of each other, namely:

$$S\left(\vec{\rho}\right) = S_{H1}\left(\vec{\rho}\right) + S_{H2}\left(\vec{\rho}\right) + S_{H3}\left(\vec{\rho}\right) + \cdots + S_B\left(\vec{\rho}\right) \tag{7.39}$$

Self-correlation functions of (7.39) belong to every stochastic field, and the totality of the fields is characterized by the $B\left(\vec{\rho}, \vec{\rho}'\right)$ correlation function that corresponds to a specified spatial spectrum. The approach of representing the total random field by a totality of the individual random fields corresponding to different scales of phase inhomogeneity is most useful for investigating turbulence in the upper atmosphere. This technique also allows enhancing the stochastic field by involving anisotropy of the field for selected scales of phase inhomogeneity.

As an illustration of the applied numerical technique, using the totality of independent stochastic fields of various spatial scales, the results of propagation of laser beam through the upper atmosphere are presented in Fig. 7.5.

In the first row of Fig. 7.5, individual realizations of the radiation intensity in the far field of the beam that has passed through the S_{H1} phase screen are shown. In the

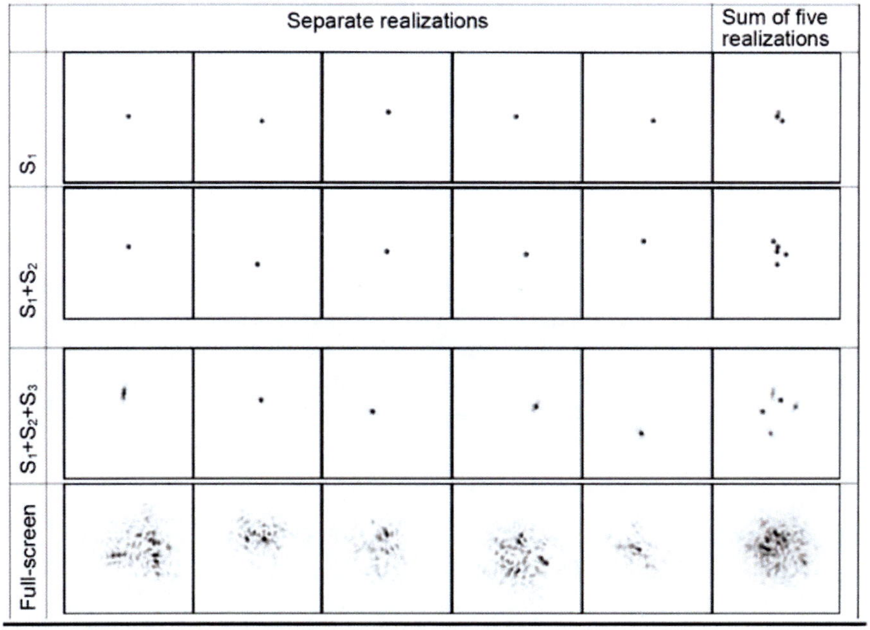

Fig. 7.5 Radiation intensity distribution in the far field of the laser beam. Initial beam diameter is of 1.5 m. Angular size of every square side is equal to 12μrad. (Model of Kolmogorov's turbulence with: $l_0 = 0{,}017$ m, $L_0 = 300$ m, and $C_n^2 = 6.0 \times 10^{-16}$ m$^{-2/3}$)

second row, realizations of the far-field beam distribution of radiation intensity, which has passed through the S_{H1} + S_{H2} screen, is shown. And consequently, far-field intensity of the radiation passed through the S_{H1} + S_{H2} + S_{H3} screens is shown in the third and fourth rows. In the fourth row, the total far-field of the beam that has passed through all of the complex phase screens is demonstrated. In the fifth column of the figure, all five realizations are presented to demonstrate the beam displacement as a whole.

It is seen that first two screens simulate such atmospheric distortions as a wedge that results in a displacement of the beam as a whole, the third screen results in speckling of the beam across aperture, and the screen simulating small-scale distortions of the atmosphere results in smoothing of the intensity distribution.

Figure 7.6 illustrates patterns of the radiation intensity distributions in a far field of the laser beam that propagates from the altitude of 10 km to the orbit of 30,000.00 km [55]. The turbulence effects are shown in a form of the intensity distribution across the beam aperture, delivered in 30,000.00 km orbit. The calculations were carried out for the case of a strong turbulence (see in previous section) and for a super Gaussian beam ($N_s = 100$) of 1 m in diameter, form-factor $Q_s = 1.006$, radiation wavelength $\lambda = 1.06\mu$m, pulse energy 100 J, and pulse width of 10^{-8} s. Zenith angle is assumed to be equal to zero.

As we see, the turbulence results not only in wandering and jitter of the laser beam but also in speckling of its far-field intensity distribution. This effect is expected to decrease the production efficiency of laser propulsion.

7.4.1 *Tentative Conclusion*

So, to predict adequately the turbulence effects on laser beam propagation in the upper atmosphere and on the efficiency of the laser power delivery to space vehicle, a numerical method based on the use of phase screen models of the turbulence has been shown to be effective. On the basis of this method, the spectral technique was used. To extend the practical fields of the spectral method applied, both the algorithm of nested grids and considered method of sub-harmonics were exploited. A special approach was also applied to create the phase screens, which considered the overall range of turbulence spatial scales corresponding adequately for solving the problem of laser energy delivery through the turbulent atmosphere.

7.5 Basic Atmospheric Effects Limiting Delivery of the Airborne Laser Power to Space Vehicle

7.5.1 *Scenario of Laser Power Delivery to a Space Vehicle*

The laser subsystems and optical elements being a part of high-power lasers are practically under the influence of various spurious factors, which increase the output

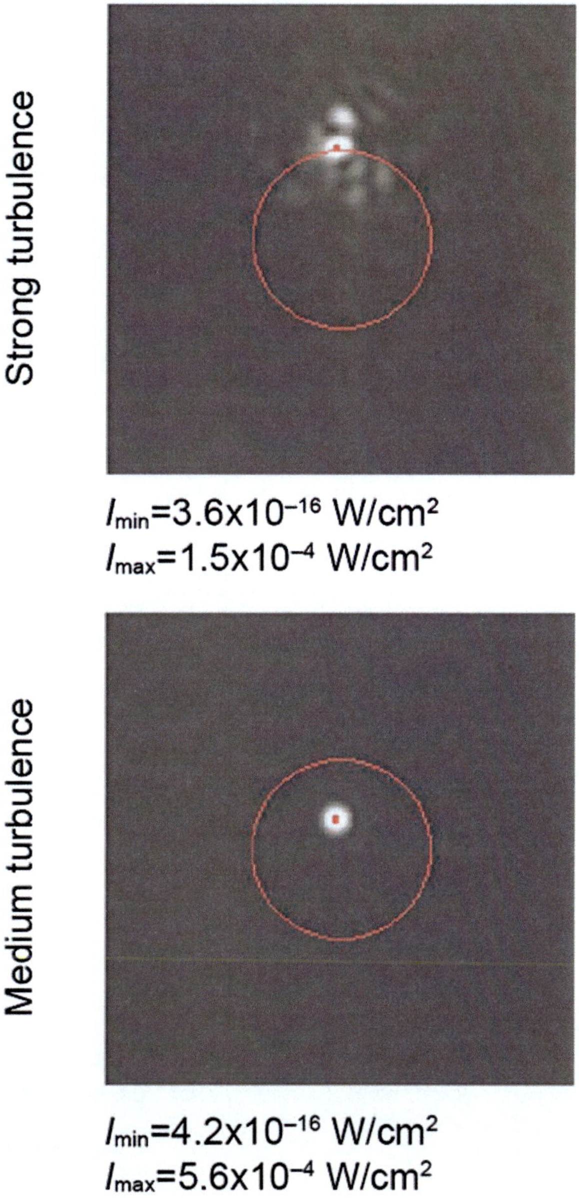

Fig. 7.6 Radiation intensity distribution in the beam delivered into the 30,000.00 km orbit for two levels of the upper atmosphere turbulence. The red circle diameter corresponds to 300 m in the orbit

beam divergence and decrease an efficiency of the laser power delivery to a space vehicle [4, 5]. These subsystems and factors are as follows:

1. Optical elements aberrations occur in both the primary and auxiliary laser subsystems

2. Optical cross coupling between the primary and auxiliary laser systems, and between the laser system and optical communication lines, and so on
3. Technological and production defects in the transmitter telescope as well as vibrations induced in the telescope structure and its thermal distortions
4. Laser beam phase distortions caused by an air surrounding the aircraft fuselage
5. Atmospheric turbulence inducing distortions in the beam
6. The so-called velocity aberrations of the laser beam, which are induced by the atmospheric turbulence because of a necessity to direct the laser beam into a lead point of space vehicle orbit for its precise illumination. The velocity aberrations are proportional to:

$$\Delta = \frac{2v_t L}{c} \tag{7.40}$$

where L is a distance between laser and vehicle, c is a speed of light, and v_t is tangential component of the vehicle velocity with respect to the laser.

In this section, we consider the basic factors limiting the laser power delivery from the airborne laser to space vehicle through the upper atmosphere, flying higher than 10 km altitude. The space vehicle orbits are considered to change from 300 km through 36,000 km. Figure 7.7 illustrates schematically the mutual positions of the aircraft and space vehicle during one of the possible scenarios of the laser and vehicle optical communication [9].

It is also assumed that the aircraft starts up from **A** point of the ground and flies to **B** point located at an altitude of $h = 10$–15 km during the time period of t_0 that is required the aircraft to arrive to B point corresponding to C point of the vehicle orbit. The vehicle orbital altitude is H, and spatial angle θ_{max} is between the OBO' surface

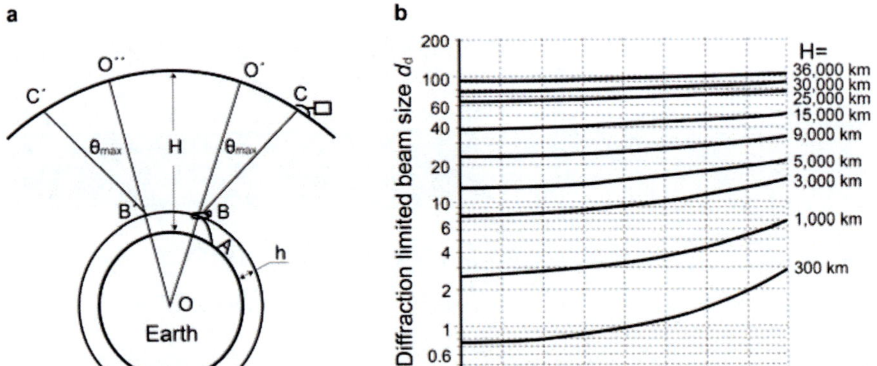

Fig. 7.7 (**a**) The general concept of propelling space vehicles by the airborne laser, and (**b**) diffractive size of the laser beam d_d delivered to the vehicle as function of zenith angle θ and vehicle orbit altitude H

normal and direction to the vehicle. This angle is determined by the time interval needed to deliver the laser power to the vehicle through the atmosphere with minimum losses.

During the aircraft flight to B point, the telescope of the onboard transmitter system is directed into the vehicle direction with an angular accuracy of $\Delta(t)$ at a specified lead angle $\theta(t)$. Then, the operation of the laser power transmission is started, which involves tracking of the vehicle position by a special optical servo system.

As for the receiver telescope arranged on the vehicle board, its aperture should be chosen to be as large as possible, taking into account the limitation imposed on the vehicle total mass. As an example, Fig. 7.7(b) illustrates the beam diameter dependencies on zenith angle for various vehicle orbit altitudes when the transmitter telescope diameter is chosen to be $D = 1$ m. As can be seen, the beam diameter changes from a diffraction size of 0.75 m at $H = 300$ km up to 90 m at GEO.

As usual, a permissible error of the laser beam guidance is assumed to cause the laser power losses to be less than 10% of the laser output power P_l. For example, this error will be $\Delta \approx \pm 0.3(2.44\ \lambda/D) = \pm 0.3 \times D_{rec}/L$, where D is the transmitter telescope diameter, D_{rec} is the receiver telescope diameter, L is the distance between these two telescopes, and 2.44 λ/D is diffractive beam angular divergence that is defined as a spatial angle within which 83% of the laser power is propagated.

Let us consider the case when the receiver telescope diameter remains constant during the entire orbital mission of the vehicle from low Earth orbit to GEO. That means that the laser beam has to be defocused to fill the entire receiver mirror surface, that is, $d = D_{rec}$ when the vehicle maneuvers at low Earth orbits. In these orbits, the permissible guidance error will be determined by the collector telescope diameter. As for intermediate vehicle orbits, corresponding to $L \approx D \times D_{rec}/(2\lambda) \approx 5000$ km, the beam diffractive diameter will be close to the receiver telescope diameter, and the guidance is allowed to be $\Delta \approx \pm 0.3 \times (2.44\ \lambda/D) = \pm\ 0.3 \times D_{rec}/L \approx \pm 3 \times 10^{-7}$ rad. And finally, the diffractive diameter of the laser beam will exceed the collector diameter at the orbits located close to GEO, and the permissible guidance error decreases with increasing of the collector's angular size.

It should be noted that the beam steering into the direction of the space vehicle with a specified angular accuracy has to be achieved simultaneously with operation of introducing a lead angle into the laser beam's pointing direction. The lead angle will change from $\Theta = 4.7 \times 10^{-5}$ rad up to 1.6×10^{-5} rad as far as the vehicle moves from LEO to GEO [10].

Thus, we can conclude that the laser beam that propels a space vehicle has to be directed to the vehicle with a guidance accuracy of $\pm 2 \times 10^{-7}$ rad that will be of order of the laser beam divergence, and many real factors such as atmosphere turbulence will disturb required guidance accuracy.

7.5.2 Turbulence Effects on a Laser Beam as Applied to High-Power Laser Propulsion

As we know, a major effect limiting efficiency of the laser power delivery to an LOTV is atmospheric turbulence even in the case when the aircraft with an onboard laser flying at 10 km altitude. Therefore, let us consider the turbulence influence on the general scenario of space vehicle missions supported by high-power laser propulsion. The laser beam is assumed to be a Gaussian with the intensity distribution $I(r) = I_o \exp\left\{-\frac{r^2}{2a^2} - ik\frac{r^2}{2F}\right\}$, where a is a radius of the beam, F is its radius of curvature, and I_o is the intensity maximum. It is also assumed that the turbulence is characterized by Kolmogorov's spatial spectrum in the following form [30]:

$$\Phi_n(\chi) = 0.033 C_n^2 \chi^{-11/3} \tag{7.41}$$

where χ and $C_n{}^2$ are the spatial frequency and structural constant of the turbulence in the atmosphere. In this case, the averaged intensity distribution $<I(z, r)>$ of a Gaussian beam after taking a vertical atmospheric path is determined as:

$$< I(z,r) >= \frac{a^2}{a_d^2} \int_0^{\infty} J_o\left(2\frac{r}{a_d}t^{1/2}\right) \exp\left\{-t - \frac{1}{2}D_s(2a)t^{5/6}\right\} dt \tag{7.42}$$

Here, $D_s(2a) = 0.182\ k^2(2a)^{5/3} \int_0^L \left(\frac{x-\varsigma}{L}\right)^{5/3} C_n^2(h + \varsigma\cos\theta)d\varsigma$ is a structural function of a spherical wave phase, L—a path length in meters; $J_0(x)$—a Bessel function; $k = 2\pi/\lambda$; a_d—radius of a beam that passes through a homogeneous medium, namely, $a_d = Lg_1/ka$, $g_1{}^2 = 1 + N^2{}_f\,(1 - L/F)^2$, N_f is the Fresnel number.

The effective radius of the Gaussian beam that passed through the upper atmosphere at various turbulence levels are shown in Fig. 7.8 [9]. Here, the beam spread function during the atmospheric path to a space vehicle is presented for strong ($C_n{}^2 \sim 10^{-15}$ m$^{-2/3}$) and weak ($C_n{}^2 \sim 10^{-17}$ m$^{-2/3}$) atmospheric turbulence when the aircraft altitude is $h = 10$ km (1) and $h = 15$ km (2), ($D = 1$ m, $\lambda = 1.06\mu$m).

As we see from the figure, the laser beam broadening will be 4–15 times in a vertical atmospheric path as compared with a diffractive beam if the aircraft flies at a 10 km altitude. But, if the aircraft is at a 15 km altitude then the turbulence effect on the beam spread will decrease significantly. We can see also the zenith angle at which the laser could operate must not exceed 40° to exclude strong turbulence effects.

In addition, such parameters of the laser radiation propagation in the atmosphere as the Freed radius and isoplanatic angle and their influence the efficiency of the laser power delivery as well [25]. Both parameters are important when considering the use of laser adaptive optical techniques to compensate for the beam wave front distortions caused by the atmospheric turbulence. The Freed radius defines spatial

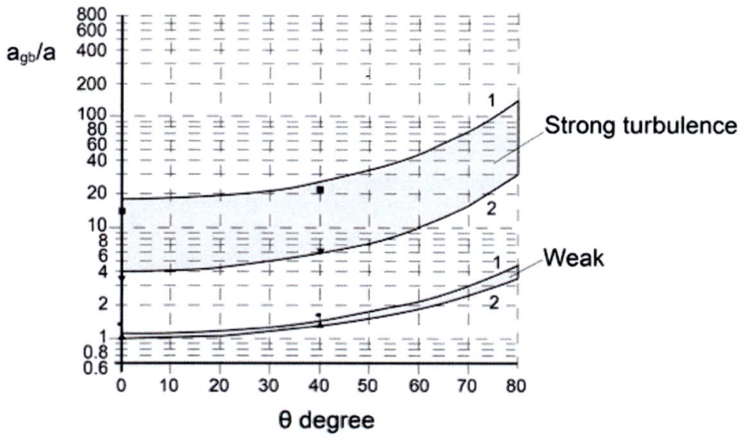

Fig. 7.8 Relative spread function of the Gaussian beam at various turbulence conditions

coherence of a spherical wave in the plane of receiver telescope, and the parameter depends on the turbulence level as follows:

$$r_o = \left[0.43k^2 \int_0^L C_n^2(h + z\cos\theta)\{(L-z)/L\}^{5/3}dz\right]^{-3/5} \tag{7.43}$$

Figure 7.9(a) [9] illustrates the Freed radius dependence on zenith angle for the effective beam diameter of $D = 1$ m.

As can be seen from the figure, the Freed radius is large and it exceeds the beam aperture if the turbulence is weak. In this case, the beam broadening because of the turbulent phase distortions of the beam wave front is small. However, if the Freed radius is smaller than the beam aperture when the turbulence is strong that causes strong phase distortions of the beam wave front and its broadening. It is also seen that the Freed radius depends on the zenith angle, and it increases with increasing θ.

The isoplanatic angle, α, depends on both the turbulent characteristics of the atmosphere and the diameter of the laser beam and radiation wavelength. Physically, the isoplanatic angle characterizes the phase distortion between two spherical waves which propagate in a turbulent atmosphere and are emitted by two point sources being mutually spaced, namely [30]:

$$\alpha = 0.528\left[k^2 \int_0^L C_n^2(h + z\cos\theta)z^{5/3}dz\right]^{-3/5} \tag{7.44}$$

That means that two laser beams which pass through the turbulence in strict opposite directions will get the phase distortions that are equal. But, insertion of a

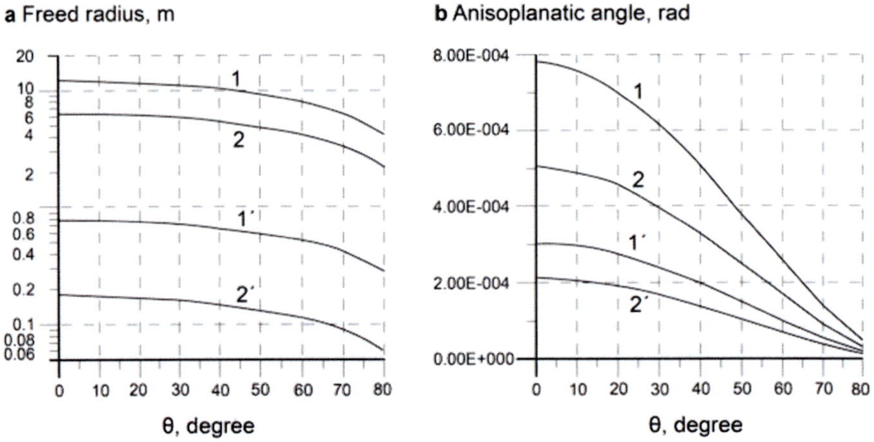

Fig. 7.9 (**a**) Freed radius and (**b**) isoplanatic angle as functions of zenith angle, θ. Turbulence: 1, 2—weak, 1′, 2′—strong; 1, 1′—$h = 15$ km, 2, 2′—$h = 10$ km, $D = 1$ m, $\lambda = 1.06\mu$m

lead angle into one of the beams will cause the difference in the beam wave fronts due to the atmospheric turbulence.

Figure 7.9(b) illustrates the isoplanatic angle dependencies on a zenith angle for following beam propagation parameters, namely, $D = 1$ m, $\lambda = 1.06\mu$m, $\alpha_{\text{dif}} = 2.44\ \lambda/D = 2.6\mu$rad. As can be seen, the angle varies over a wide range as functions of the atmospheric turbulence and on the zenith angle.

As for compensation for the high-power laser beam distortions, it is important to estimate the comparative influence of isoplanatic angle, α, and lead angle, γ, on the efficiency of laser adaptive techniques as applied to space vehicle missions, taking into account a limited nature of the speed of light when both the aircraft and vehicle move with different velocities. In this case:

$$\gamma = \frac{2\left(V_{\text{s}}^{\perp} - V_{\text{a}}^{\perp}\right)}{c} = \frac{\sqrt{\frac{gR_{\text{Earth}}^2}{R}}\cos\nu - V_{\text{a}}'\cos\theta}{c/2} \tag{7.45}$$

$$\nu = \arcsin\left(\frac{r_{\text{orb}}}{R_{\text{orb}}}\sin\theta\right)$$

Here, $V_{\text{a}}^{⊥}$ and $V_{\text{s}}^{⊥}$ are the velocity components of the aircraft and vehicle, being perpendicular to a radius vector between the center of the Earth and the vehicle, R_{Earth} is the Earth radius at an equator, c is the speed of light. As is seen from Fig. 7.10 [9] the lead angle decreases at increasing vehicle orbit altitude, H.

To estimate the laser power delivery as a function of the lead angle, the transmitter telescope magnification, $G = \frac{\langle I_{\max}\rangle}{G_{\text{DL}}}$, is required. The magnification is a ratio of an averaged radiation intensity delivered to the space vehicle to the theoretical limit of the intensity, that is, $G_{\text{DL}} = E/4\pi(D_{\text{em}}/\lambda L)^2$. Usually, this parameter is less than

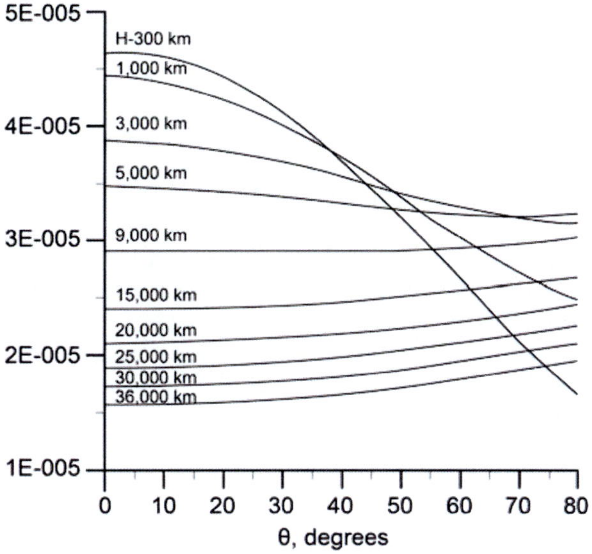

Fig. 7.10 Lead angle (radian) dependencies on zenith angle when the aircraft velocity $V^{a} = 250$ m/s

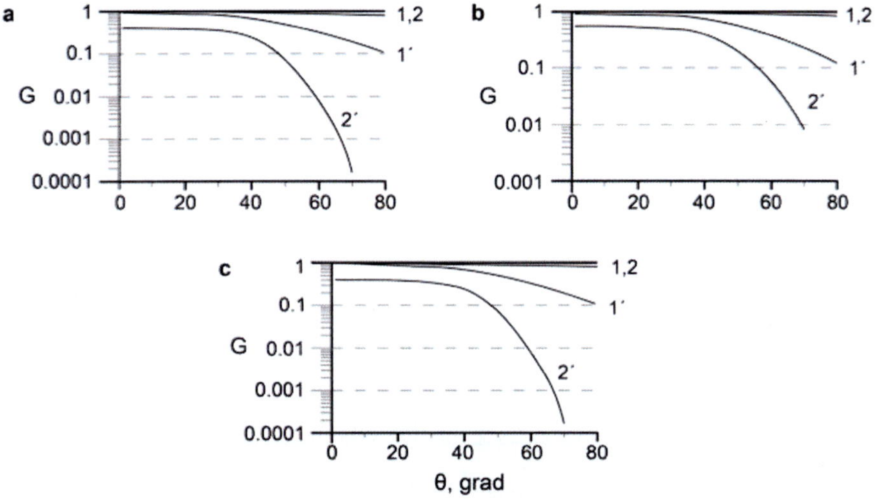

Fig. 7.11 The transmitter telescope magnification versus zenith angle for various altitudes of vehicle orbit, H, and at different turbulence. (**a**) $H = 3000$ km; (**b**) $H = 15{,}000$ km; (**c**) $H = 36{,}000$ km. 1, 2—weak turbulence; 1′, 2′—strong turbulence. 1, 1′—$h = 15$ km; 2, 2′—$h = 10$ km

unity. The results of the numerical estimations of the transmitter telescope magnification for the same parameters of the laser power are presented in Fig. 7.11.

For effective laser beam delivery to a space vehicle, it is necessary to limit a permissible range of the zenith angle to a value of $\theta \sim 30^{o}$–40° to transmit the laser power from an airborne laser to space vehicle with minimum losses.

7.6 Adaptive Laser Systems for the High-Power Laser Propulsion

7.6.1 Statement of the Problem

It follows from above that the principal problem that will limit the application of high-power laser propulsion is the Earth's atmosphere which has a strong impact on the laser beam and limits the efficiency of laser power delivery to a space vehicle. Therefore, we consider some laser adaptive optical techniques to compensate for the high-power laser beam wave front distortions, which have been developed. The main attention is paid to the airborne laser systems considered in Refs. [5, 37].

Usually, the high-power laser system includes both high-power lasers that generate near-IR radiation, a transmit telescope about 1 m in diameter, and auxiliary optic-electronic systems incorporated into the communication optical part of the system. The communication systems are used to find and track a space vehicle and to collect the information concerning phase distortions of the beam along the atmospheric path to the space vehicle. The system design stipulated that the transmitter and receiver telescope arranged on the vehicle board used one and the same principles [6].

In practice, all laser subsystems incorporated into the high-power laser system are exposed to various distortions which have to be compensated by the adaptive optical system as well. But here, we consider only those which have a strong influence on the output laser beam as it travels through the atmosphere from the laser to a vehicle. These distortions are as follows:

- Phase and amplitude distortions imparted by the atmospheric turbulence into the beam wave front, which we considered in previous sections.
- The so-called velocity aberrations of the laser beam, which are caused by a limited nature of the speed of light at pointing of space vehicle (see (7.40)).

Hence, let us consider the problem of compensation for laser beam distortions caused by the Earth's atmosphere.

As is known, adaptive optical systems are widely used to observe astronomical objects, including planets and stars, through the Earth's atmosphere to a diffractive limited quality. To get high-quality images of these objects, flexible mirrors guided by a number of actuators are applied in the adaptive optical system. The actuators operate with a 500–1000 Hz frequency that allows compensating for the image distortions. This fast speed of the adaptive optical system operation is imposed by the time constant of atmospheric turbulence. As the examples of the adaptive optical systems, we can mention the Altai telescope (Russian Federation) from Ref. [38] and the Sun telescope from Ref. [39].

One of the principal techniques of the adaptive optical system is utilization of a reference light source to measure its wave front errors induced by the Earth's atmosphere. Usually, bright stars are used as the reference sources during the first stages of the system development. Later, the Riley or sodium stars generated in the upper atmosphere by lasers are substituted for the stars. To analyze the reference source image and to form the conjugate surface profile on the flexible mirror, special Hartmann wave front detectors are used [40].

But, we must mention some of the problems concerning application of the adaptive optical technique when applied to high-power lasers. One of the problems is thermal heating of the mirrors by the high-power laser radiation. The heating results in the mirror surface deformations, which result in additional wave front distortions of the laser beam. To compensate for all the beam wave front aberrations, an internal adaptive optical loop must be developed. Additionally, to decrease heating of the mirrors, special optical coatings of its surfaces are used, which increase reflection of the laser radiation.

Another problem is related to the "velocity phase aberrations" mentioned above. The adaptive optical techniques usually use the same optical path for the reference and high-power beams that result in consequent compensation for the high-power beam aberrations. This condition is satisfied if the beam passes within the isoplanatic angular. But, for the longer beam atmosphere paths, the isoplanatic angle range is smaller. In a common case, this angular depends on the relative velocity of a space vehicle, inclination of the beam atmosphere path, and on the atmospheric turbulence level (see Fig. 7.9(b)).

Summarizing the argumentation mentioned above, one can conclude that the development of the adaptive optical system compensating for the high-power laser beam aberrations caused by passing the beam from the airborne laser to a space vehicle must include three stages, namely:

1. Analysis of the adaptive methods of compensation for laser beam distortions, which have been developed, and searching of the optimal techniques to increase the beam delivery efficiency through a turbulent atmosphere.
2. Solving the problem of heating internal optical elements of the high-power laser system, starting from the laser resonator up to the transmit telescope by using special coatings for the optics, as well as a development of an additional adaptive optical subsystems.
3. Development of the laser adaptive optical system for compensating for the beam's wave front distortions of caused by the beam atmospheric path through injection of a lead angle into the laser beam direction.

It should be also noted that the optical systems developed for tracking of space vehicles [41] and laser beam guidance systems [42] are capable of being applied to the high-power laser-propulsion systems. There is also the possibility of development of a ground-based adaptive optical system to compensate for the laser beam atmospheric aberrations.

Nevertheless, most developed laser adaptive optical system is the ABL an installed on a Boeing 747; the operation capabilities of which were demonstrated in 2010 during flight experiments in upper atmosphere [43]. The ABL system is assumed to be developed on the following principles of the beam tracking and pointing at a remote target:

- compensation for the beam wave front distortions caused by the inner and external optical paths aberrations simultaneously
- compensation for the mechanical and acoustic vibrations of the optics elements which are typical ones for the airborne laser systems
- achievement of a pointing at a remote target by the laser beam with a sub-diffractive accuracy
- detection of all aberrations in the beam path into a direction for a lead angle to compensate for "the velocity aberrations" of the beam
- development of the adaptive system with a fast response as a whole system because of fast dynamic rate of the turbulence characteristics

The following phase conjugation techniques are usually applied:

1. linear adaptive systems which include the beam wave front analyzers, fast computers, and electrically driven flexible mirrors being
2. nonlinear phase conjugation via interaction of the probe laser beam with a nonlinear optical medium in the phase conjugation mirror (PCM)
3. holography correctors which can be used instead of PCM [44]

In any case, the adaptive optical technique that is used in the laser adaptive system has to possess a wide field of view of $\pm(50–100)\ \theta_d$, where θ_d is diffractive divergence of the laser beam, at high sensitivity of PCM to laser power and to provide a high speed of response to compensate for all possible aberrations of the beam.

7.6.2 Adaptive Optical Laser Circuits and Special Equipment

The linear adaptive optical technique based on aperture sounding of the beam path to a remote target operates due to analysis of wave front phase distortions of a probe beam and generation of such conjugate beam phase distortions which result in a maximum power reflected by the target [44]. The power of the reflected beam is a criterion of conjugation of the probe beam wave front.

The aperture sounding technique possesses some advantages compared with nonlinear phase conjugation technique. First of all, there are no limitations on its application under strong atmospheric turbulence, and the technique can be easily applied to CW lasers. Some examples of the applications for solving the problems of laser power delivery to a space vehicle are presented in Ref. [43]. But, the technique is inherently limited by the number of the channels which have as interaction of these channels with each other. But, the principal disadvantage of the aperture sounding technique is a large feedback time of response that cannot allow the technique to be used for very long-distance laser power transfer.

As for the phase conjugate technique, the adaptive laser system must use a few adaptive circuits which differ from each other by both the beam wave front analyzer parameters, including number of optical detectors and their sensitivity and the

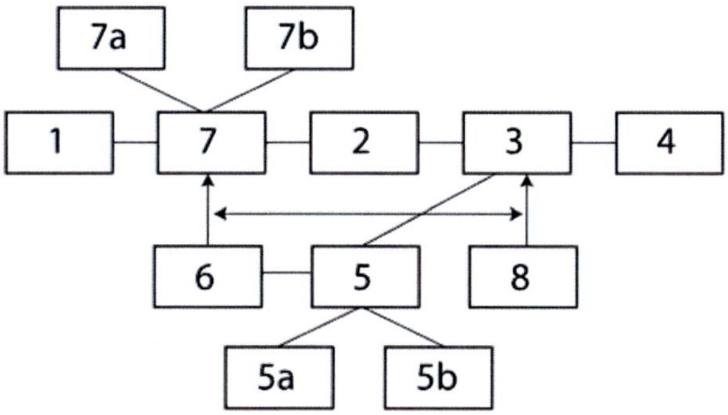

Fig. 7.12 Block diagram of a laser adaptive optical system. (1) master oscillator; (2) multi-pass power amplifier; (3) diffractive beam splitter,(4) transmit telescope; (5) beam wave front analyzer (5a—internal, **5b**—atmospheric); (6) computer; (7) phase corrector (7**a**—internal corrector, 7**b**—atmospheric corrector), (8) probe laser

drivers correcting the wave front phase, including its speed of response and number of actuators.

The typical block diagram of two-loop adaptive laser optical system is shown in Fig. 7.12 [45]. Here, laser radiation of the master oscillator (1) is directed into power amplifier (2) and then to the transmit telescope (4) by means of a few optical elements.

To control the beam wave front inside the laser, the internal adaptive circuit is applied on the basis of beam wave front detectors (5a) and beam phase correctors (7a). To control the output beam wave front, the atmosphere adaptive circuit is developed on the basis of more sensitive wave front detectors (5b) and fast correctors (7b). The beams wave front characteristics are processed by a computer (6) to drive the beam phase correctors. The optical system (5) is used as additional adaptive circuit to compensate for the wave front phase aberrations caused by heating of the system's optics.

7.6.2.1 Beam Wave Front Analyzers (BWA)

The BWA is assumed to operate for both incoherent and coherent light beams. The technique, based on the use of coherent characteristic of laser beams, is interferometry operating either by direct or heterodyne detection of light signals. As for the incoherent technique, the shack Hartmann method or its modifications are applied [46].

For the example, the MIT Lincoln Laboratory [47] as a general developer of the ABL laser adaptive systems used the Hartmann analyzer to control the high-power laser beam wave front in to real atmosphere at distances of 100 km. The Hartmann analyzer consisted of 16 $\times$ 16 sub-apertures, including a binary optical system and two CCD cameras with an array detector of 64 $\times$ 64 pixels for X and Y coordinates. The detector was characterized by a dynamical sensitive range of $\pm 1.4\ \lambda$ and phase accuracy detection of $\lambda/15$ for visible wavelength, λ, and speed of response of 240μs that corresponds to a 4 kHz dynamic frequency.

BWA of coherent laser radiation allow detecting the light signals of a small intensity by using a heterodyne technique of signal detection. This type of BWA is usually applied to develop the laser adaptive optical systems operating through super-long space distances up to 40,000 km. In this case, the probe laser is stabilized throughout the radiation frequency and power. This laser is called as heterodyne laser.

7.6.2.2 Beam Wave Front Phase Correctors

There are two different techniques that form the basis of the mirror correctors which can be used to control a laser beam wave front. The first technique is called electro-optical correctors, which are made of optical materials whose index of refraction depends on the strength of the applied electric field. The second technique uses flexible mirrors with a mechanically controlled optical surface, which is also called adaptive mirrors. The frequency range of the mirrors control is changed from a few Hz through a few kHz (Table 7.6).

The Litton/Itek Optical Systems (USA) which is a leader, starting from 1973, in the field of production of the flexible mirrors with a controlled surface. For example, LVEM mirrors from that Company with 69 and 241 control channels were used in

Table 7.6 Litton/Itek adaptive mirror correctors

Parameter	Solid piezomirror SPM	Mirrors with discrete actuators: SADM	LVEM (low voltage)	CSEM (cooled silicon)
Diameter, cm	2.5–20	16 or 25	10–100	10–20
Number of actuators	21–35	37 or 55	21–500 (>2000)	250/241
Distance between actuators, cm	0.25–1	2.75	0.7–0.8	<1
Sensitivity (nm/V)	0.4	5.5	15	15
Maximum bias, μ	±0.7	±7.5	±2.5–3.5	±2
Frequency range, kHz	10	0.3	1–10	1
Surface optical quality ($\lambda = 0.6$μm)	λ/10	λ/4	λ/15	λ/15
Maximum voltage, V	±2700	±1500	±75	±75
Actuator material	PZT (Pb(Zn, Ti)0.3)	PZT	PMN Pb (Mg1/3Nb2/3) 0.3)	PLZT Zr, Ti, Pb + La
Individual properties	Stability	Hysteresis, high supply voltage	Without hysteresis and drift	Large amplitude of deformation

the experiments on compensation for the thermal blooming atmospheric effect both in the laboratory and real atmospheres, which were carried out by MIT Lincoln Laboratory [48]. Moreover, the flexible mirrors consisting of 241 actuators with a rate of response less than 300μs were developed in the Laboratory.

7.6.3 *Laser Adaptive Optical Systems as Applied to Beaming a Remote Target*

As examples, here we present two typical adaptive systems developed for high-power lasers as applied to transferring laser radiation through the Earth's atmosphere. These systems differ from each other via the beam wave front correctors used in the systems.

7.6.3.1 Linear Adaptive Laser Systems

One of the first laser adaptive optical systems developed was to transfer CO_2 laser radiation of 1 MW power through a vertical atmosphere path, it is shown in Fig. 7.13, in terms of a block schematic diagram of a multi-circuit single-channel system [45]. The adaptive loops of the system are based on a wave front analyzer (WFA) and heterodyne analyzer of a beam wave front (HWFA).

The high-power laser operates as follows: laser radiation that is generated in a master oscillator (MO) and amplified by laser power amplifier (A) comes to the diffractive splitter DS1 that redirects a part of the radiation to the Hartmann wave front analyzer (HWFA). The analyzer processes the beam wave front to define the beam phase aberrations such as wedge, sphere, and aberrations of higher orders and generates conjugate signals for driving the flexible mirror (PC). The laser adaptive optical system can embody a few adaptive loops (circuits) independent of the internal optical path length, the number of power amplifiers, and so on. The phase-corrected laser beam is then directed to transmitter (T) that sends the beam to a remote target.

The presented adaptive optical laser system consists of two adaptive loops, including (1) an adaptive loop to correct aberrations of the transmit telescope, and (2) an adaptive loop to correct the laser beam aberrations which are caused by the Earth's atmosphere. To develop the second adaptive loop, a laser beacon radiation is passed from a remote target through the atmosphere and collected by the laser telescope. After the telescope, the beacon radiation reflected from the beam splitter (BS2) is directed to a high sensitive wave front analyzer (WFAL) using heterodyne detection techniques. A multiple-actuators flexible mirror is used to correct the beam wave front for atmosphere-induced aberrations, which is installed between MO and PA. To steer the laser beam pointing direction, an intracavity spatial light modulator (SLM) is used in the adaptive system.

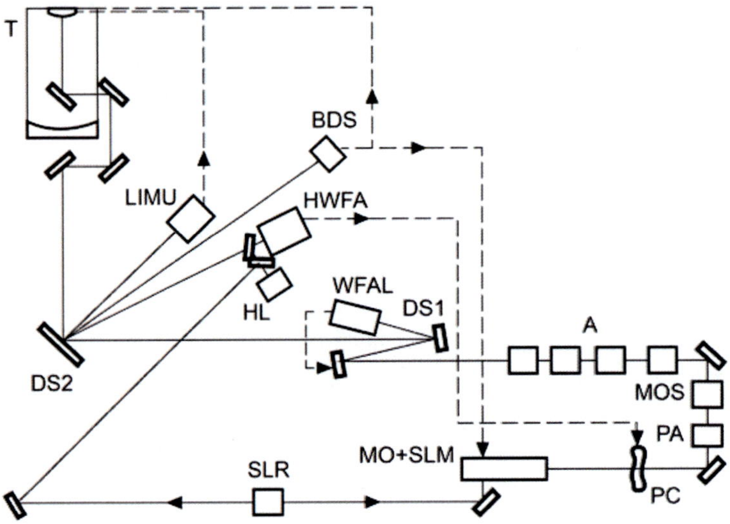

Fig. 7.13 Optical block scheme of the adaptive single-channel high-power laser [3]. *T* is transmitting telescope, *MO* is master oscillator, *A* is laser power amplifier, *WFAL* is wave front analyzer, *HWFA* is the heterodyne analyzer of a wave front, *SRL* is probe stabilized laser, *DS1* and *DS2* are diffractive splitters, *BDS* is beam direction sensor, *LIMU* are laser interference measuring unit, *MOS* is matching optical system, *PC* is phase conjugate mirror, and *SLM* is spatial light modulator

This laser adaptive system was tested experimentally by using a 100 kW CO_2 laser [49] with 3 kJ pulse energy at 100 Hz pulse repetition rate (see Fig. 7.14). The output laser beam was directed by using water cooled metal optic to an adaptive telescope consisting of 18 sub-apertures of 1.8 m in total diameter through a horizontal optical tube. The telescope magnification was $M = 5.6^{\times}$. After the telescope, the laser beam was focused at a distance of 0.5 km along a horizontal atmosphere range.

In the experiments, the adaptive laser system provided optimal focusing of a high-power laser beam at a 0.5 km distance by using a signal from a laser beacon installed at the end of the range. The beacon radiation of a 0.63μm wavelength "read out" the atmospheric aberrations distributed along the range, and BWA analyzer transformed it to control output laser beam wave front. The experiments proved the operating capability of the laser adaptive optical system developed.

Further improvements of the laser adaptive system were directed to phasing of sub-apertures of the transmitter telescope primary mirror.

Another approach to the development of an airborne adaptive optical laser system was developed during the American AirBorne Laser program [4, 5]. The ABL adaptive optical system is based on a three-circuit adaptive system that incorporates: (a) a two-loop system of the active illumination of vehicle flying in the atmosphere and (b) a loop of high-power laser beaming the vehicle with a high spatial accuracy

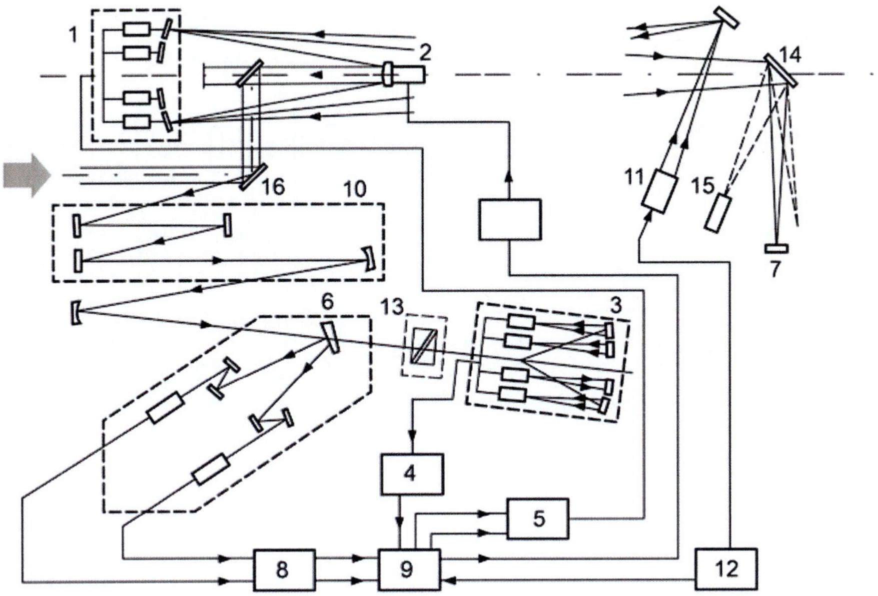

Fig. 7.14 Schematic diagrams of the laser adaptive telescope and laser beam control system. In the experiments, analysis of the laser beam wave front phase distortions was carried out by using 18-channals Hartmann analyzer. (1) 18 sub-apertures primary mirror; (2) secondary mirror with actuator to control its longitudinal displacement; (3, 4) 18-channels BWA; (5) PM drivers control unit; (6, 7) beam focus detectors; (8) driver control unit for the secondary mirror; (9) signals processing unit; (10) optical units; (11, 12) simulator of reflected signals; (13) simulator of beam wave front tilts; (14, 16) diffraction beam splitter; (15) matrix detector

by using the linear adaptive optical circuit compensating for the laser beam wave front aberrations caused by the atmosphere.

The physical scale model of first two-circuit part (a) of the ABL system, which was tested at the MIT Lincoln Laboratory real atmospheric test range, is shown in Fig. 7.15. The application of the nine-beam illuminator allowed the collection of lengthy image data of the illuminated rocket, which allow selecting of the part of the rocket to be illuminated by a four-beam illuminator. The laser radiation of the four-beam illuminator that is reflected by a moving rocket is used as a reference laser beam in the adaptive optical laser circuit. The reflected reference beam is used to form a phase shape of the high-power laser beam wave front.

The red lines in the figure correspond to the alignment of laser radiation simulating propagation of high-power laser beam through the system after conjugation procedure of the reference beam wave front (Table 7.7).

One of the principal results of the test experiments is shown in Fig. 7.16, illustrating the efficiency of the ABL adaptive optical system to compensate for the atmospheric turbulence distortions of high-power laser beam. The efficiency is estimated by a Strell number that is a ratio of the maximum intensity of the beam that passed through the turbulent atmosphere to the maximum intensity of the beam at

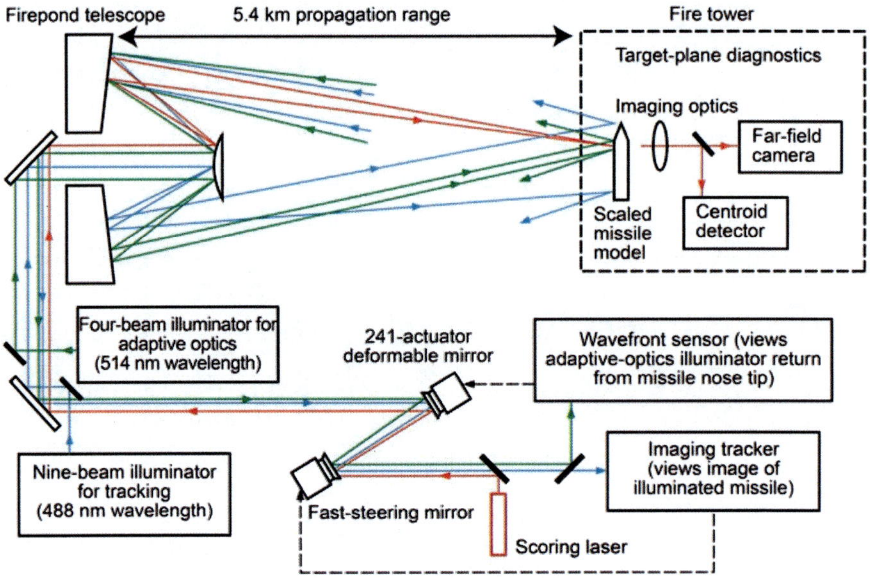

Fig. 7.15 Schematic diagram of the ABL adaptive two-circuit system

Table 7.7 The experimental equipment parameters of the ABL laser adaptive system

Flexible mirror		Beam wave front detector	
Number of actuators	341	Type	Hartmann's
Geometry	21 × 21	Geometry	218 by X axis 218 by Y axis
Shift amplitude	4μm	Detector	64 × 64 CCD array with a 3 kHz frame rate

Fig. 7.16 Results of the experimental tests of the ABL adaptive systems at the atmospheric path of a 5.4 km length

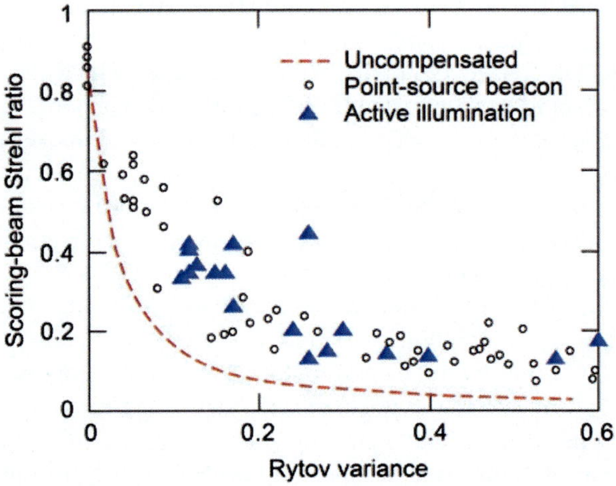

the diffraction approach. A characteristic parameter of the atmospheric test range, the Rytov variance is used, which is defined by both the turbulence, length of the atmospheric path, and the radiation wavelength [50].

As one can see, the beam aberrations caused by a strong turbulence (Rytov parameter is of more than 0.2) cannot be compensated completely by the developed adaptive optical system. There are a few reasons for that. One of these is operational speed of the system as a whole, or the system response time rate, which has to be less than the turbulence "freeze time." Another reason may be inaccuracy of phase conjugation of the wave fronts of reference beams, and so on.

7.6.3.2 Nonlinear Adaptive System Based on the Interaction of Laser Radiations with a Nonlinear Optical Medium

There are a number of nonlinear phase conjugation techniques that have been developed and applied to both RP and CW high-power laser systems. All of these techniques are based on the generation of a beam with a conjugated wave front via the interaction of a probe (reference) and signal beams in nonlinear optical mediums. A number of techniques such as four-wave mixing, Brillion scattering, and so on can be used [51, 52].

Four-wave mixing is a process similar to the dynamic holography phase corrector being written in an optically addressed liquid crystal modulators of light [53]. At the present time, the crystals enable writing the hologram gratings with spatial frequency of 100–200 lines per mm as for visible and for IR wavelengths. Other nonlinear processes such as thermal gratings, photo refraction, saturation of amplification in a resonant medium, and so on can serve as phase conjugation mirrors, too. Resonance media are preferable for that use because they possess a low intensity threshold to write a hologram, no limitations on the time of the beam interaction, the possibility of obtaining a reflection coefficient of 100%.

Degenerate four-wave mixing [52] increases the generation efficiency of the conjugated wave in the case of a weak reference beam. In this case, the PCM medium itself can be used to generate a strong conjugated wave. The process of DFWM is also achieved in lasers active media in a wide spectral range starting from UV to the far IR, including Nd:YAG, Nd:KGW, Nd:Cr:GSGG, Nd:YVO_4, and CO_2 lasers.

One of most interesting results was obtained by using an Nd:YVO_4 crystal as a nonlinear medium for a PCM [54]. Among the basic advantages of this crystal are a large cross section of resonance transmittance $\sigma_0 \approx 18 \times 10^{-19}$ cm^2 that is five times larger than Nd:YAG, a high absorption coefficient of ~30 cm^{-1} for 1-at %-doped Nd^{3+} at laser diode pumping wavelength, and a long excited state of Nd ($\tau \approx 70\mu s$). Maximum coefficient of transformation of pump beam into the conjugate reaches 10% for the two-pass PCM and 170% for a four-pass mirror.

The experimental results on phase conjugation which were obtained by using DFFW mirrors are assumed to be applied to transfer high-power laser radiation to a space vehicle by using various artificial sources for the reference beam, including "laser-guided stars," stimulated backscattering of high-power radiation by the atmosphere aerosols and molecules, and so on. These techniques are considered as applied to propelling of space vehicles for its orbital missions [55] as well as illuminating of satellites by high-power CW lasers [56, 57], and so on [58].

Here, we would like also to present the original laser adaptive optical technique to propel space vehicles (LOTV) by using high-power lasers, which is based on phase conjugation of a reference beam by using "artificial source of a coherent reference radiation" and dynamic holography correctors [55].

The general objective of the algorithm is solving the problem of a non-reciprocal propagation of the probe and high-power laser beams through the Earth's atmosphere as applied to propelling space vehicles with laser propulsion. The basic idea of the algorithm is a functional separation of two processes being responsible for generation of a phase-conjugated beam, namely, recording the atmospheric optical distortions in the direction of a lead point of the vehicle orbit, and generation of high-power laser beam with a conjugated wave front. The algorithm is proposed to be developed in the form of a two-circuit adaptive laser system as it is shown in Fig. 7.17.

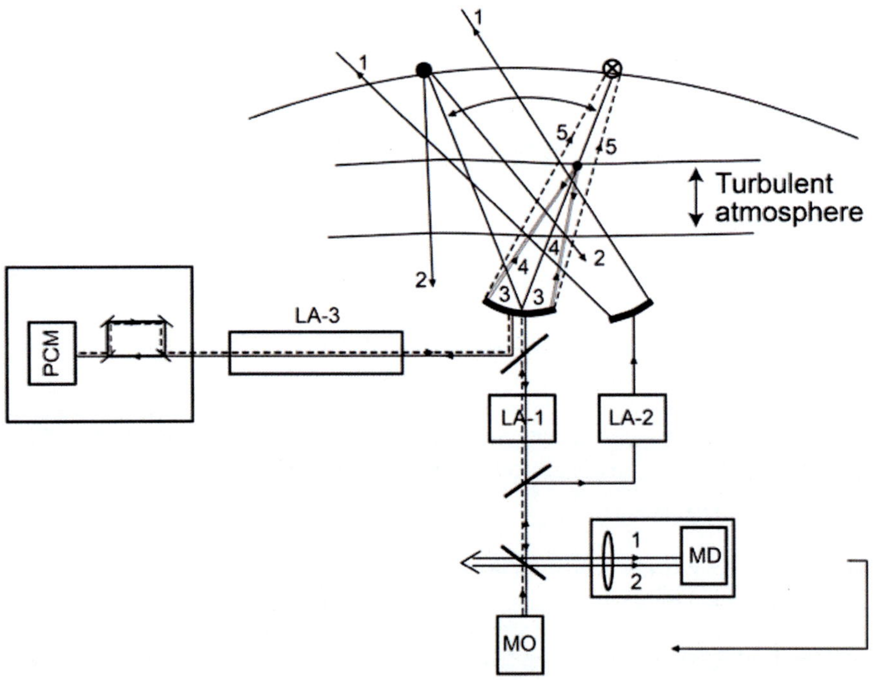

Fig. 7.17 Circuit diagram of beam-addressing algorithm on the basis of dynamic holography correctors. (1) first probe beam; (2) reflected beam; (3) second probe beam; (4) stimulated Raman scattering; (5) conjugated high-power laser beam. *(LA-1)* first power amplifier, *(LA-2)* second power amplifier, *(LA-3)* final power amplifier. *PCM* phase conjugation mirror, *MO* master oscillator, *MD* detector array

The first adaptive circuit is designed to generate an auxiliary source of a reference beam in the direction of the lead point of vehicle orbit. This source is assumed to originate due to counter-propagating stimulated Brillouin scattering (SBS) of an auxiliary laser beam with a wavelength corresponding the wavelength of high-power laser. The second adaptive circuit is designed on the basis of conventional PC mirror to generate a high-power laser beam with a conjugated wave front. To obtain the conjugated wave front of the beam, a hologram-corrector is written in a nonlinear optical medium by means of the auxiliary reference and probe beams.

The adaptive laser optical system operates as follows.

At first operation stage, the master oscillator pulse is amplified by the first laser amplifier (LA-1) and irradiates the space vehicle in the angular direction specified by the system of preliminary guidance of the laser adaptive optical system. This beam is transmitted by using an auxiliary telescope system that possesses the guidance precision of one tenth of the diffraction limits of the beam. The angular direction of the master oscillator beam is monitored by a detector array that also records the satellite image. The reference radiation pulse reflected by the vehicle is collected by the second receiver-transmitter telescope and then is directed to the same detector array that produces a signal of angular deflection between the master oscillator beam and image of the satellite. The angular deflection signal is then used at the second stage of the algorithm operation to form artificial reference source in the upper atmosphere.

The second stage, the master oscillator beam is amplified by a second power amplifier (LA-2), and the beam is focused at an altitude of 30 km in a lead angular direction corresponding to a predicted angular position of the satellite. In this process, the atmospheric aberrations registered at the first operational stage are taken into account by inserting it into the master oscillator beam wave front. To produce an artificial reference source in the atmosphere, stimulated Raman scattering of the master oscillator beam at a specified altitude is assumed to be generated. It is known that the Raman scattering of radiation is a threshold effect that originates in a spatial point of maximum radiation intensity. That is why the angular size of the reference source will appear as a point source “seen” by the receiver-transmitter telescope. The angular position of the reference source defines the high-power laser beam direction, and the reference radiation is used to generate a conjugated wave front of the high-power beam.

At third stage, the beam of the reference source that was collected by the receiver-transmitter telescope is amplified by a third amplifier (LA-3) and is conjugated by a PC mirror. Then the conjugated beam passes back by a non-reciprocal optical circuit of the system to a final amplifier (LA-3) once more. After that, the high-power laser beam is directed to a satellite.

One of the basic problems of the laser adaptive optical systems is the generation of the artificial reference source as well as registration of phase distortions of the atmosphere. The stimulated Raman scattering and Brillion scattering of the high-power laser pulses are assumed to be used for that purpose. It may be possible to use aerosol’s scattering of high-peak power beam in the atmosphere, which is theoretically predicted in Ref. [57] for the adaptive optical system, based on Hartmann wave front analyzers.

7.6.4 Principal Outcomes

It must be emphasized that one of the key technologies defining a possibility of the high-power laser-propulsion creation is a development of a laser adaptive optical system that would allow to deliver the laser power to a space vehicle through the Earth's atmosphere. The optical system has to allow to compensate for the beam atmospheric distortions of a wide spatial spectrum range, starting from the Kolmogorov's turbulence for the altitudes of $h < 20$ km up to the two-component spectrum of refraction index fluctuations in a stratosphere ($h > 20$ km).

The proposed laser adaptive optical system is based on the algorithm of compensation for the atmospheric optical distortion, which includes the following principal operation procedure, namely:

1. generation of a reference coherent radiation source in an upper atmosphere to record the atmospheric distortions in the laser beam propagation path from the laser to a lead point
2. generation of the high-power laser beam with a conjugated beam wave front by using a dynamic hologram-corrector that was recorded with the use of the reference source beam

Numerical simulation of the reference source algorithm operation demonstrated a feasibility of the compensation for the laser beam distortions, increasing a delivery of the laser power to a 3000 km orbit up to 65% of the laser initial power, for example.

References

1. Lukin, V.P.: Forming of optical beams and images on the basis of adaptive optical systems. Adv. Phys. Sci. **184**, 599–640 (2014)
2. Asanov, S.V., et al.: Prediction of high-power laser propagation of the near-IR and IR spectral ranges along inclined paths in the atmosphere. Opt. Atm. Ocean (Rus.). **29**, 167–176 (2016)
3. Sprangle, P., et al.: High-power lasers for directed-energy applications. Appl. Opt. **54**, F200 (2015). https://doi.org/10.1364/AO.56.004825
4. Lamberson, S., Schall, H., Shattuck, P.: The airborne laser. Proc. SPIE. **6346**, 63461M (2001). https://doi.org/10.1117/12.738802
5. Biliman, K.W., Horwitz, B.A., Shattuck, P.L.: Airborne laser system common path/common mode design approach. Proc. SPIE. **3706**, 196–203 (1999). https://doi.org/10.1117/12.356958
6. Mak, A.A., et al.: State-of-the-art optical technologies as applied to the problem of laser power transfer at space distances. J. Opt.Techn. (Rus.). **65**, 52–61 (1998)
7. Avdeev, A.V., et al.: Construction arrangement of HF-DF-laser facility arranged on space vehicle board. Proc. MAI **71** (2013). http://www.mai.ru/science/trudy/published.php
8. Myrabo, L.N., Lewis, J.S.: Lightcraft. Flight handbook. (Hypersonic flight transport for an era beyond oil). Apogee Books (2009)
9. Sherstobitov, V.E., et al.: On a possibility of laser beam control in LOTV mission by means of nonlinear and coherent optics techniques. AIP Conf. Proc. **702**, 296–309 (2004). https://doi.org/10.1063/1.1721009

10. Sherstobitov, V.E., et al.: Computer simulation of a solid-state laser system for propulsion of a space "tug-boat" from LEO to GEO. AIP Conf. Proc. **766**, 347–360 (2005). https://doi.org/10.1063/1.1925156
11. Krekov, M.G., Zvenigorodsky, S.G.: Optical model of middle atmosphere. Nauka, Novosibirsk (1990)
12. McCormic, M.P., et al.: Satellite studies of the stratospheric aerosol. Bull. Amer. Meteor. Soc. **60**, 1038–1046 (1979)
13. Yue, G.K., et al.: Comparative studies of aerosol extinction measurements made by the SAM II and SAGE-II satellite experiments. J. Geophys. Res. **94**, 8412–8424 (1989)
14. Selby, J.E.A., McClatchey, R.A.: Atmospheric Transmittance from 0,25 to 28,5 μm. Computer code LOWTRAN-2, AFCRL–TR–72-0745, AD A763721 (1972)
15. Kneizys, F.X., et al.: Computer code LOWTRAN-7, (AFGL). Hanscom, Massachusetts, environmental research papers, N1010 (1988)
16. Polikov, A.V., Timofeev, Y.M., Poberovsky, A.V.: Reconstruction of vertical distribution of aerosol extinction in stratosphere, using the results of "Ozon-Mir" instrumentation measurements. Proc. Rus. Acad. Sci. Atm. Ocean Phys. **35**, 312–321 (1999)
17. McClatchey, R.A., Fenn, R.W., Selby, I.E.A.: Optical properties of the atmosphere. AFGL-70-0527, N331 (1982)
18. Anderson, G.P., et al.: AFGL atmospheric constituent profiles (0–120 km). Env. Res. Pap., N954 AD A175173 (1986)
19. Rothman, L.S., et al.: The HITRAN molecular data base: edition 1991 and 1992. J. Quant. Spectr. Rad. Trans. **48**, 469–507 (1992). https://doi.org/10.1063/1.4815858
20. Osipov, V.V., Borisova, N.F., Tsukanov, V.V.: Structure, software and validation of a database UVACS (UltraViolet Absorption Cross Section). Abstract International Radiation Symposium (IRS 2000), St. Petersburg State University E30, 139 (2000)
21. Osipov, V.M., Borisova, N.F.: Attenuation of UV radiation at atmosphere ranges. Opt. Atm. Ocean. **11**, 440–444 (1998)
22. Vorob'ev, V.V.: Thermal blooming of laser beams in the atmosphere. Progr. Quant. Electron. **15** (1991). https://doi.org/10.1016/0079-6727(91)90003-Z
23. Sheremetieva, T.A., Phillpov, G.N.: Method of simulation of random fluctuations of wave front with a wide range of the fluctuations. Atm. Ocean Opt. **13**, 529–533 (2003)
24. Shen, I.: Electrostriction, Kerr optical effect, and self-focusing of laser beams. In: Raizer, Yu.P. (ed.) Effects of laser radiation. Mir, Moscow (1968)
25. Aksenov, V.P., et al.: High-power laser beams in the random-inhomogeneous atmosphere. Nauka, Novosibirsk (1998)
26. Vorontsov, M.A., et al.: Characterization of atmospheric turbulence effects over 149 km propagation path using multi-wavelength laser beacons. In: Proc. of the 2010 AMOS Conference. (2010)
27. Vorontsov, M.A., et al.: Deep turbulence effects compensation experiments with a cascaded adaptive optics system using a 3.63 m telescope. Appl. Opt. **48**, A47–A57 (2009). https://doi.org/10.1364/AO.48.000A47
28. Galkin, A.A., et al.: Imaging of space vehicles by using Altai optic-laser center adaptive telescope. Space expl. (Rus.). **46**, 201–205 (2008)
29. Lukin, V.P., et al.: Capabilities of adaptive optics as applied to solar telescopes. Opt. Atm. Ocean. **22**, 499–511 (2009)
30. Tatarskii, V.I.: The effects of the turbulent atmosphere on wave propagation. Nat. Tech. Inf. Serv, Springfield, VA (1971)
31. Fritts, D.C.: A review of gravity waves saturated processes, effects, and variability in the middle atmosphere. Pure Appl. Geophys. **130**, 343–371 (1989). https://doi.org/10.1007/BF00874464
32. Gurvich, A.S., et al.: Laser radiation in the turbulent atmosphere. Nauka, Moscow (1976)
33. Gracheva, M.E., Gurvich, A.S.: Simple model for calculation of turbulent noise in optical devices. Izv. Atm. Ocean. Phys. **16**, 1107–1111 (1980)

34. Grechko, G.M., et al.: Observations of atmospheric turbulence at altitudes of 20–70 km. Trans. Russ. Acad. Sci., Earth Sci. Sect. **357A**, 1382–1385 (1997)
35. Smith, S.A., Fritts, D.C., Van Zandt, T.E.: Evidence for a saturated spectrum of atmospheric gravity waves. J. Atm. Sci. **44**, 1404–1410 (1987). https://doi.org/10.1175/1520-0469(1987)044<1404:EFASSO>2.0.CO;2
36. Fritts, D.C., Van Zandt, T.E.: Spectral estimates of gravity wave energy and momentum fluxes, Part I: energy dissipation, acceleration, and constraints. J. Atm. Sci. **50**, 3685–3694 (1993)
37. Lamberson, S., Schall, H., Shattuck, P.: The airborne laser. Proc. SPIE. **6346**, 63461M (2007). https://doi.org/10.1117/12.482116
38. Galkin, A.A., et al.: Imaging of space vehicles by using Altai's optic-laser center telescope designed with adaptive optics. Space expl. **46**, 201–205 (2008)
39. Lukin, V.P., et al.: Application abilities of adaptive optics for Sun telescopes. Atm. Ocean Opt. **22**, 499–511 (2009)
40. Lukin, V.P.: Adaptive optics in the formation of optical beams and images. Adv. Phys. Sci. **184**, 599–640 (2014). https://doi.org/10.3367/UFNe.0184.201406b.0599
41. Crawford, D.P.: Lasers for Security, Director of Laser Weapon Systems. Directed Energy Systems Northrop Grumman Space Technology. Presented at the OSA Congressional R&D Caucus Event. (2007)
42. Fenton, C.G.: Directed-energy weapons. Technology, Applications and Implications, Lexington Institute, USMC. Joint Nonlethal Weapons Directorate, February (2003)
43. Higgs, C., et al.: Atmospheric compensation and tracking using active illumination. Lincoln Lab. J. **11** (1998)
44. Vorontzov, M.A., Shmalgausen, V.Ya.: Adaptive optics principles. Nauka, Moscow (1985)
45. Belousova, I.M., et al.: On the effective transportation of laser radiation through the atmosphere. Proc. SPIE. **2771**, 252–262 (1996). https://doi.org/10.1117/12.138073
46. Belousova, I.M., et al.: Laser beam wavefront analyzer for adaptive optical systems. J. Opt. Techn. **6**, 33–44 (1992)
47. Primmerman, C.A., et al.: Atmospheric-compensation experiments in strong-scintillation conditions. Appl. Opt. **34**, 2081–2088 (1995). https://doi.org/10.1364/AO.34.002081
48. MIT Lincoln Laboratory Annual report (2010)
49. Bulaev, V.D., et al.: The experimental laser facility based on the high-power repetitively-pulsed E-beam sustained CO_2 laser. AIP Conf. Proc. **766**, 361–372 (2004). https://doi.org/10.1063/1.1925157
50. Higgs, C., et al.: Atmosphere compensation and tracking using active illumination. Lincoln Lab. J. **11** (1998)
51. Zel'dovich, B.Y.: Phase conjugation methods. Nauka. Moscow. (1985)
52. Ageichik, A.A., et al.: Phase conjugation in a high-power E-beam-sustained CO_2 laser. Proc. SPIE. **1841**, 181–189 (1992)
53. Vasil'ev, A.A., et al.: Spatial light modulators. Radio i Svyaz, Moscow (1987)
54. Brignon, A., Huignard, R., Rajbenbach, J.-P.: Phase conjugation by saturable-gain degenerate four-wave mixing in solid state laser medium. Proc. SPIE. **2771**, 64–74 (1996). https://doi.org/10.1117/12.238046
55. Stepanov, V.V., et al.: Laser beam control system to compensate for turbulent distortions of the laser beam at introducing a lead angle. AIP Conf. Proc. **702**, 283–295 (2004). https://doi.org/10.1063/1.1721008
56. Kingsbury, R.W.: Optical communications for small satellites. Submitted to the Department of Aeronautics and Astronautics Doctor of Philosophy in Communications and Networking at the Massachusetts Institute of Technology (2015)
57. Banakh, V.A., et al.: Compensation for wavefront of a partially coherent laser beam by using atmosphere radiation backscattering effects. Atm. Ocean Opt. **24**, 549–554 (2011). https://doi.org/10.1134/S1024856011060030
58. Sandler, D.G., et al.: Shearing interferometry for laser-guide star atmospheric correction at large D/r0. J. Opt. Soc. Amer. **11**, 858–873 (1994). https://doi.org/10.1364/JOSAA.11.000858

Conclusion

In this book, we tried to follow the task that was formulated by Russian Scientist K. E. Tsiolkovsky in the beginning of twentieth century and which is formulated as the development of "the method of obtaining a speed with the energy transmitted from the planet in the form of a parallel bundle of electromagnetic beams." After many years, this method has got a name of laser propulsion, in which the parallel bundle of electromagnetic beams is produced by lasers. But, those transportation systems which could be based on the use of both high-power lasers and vehicles with a laser-propulsion engine (LPE) have not been created till now in spite of the expectations that these systems would be competitive with respect to those which are based on the use of rockets and aircrafts. To disclose the technological and engineering problems of high-power laser propulsion to be developed, we tried to reveal the principal processes and phenomena of the laser propulsion, which are on the basis of these problems. And the following are among them:

1. Efficiency of a laser power conversion into a thrust both in the pulsejet and in ramjet modes of laser-propulsion production is limited by a value of 40% in maximum. The reason of that is the screening of an incoming laser radiation by a plasma that is ignited in a propellant under the action of the laser radiation.
2. In the ramjet mode, the produced thrust is also limited in a magnitude by a surging effect of the propulsion engine inlet, which is induced due to strong shock waves generated in the engine nozzle by high-peak power laser pulses.
3. A laser ablation propulsion (LAP) has some advantages as compared to a pulsejet laser propulsion due to production of a thrust with a higher specific impulse. But in this case, a higher power laser radiation is required to produce the thrust of the same value as for the pulsejet propulsion. It must be noted that both types of the laser propulsion are distinguished by the maximum efficiency of the thrust production of 40%.

   In that sense, the LAP using CHO-polymers with energy storage as propellants on a large scale is a promising technology to achieve efficient thrust production

Y. A. Rezunkov, *High Power Laser Propulsion*, Springer Series on Atomic, Optical, and Plasma Physics 116, https://doi.org/10.1007/978-3-030-79693-8

due to release of the stored energy under the action of a laser radiation. Many experiments show that the use of the CHO-polymers allows achieving the LAP efficiency of more than 50%. To develop prospective high-power laser propulsion systems, the LAP is assumed to achieve the production efficiency of 70%. The choice and/or the construction of the CHO-polymer compositions is one of the principle tasks that is imposed at the present time. We hope that the scientific experience accumulated by now in the study of the LAP will allow solving this task in the near future.

4. The other problem of the high-power laser propulsion is thrust production during transition of a subsonic gas flow into a supersonic one in a laser-propulsion engine. In this book, we denote that the ramjet supersonic laser propulsion is restricted by a few gas-plasma-dynamic negative effects which are developed via the interaction of laser pulses with supersonic gas flows. To overcome these difficulties, we propose to use a laser-ablated jet that is generated in the engine nozzle to accelerate additionally the incoming supersonic flow due to the interaction of the jet with the flow. It is expected that the momentum coupling coefficient C_m of 10^{-3} N/W would be achieved in this case.

   Unfortunately, an experimental justification difficulty with this proposed laser-propulsion technology is the lack of relevant experimental equipment that is required.

5. There are only two principal designs of vehicles with a laser propulsion, which were developed to the present time and to be used as a part of new transportation systems based on the high-power laser propulsion. These designs are the Lightcraft technology demonstrator (LTD) developed by Leik N. Myrabo and the space mini-vehicle based on the aerospace laser-propulsion engine (ASLPE) developed in the NII OEP. Both of these technologies satisfy the conditions of thrust production both in the subsonic and supersonic modes of a laser-propulsion production. The LTD is most adapted for launching of vehicles to low Earth orbits, and the vehicle with the ASLPE is adapted for orbital missions in a near-Earth space. But in every case, specially constructed propellants have to be stored on a board to produce a thrust in the mode of a supersonic LAP.

6. And finally, the difficulty with the high-power laser propulsion is a precise beaming of a flying vehicle through the Earth's atmosphere. In any case, the specular laser adaptive systems to compensate for the laser beam wave front distortions, which are caused by the atmospheric turbulence and by a thermal blooming effect, have to be developed.

We hope that all of these restrictions of the high-power laser propulsion to be developed will be eliminated in the near future if we examine them in the context of an applied laser sciences progress. Partially, the development of a new class of diode-pumped alkaline lasers (DPAL) around the world is one of the examples of this progress. The DPAL possess such key features as a high efficiency of the electric power conversion into the laser radiation power, a low laser beam divergence that is close to a diffractive limit, a short radiation wavelength of 0.8μm, as well as small "size and mass dimensions." At the present time, the DPAL power reaches 100 kW that is well enough to be used for launching space mini-vehicles with laser propulsion.

Moreover, some of the works on LAP, as well as a laser power conversion into a plasma temperature, can be used at the development of the onboard laser mini- and micro-thrusters at present.

For example, the application of the high-power laser propulsion to solve some problems of near-Earth space exploitation can be considered. One of these problems is the necessity of space debris removal out from both the low Earth orbits and from GEO. This debris consists of the massive worked out space vehicles and their destroyed components. All of these elements impose a hazard for future missions of the space stations and manned spacecrafts in the near-Earth space. In spite of a complexity of the considered trash-collector concept, we predict that this type of the space transportation systems which are based on laser propulsion will be realized immediately after the development of the air-borne high-power lasers of a 100 kW power and more. Unfortunately, the use of the high-power lasers in a space depends not only on the progress in a laser science but also on the political initiatives of leading countries which will be interested in the peaceful use of the near-Earth space.

And finally, the development of the high-power laser-propulsion systems has to be supported by the progress in (a) an optic-electronic equipment to control the high-power laser radiation for a precise beaming of the space vehicles with laser propulsion, and (b) new optical design of the space vehicle device.

Index

A

Aberration
- coma, 138
- spherical, 138
- phase aberrations of the beam wave front, 251
- velocity aberrations of the laser beam, 262, 268, 269

Ablated mass rate, 24, 94, 121, 192, 196, 204
"Ablation jet-main stream" interaction, 194
Ablation of solid materials
- "closed" laser ablation, 20
- combined ablation, 17
- direct ablation, 119
- laser ablation, 94

Ablation pressure, 120
"Absorption explosion" model of plasma ignition, 99–102
Absorption of laser power
- multi-photon absorption, 86

Adaptive optics, 6, 10
Aerodynamic drag coefficient, 214
Aerodynamic drag of the upper atmosphere, 211
Aerosols
- AFGL model of aerosols, 244
- Lowtran aerosols models, 244
- solid and gaseous aerosols of the atmosphere, 242

Aerosol's absorption coefficient, 245
Aerosol's extinction coefficient, 246
Aerospace laser propulsion engine (ASLPE), 134–139, 145, 147, 149–169, 229, 234, 235
Afterbody
- off-axial parabolic afterbody, 197

Afterburning of detonation products, 120
Airborne laser (ABL), 29, 218, 242, 254, 260, 262–267, 269, 271, 274–276
Airborne laser systems, 268, 270
Air-breathing engine, 8
Air-breathing gas-dynamics, 44
Air Force Research Propulsion Laboratory (AFRPL), 5
Alpha Centauri star, vi
Apex angle of cone nozzle, 47
Apollo Lightcraft, vi
ASLPE beam concentrators, 137, 139, 145, 146, 159, 162, 166, 167
Atmosphere turbulence, 251–256, 258, 261, 263
Atom ground state, 64
Auxiliary laser rangefinder, 232
Averaged ionization potential, 127
Avogadro number, 96

B

Ballistic missiles, 5
Beam broadening, 264, 265
Beam concentrator
- diffractive-limited beam, 146
- matrix, 10, 20, 21, 144
- two-mirror, 141, 143, 147, 149

Beam concentrator mirrors as a whole system, 138
Beamed energy propulsion (BEP), 2
Beam incident angle, 36, 193

Y. A. Rezunkov, *High Power Laser Propulsion*, Springer Series on Atomic, Optical, and Plasma Physics 116, https://doi.org/10.1007/978-3-030-79693-8

Beaming of a vehicle, 39
Beam pressure, 2
Beam spread function, 264
Beam steering, 263
Beam thermal defocusing, 250
Beam wave front, 39, 139, 143, 250, 251, 258, 264, 265, 268–271, 273, 275, 280
Beam wave front analyzers (BWA), 271–272, 274, 275
Beam wave front phase corrector, 272–273
Beam wave front phase inclination, 250
Benzene ring links, 115
Bernoulli equation, 98
Bessel function of a zero order, 146
Biberman-Norman formula, 70
Block construction of the laser-propulsion engines, 234
Boiling point of the material, 97
Boltzmann distribution function, 66
Boltzmann equation, 67
Bouguer-Lambert-Beer absorption law, 126
Breakdown region, 46, 53, 72, 73, 80, 88, 172, 173, 180, 183, 186
Breakthrough Starshot project, 12
Brenta frequency, 253
Brillion scattering, 277, 279
Broadening of the beam, 251

C

"Cannon-ball" effect, 108
Chemical oxidation of the vapor components, 95
Chocking effect in the nozzle inlet, 180
"Cluster constellation" on the basis of micro- and nano-vehicles, 208
C-O bond energy of the molecular, 126
C-O-C stretching vibrations, 117
Coefficient of radiation absorption, 97
Coefficient of thermal diffusivity, 97
Combustion chamber, 7, 8, 16, 18, 22, 95, 127, 135, 156–158, 186, 187
Complete thermodynamic equilibrium, 65
Complex parameter of local explosive theory, 49
Compensation for the beam wave front distortions, 270
Component molecular mass, 85
Concentrator of laser beam, 44
Concept of "averaged" medium, 82
Conical beam transformer, 137
Conservation law of momentum fluxes, 108
Continuity concept of gas flow, 79
Conversion efficiency, 62, 72–77, 101, 128
Coordinate photo detector array, 232
Corner reflectors, 232, 234
Coupling coefficient
  momentum coupling coefficient, 47, 106, 112, 114, 118–120, 128, 130, 135, 151–154, 156, 161, 163, 192, 204
Courant-Friedrichs-Lewy (CFL) criterion, 194
Critical nozzle cross section, 98

D

Degree of plasma ionization, 63, 102, 119
Delayed burning, 95, 113, 115, 117, 118
Density number of electrons, 88
Detonation wave, 3, 5, 7, 16, 34, 98, 102, 105, 111, 173, 183
Diagram
  x-t diagrams, 87, 88
Dielectric constant of plasma, 75
Diffractive beam angular divergence, 263
Diffractive limit of beam divergence, 284
Diffractive radius of the beam, 255
Direct photo-ionization rate, 64
Double laser pulses, 34
Dynamic adaptive numerical grid, 81, 162

E

Earth equatorial plane, 216, 223
Earth gravitation constant, 214
Earth orbits, 10, 12, 13, 17, 23–25, 30, 35, 36, 38, 39, 44, 94, 133, 156, 208–212, 222–229, 242, 263
Eccentricity of hyperbola, 143
Effective radius of the beam, 255, 256
Electrically driven flexible mirror, 270
Electric-mechanical deflectors, 233
Electrodynamics tethers, 210
Electron-hard particles collision, 74, 86
Electron number density, 63, 69, 70, 77
Elliptical trajectory of satellite, 216
Elvert system of equations for the plasma ionization equilibrium, 68
Energetic materials, 17, 21, 22, 111
Energy
  detonation, 14, 112, 113, 115, 118
  equilibrium vibrational, 85
  reciprocating and rotating, 85
  vibrational, 85
Energy conservation law, 95
Energy cost of launching a vehicle
  mission energy cost, 26

Energy of delayed burning, 112
Energy of electron excitation, 85
Engine
electrojet, 19, 223
jet, 19, 20, 22, 139, 220
rocket, 7, 44, 113, 129, 133, 134, 156, 158
Equilibrium constant for *i*-th reaction, 86
Equilibrium of physical-chemical reactions running under a laser pulse, 82
Evaporation heat of the target material, 96
Evaporation temperature, 100, 126, 193
Evaporation threshold of the target material, 95
Expansion ratio of the lateral jet, 190
Explosive breakdown of substance, 15
"Explosive" energy, 46

F

First ionization potential, 64, 69–72, 77
Flow
adiabatic, 58, 84, 98, 99, 185, 192
over-expanded jet flow, 123
supersonic gas, 16, 88, 172, 176, 178, 185, 188, 190, 192, 194, 199, 204
transverse, 158, 169
Forebody cone of the vehicle, 175
Formaldehyde monomer (CH_2O), 126
Formation enthalpy of polymers, 113
Formation enthalpy of reaction products, 113
Fourier spectrum, 248
Four-wave mixing
degenerate four-wave mixing, 277
Four-way mirror pyramid, 234
Freed radius, 264–266
Free electron density number, 62
Fresnel distance of the beam propagation, 235
Fuel
carbon-to-hydrogen, 113
onboard, 211
Fused quartz, 168

G

Gas-dynamic effects, 82, 121, 163, 190
Gas-dynamic equations, 5, 16, 81, 85
Gas-dynamic processes, 4, 17, 44, 77, 79, 88, 121, 162
Gas flow structure, 82, 172
Gas-kinetic physics, 96
Gas-plasma-dynamic effects, 284
Gauss functions, 69
Gaussian laser beam, 264
Gaussian profile of intensity distribution, 254
Geocentric equatorial coordinate system, 212, 213, 224
Global computer networks, 208
Godunov numerical scheme, 81, 162
Gravitation constant, 213

H

Hartmann analyzer, 271, 275
Hartmann wave front detectors, 269
Heat
specific heat capacities, 86
specific heat of combustion, 112, 113, 219
specific heat of polymer combustion, 129
Heat of the surface evaporation, 96
Henry T. Nagamatsu laboratory, 178
Heterodyne analyzer of beam wavefront, 273
High-melting point material, 94, 96
High-power laser propulsion (HPLP), 2, 3, 6, 10, 13, 15, 34–40, 51, 106, 125, 133, 154, 169, 198, 209, 210, 242, 243, 247, 252, 264–275, 277–280
High-temperature oxidation of the ablated CHO-materials, 95
HITRAN databank, 245
Holography correctors, 270, 278
Hybrid rocket engine system, 223
Hypersonic aerodynamics, 13
Hugoniot adiabatic, 102

I

Ignition intensity threshold, 62
Impulse
mechanical, 36, 46
Incident flow stagnation point, 181
Independent stochastic fields, 259
Index of refraction of the atmospheric, 247
Ingless-Teller formula, 71
Inner gravitational waves (IGW), 252–255, 257, 258
Intermediate circular focus of the beam concentrator, 137
Intermediate vehicle orbits, 263
International Symposium on Beamed Energy Propulsion (ISBEP), 12
Interplanetary human travel, 2
Inverse Breamstranstrum effect, 70
Ionization
avalanche ionization of gases, 99
avalanche ionization or cascade collisional ionization of gas molecules, 86
collisions, 64, 67, 71, 74, 85, 123

Ionization (*cont.*)
degree of plasma ionization, 63, 71, 102, 119
direct, 64, 65
twofold, 62
Ionization cross-section, 64
Ionization degree of plasma, 74
Ionization potential of atoms, 63
Isentropic stochastic phase screens, 258
Isoplanatic angle, 264–266, 269

J

Jet
ablated, 94, 98, 124, 192, 193, 201, 202
ablated vapor, 95, 103, 107, 121, 192
under-expanded vapor jet, 97
vapor-plasma, 123, 124, 127
Jet exhaust, 18, 24, 25
Juge point of Hugoniot adiabatic, 102

K

K-factor, 114, 115, 117
Knudsen number
characteristics, 79
Kolmogorov's isotropic turbulence, 252
Kolmogorov's spatial spectrum, 264

L

Laser
airborne, 28, 38, 134, 274
chemical, 2, 4, 5, 17, 21–23, 34, 85, 94, 95, 112, 115, 125, 126, 128–130
diode-pumped alkaline lasers (DPAL), 284
free-electron, 39, 102
ground based, 2, 10, 35, 38, 39, 238
high-power, 2, 4, 5, 10–12, 15, 17, 22, 35, 36, 38–39, 51, 53, 112, 130, 134, 138, 156, 163, 165, 172, 183, 210, 242–244, 247–251, 260, 266, 268, 269, 271, 273–275, 277–280
solar-pumped space-based lasers, 238
solid-state, 15, 247
solid-state disk laser, 39
Laser ablation
combined, 44, 105, 107, 108, 130
confined, 22, 105, 108–111, 129
direct, 17, 105–108, 119
Laser ablation propulsion (LAP)
combined laser ablation propulsion, 107–108
direct laser ablation propulsion, 17, 105–106, 119
Laser adaptive optical techniques, 264, 268, 278
Laser adaptive system
nonlinear adaptive laser system, 39, 277–279
Laser beam concentrator, 4, 135, 197
Laser beam distortions, 5, 242, 266, 268, 269, 280
Laser beam focusing system, 134
Laser beam guidance systems, 269
Laser beam phase distortions, 262
Laser breakdown thresholds, 52, 165
Laser chemical propulsion (LChP), 18, 22
Laser Energy Network program, 10
Laser impulse space propulsion (LISP) project, 36–38
Laser-induced breakdown of propellant, 44
Laser irradiance flux, 100
Laser optics, 5, 138
Laser orbital transfer vehicle (LOTV)
orbital missions, 10
Laser plasma, 45, 62–77, 107
Laser power conversion, 54, 74, 77, 80, 82, 84, 87, 98, 99, 101, 114, 156, 172, 185
Laser power conversion into plasma temperature, 86
Laser power fluxes, 36, 38, 80, 102, 103, 106, 121, 122, 125–127
Laser power transmission, 5, 34, 263
Laser propulsion
air-breathing laser propulsion (ABLP), 19, 20, 22, 56, 88, 134, 176, 192
CW, 16, 19, 44, 156, 158, 171, 172
explosive laser propulsion, 16
high-power laser propulsion (HPLP), v–vii, 2, 3, 6, 10, 13, 15, 34–40, 51, 106, 125, 130, 133, 134, 154, 169, 209, 242, 243, 252, 264–280, 284, 285
pulsejet supersonic propulsion, 173–183
ramjet supersonic, 172
rocket, 2, 4, 16, 23, 26, 30, 54–61, 129
space, 3, 5, 8, 10, 11, 14, 20, 23, 26, 36, 54–61, 211, 229, 234, 242
supersonic, 35, 45, 134, 156, 172, 173, 181, 183, 192, 204
Laser propulsion mechanism, 10, 16
Laser-propulsion production efficiency, 35, 51, 220, 283
Laser pulse duration, 62, 107, 108, 119, 130, 197
Laser pulse interaction with plasma, 53

Laser pulse shape, 100
Laser radiation interaction with plasma, 44, 61, 94, 192
Laser radiation wavelength, 62, 73, 80, 105, 108, 119, 138, 168, 244, 251
Laser-supported combustion wave, 5
Laser-supported detonation wave, 105
Laser-thermal propulsion (LTP), 16, 22
Launching of satellites, 23, 209
Lightcraft, 10, 12–16, 35, 55, 173–176, 178–182, 197, 199, 202, 209, 242
Lightcraft Technology Demonstrator (LTD), 10, 11, 16, 35, 134, 173
Lightsat orbital cluster, 209
Lightsat system, 209
Linear adaptive optical technique, 270
Local explosion theory, 22, 46, 47, 49, 82
Local instantaneous explosion, 47
Local thermodynamic equilibrium state, 66
Low Earth orbit (LEO), 6, 10, 23–30, 32–34, 45, 214, 215, 230, 263
Low Earth orbit satellite, 209
Lowtran software, 244

M
Mach cone, 8
Mach disk in a supersonic jet, 127
Mach numbers, 8, 81, 172, 174, 178, 181, 182, 186, 190, 192, 194, 196–200, 247
Manned vehicle flights, 7
Mass consumption rate, 28, 59, 81
Mass flux density, 95–97
Master oscillator-power amplifier (MOPA), 39, 243
Matrix of direction cosines, 144
Maxwell distribution of velocities law, 67
Merging effect, 188
Merging of shock waves, 172, 188
Middle-infrared spectral range, v
Mirror
  "active mirrors", 243
  annular, 8
  flexible, 272, 273
  generatrix of the, 139
  hyperbolic, 137, 139, 141
  phased array, 11
  plane, 139, 195, 232
  primary, 8, 10, 11, 243, 251, 275
Misalignment of the concentrator mirrors, 143
Molecular absorption of laser power, 7, 247
Multi-component medium, 82
Multilayer coatings, 168
Multilayer structured targets, 105, 108–111
Multiple-actuators flexible mirror, 273
Multi-stage ionization of atoms, 62

N
Navier-Stokes equations, 193, 194
Near-Earth orbit, 2, 3, 210
Newtonian iteration process, 144
Newton's second law, 213
Non-linear atmospheric effects developing under high-power laser radiation, 247
Nonlinear divergence of the laser beam, 250
Nonlinear optical medium, 270, 277–279
Non-linear phase conjugation, 270, 277–280
Nozzle
  axial parabolic, 55, 56, 59, 60
  conical, 16, 47, 112, 152
  jet, 3, 4, 16, 20, 22, 48, 53, 55–61, 84, 97, 98, 114, 128, 134–136, 156, 173, 190, 192, 195, 197, 199–202, 204
  laval, 16, 159
  off-axis parabola nozzle, 47, 55, 56, 59, 60, 195
  parabolic, 20, 44, 47, 48, 50–57, 59, 80–83, 112, 136, 171, 172, 195–197, 199
  plug, 197
Number density of electrons in plasma, 75
Numerical technique of phase screens, 247

O
Off-axis parabola, 20, 56
Onboard receiving optics, 7
One-and-the-same receiver-transmitter telescope, 243
Optical adaptive systems, 5
Optical continues discharge (OCD), 156–159
Optical plasmatron, 156
Optical quality of the concentrator mirrors, 144
Optical surface quality of the mirrors, 236
Optical switch, 233
Optical system to collect the laser power, 229
Optical thickness of the laser radiation path, 245
Optical turret, 231–233
Optical window, 16, 135, 156
Optic-electronic equipment, 8, 38, 39
Optics
  diffractive optics, 39
Orbit
  circular, 215, 216, 223, 224
  coplanar, 224
  elliptical, 26, 30, 211, 216, 217, 224, 225

Orbit (*cont.*)
- Hohmann elliptical transfer orbit, 30
- rendezvous, 222

Orbital debris, 36
Orbital maneuver of space vehicle, 30, 134
Orbital satellites cluster, 208

P

Parabolic shroud of the Lightcraft engine, 200
Paraboloid focus, 47
Paraboloid of revolution (rotation), 4, 173
Pendulum technique, 151
Perfect gas, 79–82, 84, 88
Phase-adaptive technique, 13
Phase conjugation mirror (PCM), 243, 270, 277, 278
Phase distortions of the incoming laser beam, 143
Photo chemical model of POM ablation, 126
Photo thermal ablation model, 126–127
Plank's constant, 64
Plasma
- equilibrium, 66–70, 76, 82–84, 88, 162, 193
- high density, 73
- ideal equilibrium, 80
- laser-ignited plasma, 69, 72, 88
- laser-induced, 72, 77, 107, 185, 188
- multiply-ionized plasma, 63–72, 74, 77
- non-equilibrium, 65, 66, 77, 80, 85–88
- optically thick, 66
- weakly ionized, 62, 63, 123

Plasma-chemical reaction, 82, 85, 88, 112–118
Plasma dielectric conductivity, 74, 75
Plasma internal energy, 103
Plasma ionization equilibrium, 65, 68
Plasma luminescence spectrum, 65
Plasma resonance frequency, 74
Plasma screening of laser radiation, 161, 192
Polymers of CHO-chemical composition, 95
Polymer with accumulated internal chemical energy:
- CHO-polymer, 95, 112–118, 120–130, 135, 219, 283, 284
- Delrin polimer, 21, 22, 112, 114, 129, 130, 153, 159, 222
- polycrystal polymer, 112
- polycarbonate, 115, 117
- polymethylmethacrylate, 129
- polyoxymethylene, 114, 116–118, 129
- polyphormaldegid, 219
- polyvinylchloride, 114, 115, 117

Potential
- ionizations, 64, 65, 67–70, 77, 100

Power sources, 2, 3, 5, 8, 12, 38, 50, 95
Prandtl number, 190
Pressure impulse receiver, 136
Primary and auxiliary laser sub-systems, 261
Prokhorov's theory, 46, 52, 96
Propellant
- gas, 6, 7, 16, 18, 44, 45, 51, 55, 56, 59, 60, 86, 112, 196
- high-energetic polymer, 112
- liquid, 14–16, 55
- low-melting, 98
- non-chemical propellant, 129
- onboard, 4, 14, 16, 28, 29, 59, 219, 224, 225, 229, 235
- "SLAVIT" propellant, 112

Propellant mass consumption
- stored propellant mass, 32, 33

Propellant's thermal energy, 44
Propulsion
- air-breathing, 16, 20, 22, 44, 51, 59, 94, 114, 133
- air-breathing laser, 19, 22, 134
- explosives, 15, 16, 44, 46, 47
- laser ablative, 24, 88
- pulsejet, 46–48, 50, 52, 53, 55, 62, 88, 133, 151, 169, 172, 173, 176, 181, 183
- pulsejet laser, 46, 49, 53, 58, 78–81, 84, 86, 163, 172
- repetitively pulsed mode of laser propulsion, 54, 185

Propulsion mechanism
- explosive mechanism of laser propulsion, 46

Pulse energy consumed to ablate one gram of the target material, 193
Pulse repetition rate, 16, 20, 37, 52, 53, 60, 111, 165, 185, 186, 189, 193, 247, 274
Pyrolysis process, 113

Q

Quantum electronics, 4

R

Radiation intensity threshold, 62, 101
Radiation transmittance, 74
Raizer's method, 63, 67–69
Ramjet engines, 186, 187, 190, 204

Rate constants of i-th direct and inverse chemical reactions, 86
Ray tracing technique, 144
Receiver optics, 8, 134, 135
Recoil impulses, 2, 36, 47, 52, 94, 103–105, 109, 111, 119, 121, 192
Recombination rates, 64
Release of combustion energy, 95
Repetitively pulsed mode of laser operation, 54, 109, 112
Reusable airspace launch systems, 13
Reynolds numbers, 178, 190
Riemann problem, 194
Riley/sodium stars generated in the upper atmosphere by lasers, 268
Rocket boosters, 30, 31, 33
Rytov parameter, 277

S
Saha's equation, 65, 66, 69
Satellite
  non-operational, 211
Satellite orbit maneuver, 216, 219, 229
Schlieren's instant pictures, 121
Screw jet of gaseous propellant, 16
Self-correlation functions, 259
Shadow patterns of the plasma, 73
Shock wave
  head shock wave, 123–, 124
  quasi-plane, 51
  quasi-stationary bow, 186, 188
Shuttle orbiter, 8
Single-stage rocket, 16
Single-stage supersonic aircraft, 13
Skin-layer depth in the substrate, 108
Slit between the first mirror of the beam concentrator and engine nozzle, 139
Solar sails, 210
Space-borne laser, vi
Space communication networks, 209
Spacecraft, 2, 14, 207
Space debris, 36, 210, 211, 223
Space debris removal, 210, 222–229
Space infrastructure servicing station, 211
Space shuttle, 13
Space transportation system, 23, 210, 211, 222
Space tugs, 224, 225, 228, 229, 231, 236
Space vehicles, 5, 12, 15, 16, 23–25, 30–34, 39, 134, 208, 209, 238, 241–280
Spatial light modulator (SLM), 273, 274
Specific ablation energy, 24, 114, 115, 118, 195
Specific energy released, 77
Specific impulse, 3–5, 10, 17–20, 25, 28–32, 34, 54, 55, 58–60, 94, 98, 111, 121, 130, 156, 158, 192, 219, 220, 222, 225
Specified lead angle, 263
Speed of sound in the atmosphere (in gases), 53, 102, 185
Spherical distortion of incoming beam, 138
Spinning of the injected gas, 157
Spiral orbit trajectory, 26
Spiral vortex in the flow, 163
Sputnik, v
Star trackers, 230
"Star Wars" program, 10
Stefan-Boltzmann constant, 103
Stochastic phase screens, 258, 259
Strain sensor, 159, 161
Strategic Defense Initiative (SDI), 5, 11, 35
Strell number, 275
Structural function of a spherical wave, 264
Subsonic laser propulsion, 45, 284
Substrates, 20, 108, 109, 168
Supersonic laser propulsion, 8–10, 22, 35, 171–204

T
Target material, 36, 96, 101, 105–108, 126, 193
Target material mass consumption, 98
Telescope
  large-aperture adaptive telescope, 5, 11
  receiver, 10, 31, 32, 34, 39, 223, 229–233, 235–237, 255, 256, 258, 263, 268
  transmitter, 31, 37, 51, 229, 231, 232, 243, 250, 262, 263, 266–268, 274
Telescope primary mirror, 31, 274
Telescope system, 12, 39, 236, 237, 279
Thermal blooming effect, 250
Thermal deformation of the concentrator mirrors, 143
Thermal loads of the mirrors, 149
Thermo-dynamic functions, 82
Thrust
  jet, 3, 4, 6, 14, 15, 18, 19, 44, 45, 59, 79, 84, 94, 97, 98, 105, 107, 121, 127, 130, 156, 191, 192, 197, 204, 219
  low, 30, 31, 38, 44, 115, 156, 229
Thrust coefficient, 128
Thruster, 9, 210
Thrust production
  "air-breathing" mode of thrust production, 45

Thrust production (*cont.*)
efficiencies, 10, 18, 38, 45, 48, 50, 59, 61, 80, 94, 101, 114, 127, 147, 164, 176, 180, 197, 204, 234
"pulsejet" mode of thrust production, 45
"rocket" mode of thrust production, 45
Time interval of shock wave dissipation, 47
Toroidal laser beam, 145
Toroidal ring whirlwinds, 183
Total evaporated mass consumption, 193
Tractor beam conception, 134
Transfer period, 31, 32
Trash-collectors, 210, 222
Trash-collector system, 211
Turbulence "freeze time", 277
Turbulent phase distortions of laser beam, 39
Two-loop adaptive laser optical system, 271
Two-particle photo-recombination, 64

U
Unstable resonator, 159

V
Vehicle
bell type, 53
laser-propelled, 197, 209
laser space mini-vehicle, 15, 209–240, 288, 289
single-stage airspace, 11
space micro-vehicle, 12
space vehicle with laser propulsion, 209
Vehicle forebody, 172, 180, 181
Vehicle payload mass, 24
Vehicle velocity increment, 24, 25

W
Wandering of the laser beam, 138, 247, 252, 260
Wave
blast, 176, 179–181
combustion, 5, 7, 35
fast ionization, 183
laser detonation, 98, 183
laser-sustained detonation (LSD), 9, 102, 103, 105, 173
laser-sustained radiative, 102–103
laser-sustained shock, 107, 173
laser-sustained subsonic radiative (LSR), 103
Wavelengths, 2, 10, 17, 24, 31, 32, 37, 39, 65, 66, 69, 71, 74, 94, 99, 106, 107, 111, 122, 146, 168, 231, 236, 237, 243–246, 254, 260, 265, 272, 274, 277, 279
White Sands Missile Range, 11
Working medium (propellant), 45

Z
Zenith angles, 250, 257, 258, 260, 262–267
ZEMAX-specialized software, 145